Acoustic and Vibrational Enhanced Oil Recovery

Scrivener Publishing
100 Cummings Center, Suite 541J
Beverly, MA 01915-6106

Publishers at Scrivener
Martin Scrivener (martin@scrivenerpublishing.com)
Phillip Carmical (pcarmical@scrivenerpublishing.com)

Acoustic and Vibrational Enhanced Oil Recovery

George V. Chilingar
Kazem Majid Sadeghi
and
Oleg Leonidovich Kuznetsov

Scrivener Publishing

WILEY

This edition first published 2022 by John Wiley & Sons, Inc., 111 River Street, Hoboken, NJ 07030, USA and Scrivener Publishing LLC, 100 Cummings Center, Suite 541J, Beverly, MA 01915, USA

For more information about Scrivener publications please visit www.scrivenerpublishing.com.

Wiley Global Headquarters
111 River Street, Hoboken, NJ 07030, USA

For details of our global editorial offices, customer services, and more information about Wiley products visit us at www.wiley.com.

Library of Congress Cataloging-in-Publication Data

ISBN 9781119760153

Cover image: Pixabay.Com
Cover design by Russell Richardson

Set in size of 11pt and Minion Pro by Manila Typesetting Company, Makati, Philippines

Printed in the USA

10 9 8 7 6 5 4 3 2 1

Dedication

The authors are greatly indebted to Dr. Henry Chuang and Nina Chuang, for all of their tireless and generous help in accomplishing the completion of this book. Dr. Chuang was former student and teaching assistant in Petroleum Engineering at the University of Southern California through the guidance of Professor George V. Chilingar. The authors wish to thank them for everything and send them the very best wishes for their health and happiness.

Contents

List of Contributors

List of Major Contributors

Donald G. Hill
Valery P. Dyblenko
Rishard Ya. Sharifullin
E. Y. Marchukov
Ernst M. Simkin
I. A. Chirkin
Muhammad Haroun
M. L. Surguchev
Mehdi Shamsi

Other Contributors

Michael Gorfunkel
Julia V. Korobeynikova
Nurzhan Igissinov
Rustem Akimov
Mark S. Chilingar

1
Introduction

Physical fundamentals of vibration and acoustic actions on the reservoir for improving the oil yield and increasing the development tempo of depleted and water-encroached oil fields are presented in this book. It is shown that the area of most efficient application is reservoirs with crudes characterized by a high water content and heterogeneity, low permeability, high shaliness, etc. The developed techniques are ecologically neutral and do not cause damage to the well-designed elements. The book may be useful to students, scientists, and engineering-technical personnel engaged in geophysics and development of petroleum fields.

The book is devoted to the theoretical and applied aspects of physics of the porous media saturated with oil, gas, and (or) water. A special attention is devoted to the hydro- and thermodynamic phenomena emerging in such media upon spreading within them of elastic waves with frequencies between infrasound and ultrasound. For expert analysis of the reviewed phenomena, the authors expended years of painstaking work (creating theoretical models and conducting specialized physical experiments on the reservoir models in the in situ conditions). In the 1970s to 1990s, the authors designed a complicated special equipment for the laboratory and field experiments and then conducted experimental and industrial testing. Beginning in 1990s, the interest increased in the vibration and acoustic technologies to an increasing oil-gas production in Russia, USA, France, Brazil, and China. The authors believe that the results achieved will be useful first of all for specialists designing new methods of improving well productivity. A useful information, however, will be found here also by geophysicists creating new technologies of rock diagnostics and of seismic and acoustic identification of commercial hydrocarbon accumulations.

Having read the book, the reader will get the idea of major aspects in theory and practice of vibration and acoustic effect on the porous media. It is important that the reviewed technologies are nondamaging to the environment. The authors believe that the book will initiate a burst of interest not only in the hydraulics and flow physics in the field of vibration but also in the nonlinear geophysics. The most important is the synergetic

George V. Chilingar, Kazem Majid Sadeghi, and Oleg Leonidovich Kuznetsov. Acoustic and Vibrational Enhanced Oil Recovery, (1–18) © 2022 Scrivener Publishing LLC

physicochemical effects forming the foundation of seismo-geochemistry and the study of the Earth's gas respiration in the process of its oscillatory motions.

Research presented in this book was initiated partly because of the recorded dependence of the rate of oil production on earthquake occurrences in seismically active areas (Surguchev *et al.*, 1975 [33]). It was noticed that several days after the occurrence of an earthquake with the epicenter located in the vicinity of the oil-producing oil field, the rate of oil production increased and remained higher than the pre-earthquake level for a considerable period of time. It was also noticed in the soil remediation studies (e.g., Sadeghi *et al.*, 1992 [28]) that sonic energy applied to the contaminated soil increases the rate of hydrocarbon removal and decreases the percentage of residual hydrocarbons. Modern world oil and gas-producing industry is dominated by the application of artificial methods of affecting oil reservoirs. The application of diverse action systems and methods enabled an implementation of the intensive oil fields' development at high oil production tempo. The major accrual in the oil reserves in the future is anticipated in the areas of ever more complex geologic and geographic environment, significantly distant from the areas of oil consumption.

Thus, developing novel efficient enhanced oil recovery technology [e.g., Electrical Enhanced Oil Recovery (EEOR)] is imperative. At present, after a field development is ended, more than half of oil reserves remain subsurface. This means that, currently, the fields with residual oil reserves of over 0.5 BT (billion tons) are written off annually. If such amount of the oil loss is maintained, then a stable increase of oil production cannot be guaranteed for an extended period of time due to limited oil reserves.

The following major reasons may be identified for a decreased oil yield:

- The operation of capillary forces preventing oil displacement from smaller pores of a micro-nonuniform porous medium;
- Unfavorable interrelation between the mobilities of the displaced and displacing liquids;
- Geological heterogeneity of the productive reservoir.

Almost 90% of oil in Russia, for example, is recovered from the fields which experienced poor sweep efficiency during the waterflooding.

Another important trend is development of the techniques improving interrelation of the mobilities between the oil and water. For the application of one or another action method on the oil and gas reservoirs, it is important to know where exactly the oil is remaining after flooding. Due to a low sweep efficiency by water, the non-recovered oil in the flooding process, will constitute 60% to 90% of the entire reserve. About 10% to 40% of the oil is not recovered due to a low sweep efficiency by waterflooding.

In the development of low-viscosity oil fields with productive reservoirs of moderate heterogeneity, the major cause of incomplete oil recovery is associated with the capillary forces. Oil recovery improving methods in such fields must be the elimination (total or partial) of capillary forces causing the problem.

If a field under development is characterized by a strong geological heterogeneity of productive reservoirs, then the oil yield improving methods should first of all include increase in the sweep efficiency by the water in waterflooding.

At the development of high-viscosity oil fields, a substantial fraction of non-recovered oil is associated with unfavorable mobility ratio. That is why the means of increasing the oil yield must be directed first of all to increasing viscosity of the displacing fluid and decreasing viscosity of the displaced oil.

Most of the known methods of increasing the oil yield includes the means of eliminating, in part or in full, the manifestation of one of the three aforementioned major causes of lowering efficiency of oil displacement from the productive reservoirs. However, the application of these means in the late development stages gives disappointing results. One of the major causes of the low efficiency of the methods is that they ignore natural tendency of hydrocarbon fluids to move when subjected to the action of gravity and capillary forces. This explains the attention of petroleum engineers and geologists to vibration and acoustic technologies capable of taking these tendencies into account.

Various options are created of base technologies and technical solutions for industrial implementation. Depending on technology and the applied technical means, vibration and acoustic methods may be used for the solution of the following tasks:

- Increasing the productivity of producing wells in which the application of conventional methods turned out impossible or of low efficiency;
- Increasing the oil and gas yield from the water-flooded low-productivity reservoirs;
- Increasing sweep efficiency during waterflooding.

Vibration technologies of increasing well productivity are most popular as a result of their relative simplicity and low cost. They are based on various techniques of passing the energy from the borehole vibration source to the reservoir by way of the borehole fluid. Due to a strong fading of vibrations in a liquid, this way of the energy transfer results in vibrations fading already at 1 m from the borehole walls. However, this is quite sufficient for efficient cleaning up of the borehole walls and the bottomhole zone from dirt and colmatage matter (formation damage). Vibration eliminates the blocking effect of residual gas, oil and water phases. Vibration initiates the fluid filtration in low-permeability zones. In low-permeability reservoirs, on achieving sufficiently strong pressure impulses, even a hydrofracking of the reservoir is possible. In the territory of the former Soviet Union (FSU), the application of action methods to bottomhole zones resulted in the production of over 20 million tons of oil.

One can expand the vibrations to greater distances from the vibration source. The effective encompassment of the productive reservoir around the initiating borehole may reach 12 km^2. The number of wells simultaneously encompassed by the action is 25 to 50 (depending on the development grid).

The development technique of increasing the reservoir's oil-gas yield, created by the authors, is the unique "vibro-seismic" method. It is cyclical areal action in the reservoir by the low-frequency vibrations in the frequency range corresponding to the reservoir resonance. The annual oil production from the test areas as a result of vibro-seismic action increased more than 60%. The effect lasted for 6 to 18 months and longer. An increase of the productive reservoir encompassment by the thickness was 30% to 35% and more. In some cases, the wells which produced earlier by a rod-operated plunger pump switched to a durable gushing regime with almost tenfold increase in the production rate. The action efficiency was defined not only by an increase of the total production but also by a decrease of water cut. In some wells, the water cut was decreased by 30% to 40%. As a result of applying this technique, more than 500 thousand tons of oil were produced.

Figure 1.1 shows the dependence of water cut in wells located at three different oil fields in Northern Caucasus several days after an earthquake. The earthquake epicenter was located at a distance of 130 to 170 km from the producing wells. In all three cases, a decrease in water cut and a respective increase in oil production was recorded in about a week after the earthquake. Nevertheless, on the average, decrease in the water cut was about 40% for all three wells for a considerable period of time. This and other similar facts led to extensive laboratory and field studies of artificial vibro-seismic enhancement of oil production. Theoretical and experimental aspects of this phenomenon and the mechanism of enhanced oil recovery due to the presence of elastic waves generated by surface-based vibrators have been studied during the last 40 years by Surguchev *et al.*

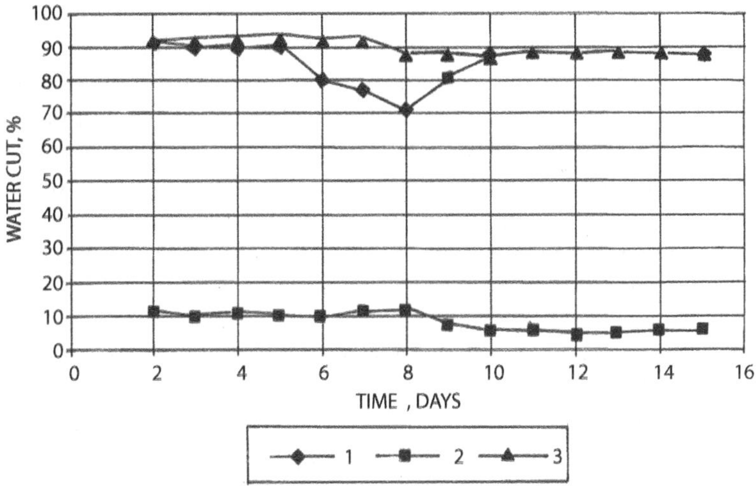

Figure 1.1 Decrease in the water cut in three wells after an earthquake: *1* = Makhachkalinskoe field, 170 km from the earthquake epicenter; *2* = Shamkal-Bulakskoe field, 130 km from the epicenter; *3* = Novo-Groznenskoe field, 130 km from the epicenter. All fields are located in the northern Caucasus.

(1984) [34], Simkin and Surguchev (1991) [31], Simkin (1992) [30], and Kouznetsov and Simkin (1994) [21].

In 1987, at one of Professor George V. Chilingar's Petroleum Engineering Classes at the University of Southern California, his star student from Iran, K. Majid Sadeghi gave a lecture on his research on vibrational dewatering of contaminated muds. Professor Chilingar asked Sadeghi: "Now that we can dewater muds, can we use this technology to increase the production of oil from the oil sands? Sadeghi answered: "Absolutely."

1.1 Origin and Migration of Oil

Under the influence of increasing overburden pressure and geothermal temperature, kerogen in argillaceous sediments will generate petroleum hydrocarbons by thermal decomposition. The chemical process of generation of oil has become well established (Welte, 1972 [37]). Many geologists believe that carrier water is necessary for the primary migration of oil (Hedberg, 1964 [16]). They consider that carrier water and oil migrate in the form of solution and/or emulsion. Interlayer water of montmorillonite released by transformation to illite during the late stage of diagenesis was considered to be essential for petroleum migration (Powers, 1967 [25]; Burst, 1969

[8]; Perry and Hower, 1972 [24]). On the other hand, Aoyagi and Asakawa (1980) [2] concluded that both the interlayer and interstitial water expelled during the middle stage of diagenesis were responsible for oil migration. On the other hand, McAuliffe (1966) [23] has objected to these opinions on the basis that oil is only very slightly soluble in water. Also, some geologists have favored the migration theories based on the movement of oil and gas due to the capillary phenomena, buoyancy effect, and gas expansion, which are generally independent of the movement of water (Dickey, 1975 [12]). Some migration of oil could also occur in a gaseous form (Chilingar and Adamson, 1964 [9]).

Based on extensive observations, shales composed of non-expandable clays such as kaolinite and illite did not act as source rocks, because of the absence of water necessary to push out the oil (Chilingar and Knight, 1960 [11]; Aoyagi *et al.*, 1975 [4]). Also, many undercompacted (overpressed) shales did not act as source rocks, because compaction mechanisms were not operative to squeeze the oil into the reservoir rocks.

It is necessary to establish what conditions existed during the primary migration of oil (from the source rocks to reservoir rocks) compared to those during the secondary migration of oil (during production) to make the former so much more efficient. Several possible explanations are presented below. Among them are (1) seismic activity (earthquakes), (2) intense electrokinetics, (3) Earth tides, (4) compaction, and (5) migration in a gaseous form.

1.1.1 Seismicity

According to the site http://www.iris.washington.edu/SeismiQuery/events.htm, 66 earthquakes with magnitudes 2 or higher occurred over the globe on average per night for the period from 01.01.2000 to 01.01.2010.

Assuming that the seismic activity of the planet Earth was the same as at the beginning of the 21st century, one can calculate how many earthquakes occurred during the Paleozoic, Mesozoic, and Cenozoic times:

- In the Paleozoic 69,507,200 million
- In the Mesozoic 27,429,750 million
- During the Cenozoic 7,924,150 million

On considering precursor activity of earthquakes, the minimum frequency oscillations in the number of earthquakes, and gravity attraction of the Sun and the Moon, then the volume of fluid movements due to dynamic forces in the migration routes in the geological past was immense (Fedin *et al.*, 2013 [14]).

The seismic vibrations (elastic oscillations) eliminate the blocking effects of the residual oil, gas, and water phases, enable movement through the low-permeability zones, and increase the areal and vertical sweep efficiency.

The physical theory behind the application of seismic vibration to increase production was discussed in detail by Beresnev and Johnson (1994) [5], Beresnev (2006) [6], Kouznetsov *et al.* (1998) [18], (2001) [19], (2002) [20], Pride *et al.* (2008) [26], and Fedin *et al.*, (2013) [14].

Hydraulic fracturing also can occur with high-pressure impulses within the low-permeability zones (Kouznetsov *et al.* (2002) [20]). In the FSU, the use of various vibration techniques resulted in an incremental oil production of 200 million bbls (see Kouznetsov *et al.* (2002) [20], Figure 1.3). Fedin *et al.* (2013) [14] reported additional recovery of millions of bbls of oil by using explosions (60-kg explosives) in producing wells.

1.1.2 Electrokinetics

Electric currents are prevalent in the Earth's crust and probably were amplified in the geologic past (Serruya *et al.*, 1967 [29]). Upon application of DC current, for example, the flow rate can increase many fold both in clays and sands (Amba *et al.*, 1965 [1]; Chilingar *et al.*, 1970 [10]). Thus, the presence of intense electric currents in the geologic past possibly facilitated primary migration of oil. Figure 1.2 shows the schematic diagram of electrokinetic double layer.

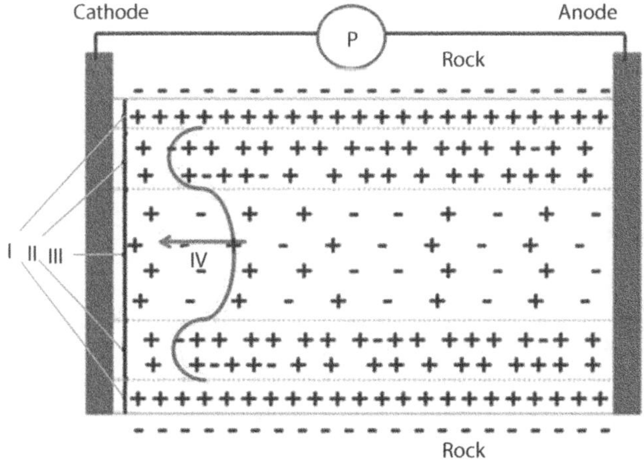

Figure 1.2 Schematic diagram of electrokinetic double layer (I: Immobile layer, II: Mobile double layer, III: Free water, IV: Velocity profile as envisioned by Dr. G. V. Chilingar. Solid curved line: velocity profile in a capillary. P-D.C., current power supply).

Theory of the electrokinetic flow is based on a double-layer theory as illustrated in Figure 1.2. According to Figure 1.2, the negatively charged surface of the clays attracts the positive ions of the aqueous medium, forming the immobile double layer. This immobile double layer is followed by a thick mobile layer with a predominance of positively charged ions (cations) with a few negatively charged ions (anions).

Upon application of direct electric current, the mobile double-layer moves toward the cathode, dragging the free water. A schematic diagram of EEOR field setup, as used by Wittle *et al.* (2008) [38] and (2011) [39] in Canada and California oil fields, is presented in Figure 1.3.

In case where imposed electric potential gradient, E, is in the same direction as the pressure drop (Δp), the total flow rate (q_t) upon application of D.C. current is as follows:

$$q_t = \frac{A \times k \times \Delta p}{\mu \times L} + \frac{A \times k_e \times E}{\mu \times L} \tag{1.1}$$

where k is the hydrodynamic permeability; $k_e = \frac{D \times \varsigma}{4 \times \pi \times F}$ is the electrokinetic permeability; Δp is the pressure drop; μ is the viscosity; L is the length

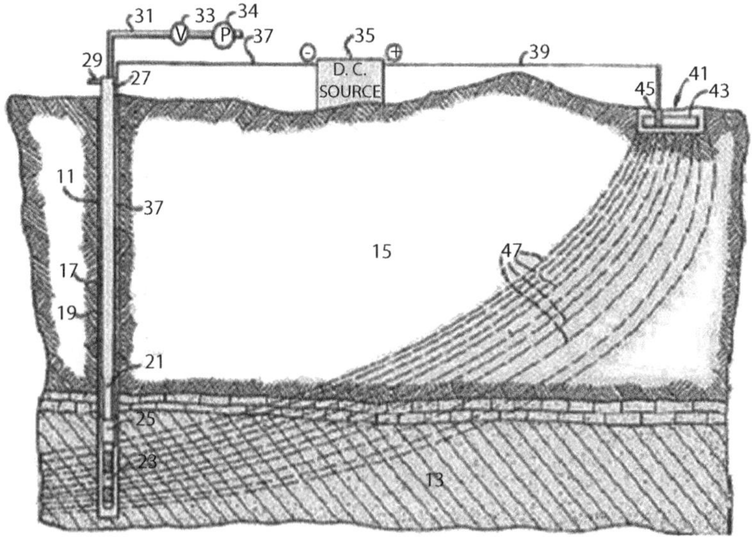

Figure 1.3 Schematic diagram of EEOR (Electrical Enhanced Oil Recovery) electrode arrangement in oil fields as used by Wittle *et al.* (2008) [38] and (2011) [39], Ambah *et al.* (1965) [1], and Titus *et al.* (1985) [35].

of porous media; A is the cross sectional area of porous media; ζ is the zeta potential; F is the formation resistivity factor (Archie's); E is the electrical potential; and D is the dielectric constant.

Equation (1.1) can be presented in a dimensionless form by normalizing the flow rates and, thus, eliminating μ, A, and L terms (Chilingar *et al.*, 1970 [10]):

$$\frac{q_t}{q_i} = 1 + \frac{k_e \times E}{k \times \Delta p} \tag{1.2}$$

and

$$\frac{q_t - q_i}{q_i} = \frac{k_e \times E}{k \times \Delta p} \tag{1.3}$$

where q_i is the initial hydrodynamic stabilized flow rate:

$$q_i = \frac{A \times k \times \Delta p}{\mu \times L} \tag{1.4}$$

Thus, the possible existence of strong D.C. current in the rocks during the geologic past could have facilitated the movement of fluids (Ambah *et al.*, 1965 [1]).

1.1.3 Earth Tides

In the classic book on wettability, Donaldson and Alam (2013) [13] pointed out that Earth tides activate the motion of fluids by continued expansion and contraction as simple harmonic oscillation resulting from diurnal tides. This effect was much greater earlier in geologic history because the moon was much closer to the Earth. Currently, the moon is moving away from the Earth at ≈3 cm/yr. Thus, the Earth tides were more rapid and energetic, resulting in greater dilation and the compression of rocks. According to Donaldson and Alam, in a 10-m-thick reservoir, with a radius of 100 m, an elastic change of porosity of 0.1% will result in an oscillatory motion of 294 bbl of fluid. Such motions could reduce the oil saturation to zero over geologic time, leaving only a trace amount of oil behind.

1.1.4 Compaction

In order to investigate some problems involved in the primary migration of oil, Aoyagi *et al.* (1985) [3] compacted Na-monmorillonite clay mixed with seawater and crude oil for 25 days under a pressure of 1,000 kg/cm^2 and a temperature of 60°C. The proportion of oil in the expelled liquid increased with time. The porosity of the compacted sample decreased from 81% to ≈26%. These authors concluded that the primary migration of oil from source rocks to reservoir rocks occurred chiefly during the late compaction state (ϕ = 10%–30%). Considering the fact that tectonic activity during the primary migration of oil was intense, the oil was simply squeezed out (expelled) from the source rock.

1.1.5 Migration in a Gaseous Form

Kapelyushnikov (1954) [17], Gerber and Dvali (1961) [15], Zaks (1952 [40], 1955 [41]), Zhuze and Yushkevich (1959) [42], and Chilingar and Adamson (1964) [9] studied the possibility of oil migration in a gaseous form. Based on some experimental work, Chilingar and Adamson (1964) [9] concluded that some migration of petroleum could have occurred in a gaseous form in the geologic past at high temperatures.

It is important to remember that with increasing polarity of oil, the relative permeability to oil increases and that to water decreases (water cut decreases) (Sinnokrot and Chilingar, 1961 [32]).

Gerber and Dvali (1961) [15] subjected to extraction four samples of rocks with CO_2 at pressures ranging from 100 to 400 kg/cm^2 and temperatures of 40°C to 74°–90°C. Their findings can be summarized as follows:

1. The bitumens dispersed in rocks and, having composition related to crude oil, can dissolve in compressed gases and migrate together with them.
2. At 200 to 400 kg/cm^2 pressure and temperature of 40°C, in dynamic conditions, CO_2 can extract not only oils and tars but also asphaltenes and porphyrins from bituminous shales of Ukhta (Russia).
3. The kerosene and the main oil fraction of bitumens can be extracted from the rocks at pressures of 100 to 200 kg/cm^2 and temperature of 40°C.
4. Mobile, syngenetic bitumens in the source rocks can dissolve just as easily in the compressed CO_2 gas as does petroleum from oil-saturated rocks.

Kapelyushnikov (1954) [17], who studied the P-V-T relationship of oil, gas, and water at pressures up to 500 atmospheres and temperatures up to 100°C, concluded that gaseous oil/gas mixtures move toward the low-pressure areas, dropping the tars and asphaltenes first, followed by medium and lighter components. He also found that at critical pressures, the water and salts present in the oil-bearing strata are also transported together with the gas toward the low pressure areas.

1.2 Seismic Vibration Techniques

The use of various seismic vibration techniques in Russia resulted in incremental oil production of millions of tons. There are several seismic vibration techniques for transmitting energy into the reservoir, with oscillations over long distance. The area of productive reservoir around the well-being stimulated may be as great as 12 km^2. The number of wells simultaneously subjected to the treatment can range up to 50 (depending on the well spacing).

The technique for increasing the oil yield of high water cut and low oil production reservoirs consists of subjecting the reservoir area to cyclic low-frequency (5 to 90 HZ) elastic oscillations within the frequency range corresponding to the reservoir resonance (Kouznetsov *et al.*, 1998 [18]). The results obtained for during seven years of using this seismic vibration technique in different areas in Russia, on the average, annual oil production increased by more than 60%. The duration of the seismic vibration effect was 6 to 18 months and sometimes longer. The vertical reservoir sweep increased by 35%. In some cases, the wells that were producing by artificial lift (sucker rod pumps) retuned to natural flow with almost a tenfold rate increase. The effect was evaluated not only by an increase in the total production rate but also by a decrease in the waste cut. In some wells, the water cut decreased by 30% to 40% (Kouznetsov *et al.*, 1998 [18]; Simkin and Surguchev, 1991 [31]).

1.2.1 Producing Well Experiments

Field experiments were conducted using vibrators located at the Earth's surface in the vicinity of oil-producing wells. Vibro-energy generated by vibrators traveled down as elastic waves to the oil-bearing formation. The goal of the experiments was to study the effect of elastic waves on the relative permeability to oil and to water, water saturation, oil production rate, and the rate of oil displacement by water. Generally, it was found that the presence of elastic waves increased the relative permeability to oil, decreased the relative permeability to water, the decreased the water cut, and enhanced the oil production. In terms of industry application, the use of vibroenergy for enhancement of oil recovery is acceptable only if the amount of generated vibro-energy is smaller than the amount of energy contained in the additionally recovered oil.

The authors studied the energy balance between the generated vibro-energy and energy contained in the additional oil produced by using vibrators. Calculations show that the amount of vibro-energy applied to one unit of rock volume does not exceed 1.5% of the energy contained in the oil present in this unit volume calculations were based on the fact that the energy content of 1 kg of oil is 4.2 = 107 J. Therefore, for this process to be economically feasible, it is necessary to increase the oil production by a minimum of 1.5% (Vahitov and Simkin, 1985 [36]). To study feasibility of enhanced oil recovery based on the use of vibro-energy, field experiments were conducted in several old oil field in Russia, Uzbekistan, and Kirgizstan.

1.2.2 Mechanisms of Interaction of Fluid Flow With the Vibro-Energy in Porous Media

Laboratory and field experiments demonstrate the dependence of fluid flow and character of oil displacement by water on application of vibro-energy. Results of experiments demonstrate that k_{ro}/k_{rw} and the rate of water displacement by water increase on application of vibro-energy. The effect of vibro-energy on fluid flow the in porous media is defined by a number of nonlinear factors. Application of vibro-energy causes periodic or quasi-periodic movements of oil and water phases in pore channels with periodically changing directions. Due to these periodic movements, molecules of oil and water stick to a lesser degree to the solid phase. Accelerations of the oil phase and water phase are related to each other as (Kouznetsov and Simkin, 1994 [21]):

$$\frac{\rho_o}{\rho_w}\frac{\partial^2 x_o}{\partial t^2} = \frac{\partial^2 x_w}{\partial t^2}, \tag{1.5}$$

where ρ_o and ρ_w are densities of oil and water, respectively; x_o and x_w are distances traveled by the oil droplet and water droplet, respectively; and t is the time.

Equation (1.5) shows that the acceleration of the oil phase, which has a lower density than water, will be larger than the acceleration of the water phase. The lower the density of oil, the higher will be the ratio of acceleration of oil to that of water due to applied vibro-energy. Thus, the time of possible static contact of oil molecules with a solid phase will decrease with decreasing density of oil. In turn, this will lead to a decrease of adhesion of oil molecules to the walls of the pore channels and better mobility of the oil phase. Mobility of the water will also increase, but to a lesser degree than that of the oil.

Periodic movements of the oil and water in pore throats and variable pressure gradient caused by the applied vibro-energy lead to the destruction of water films blocking fluid flow through narrow pore throats. This leads to a general increase of the relative permeability both to oil and to water. The molecules of oil are much larger than those of water. Therefore, destruction of water films sealing pore throats and increasing the size of the open part of pore throats should produce a larger effect on the relative permeability of rock to oil than that to water.

The relative permeability to oil also increases due to reduction in the interfacial tension and contact angle between oil and water on application of vibroenergy. As a result, the size of oil globules decreases.

The results of vibro-stimulation tests for enhanced oil recovery, using powerful surface-based vibro seismic sources, indicate that the rate of displacement of oil by water increases and percentage of non-recoverable residual oil decreases if the vibroenergy is applied to the porous medium containing oil (Kuznetsov *et al.*, 2021 [68]). Tests on sandpacks showed an increase of degassing rate due to application of vibro-energy. Results of both laboratory and field tests of the proposed enhanced oil recovery method showed an increase in the recovery of oil and oil-water ratio. Decrease in water cut is caused by the reduction in the oil-water interfacial tension and increase in the relative permeability to oil and to water.

This proposed vibro-seismic methodology will serve as an additional enhanced oil production recovery technique. Vibro-energy reduces interfacial tension, increases the relative permeability to oil, and increases the produced oil-water ratio. Possibly, this technique can be used in conjunction with other methods, such as thermal recovery, application of direct electric current, and chemical floods.

In the, 1980s, Donaldson, Chilingar, and Yen published two books on Enhanced Oil Recovery (I: Fundamental and Analyses and II: Processes and Operations) (Donaldson *et al.*, 1985 [57]; Donaldson *et al.*, 1989 [58]).

Inasmuch as both acoustic/vibrational and EEOR technologies are used by the authors in revitalizing abandoned oil fields, decision was made to invite the foremost experts on EEOR (Professor's Donald Hill and Muhammad Haroun) to make contribution to this book.

References and Bibliography

1. Ambah, S.A., Chilingar, G.V., Beeson, C.M., Application of electrokinetic phenomena in civil and petroleum engineering. *N Y Acad. Sci.*, 118, 585–602, 1965.

2. Aoyagi, K. and Asakawa, T., Primary migration theory of petroleum and its application to petroleum exploration. *Org. Geochem.*, 2, 33–43, 47–56, 1980.
3. Aoyagi, K., Kazama, T., Sekiguchi, K., Chilingarian, G.V., Experimental compaction of Na-Montmorillonite clay mixed with crude oil and seawater. *Chem. Geol.*, 49, 385–392, 1985.
4. Aoyagi, K., Kobayashi, N., Kazama, T., Sawa, T., Sasaki, S., Compaction of clays under high pressure and programming temperature. *J. Jpn. Assoc. Pet. Technol.*, 40, 111–116 (in Japanese), 1975.
5. Beresnev, I. and Johnson, P., Elastic-wave stimulation of oil production: A review of methods and results. *Geophysics*, 59, 1000, 1994.
6. Beresnev, I., Theory of vibratory mobilization of nonwetting fluids entrapped in pore constrictions. *Geophysics*, 71, 2006.
7. Buryakovsky, L., Chilingar, G.V., Rieke, H.H., Shin, S., *Fundamentals of the Petrophysics of Oil and Gas Reservoirs*, p. 374, Wiley- Scrivener, Salem, MA, 2012.
8. Burst, J.F., Diagenesis of Gulf Coast clayey sediments and its possible relation to petroleum imigration. *Am. Assoc. Pet. Geol. Bull.*, 53, 73–93, 1969.
9. Chilingar, G.V. and Adamson, L.G., Does some migration of oil occur in a gaseous form? *Int. Geol. Congr. Sec. 1, Part 1*, 64–70, 1964.
10. Chilingar, G.V., El-Nassir, A., Stevens, R.G., Effect of direct electrical current on permeability of sandstone cores. *J. Pet. Technol.*, 22, 830–836, 1970.
11. Chilingar, G.V. and Knight, L., Relationship between pressure and moisture content of kaolinite, illite and montmorillonite clays. *Am. Assoc. Pet. Geol. Bull.*, 56, 796–802, 1960.
12. Dickey, P.A., Possible primary migration of oil from source rocks in oil phase. *Am. Assoc. Pet. Geol. Bull.*, 59, 345–377, 1975.
13. Donaldson, E.C. and Alam, W., *Wettability*, p. 336, Gulf Publishing Company, Huston, TX, 2013.
14. Fedin, L.M., Fedin, K.L., Fedin, A.K., *Basics of Increasing Heavy Oil Recovery*, p. 112, Dolyam, Simpheropol, 2013.
15. Gerber, M.I. and Dvali, M.F., *Natural Compressed Gases as a Probable Factor in Migration of Oil from Source Rock*, p. 83, GosTopTechIzdat, Leningrad, 1961.
16. Hedberg, H.D., Geological aspects of origin of petroleum. *Am. Assoc. Pet. Geol. Bull.*, 48, 1755–1803, 1964.
17. Kapelyushnikov, M.A., Migration and accumulation of dispersed petroleum in sedimentary rocks. *Dokl. Akad. Nauk USSR*, 99, 078, 1954.
18. Kouznetsov, O.L., Simkin, E.M., Chilingar, G.V., Katz, S., Improved oil recovery by application of vibro-energy to waterflooded sandstones. *J. Petrol. Sci. Eng.*, 19, 191–200, 1998.
19. Kouznetsov, O.L., Simkin, E.M., Chilingar, G., *Physical Bases of Acoustic and Vibrational Effects on Oil and Gas Deposits*, p. 260, Izd. Mir, Moskow, 2001.
20. Kouznetsov, O.L., Simkin, E.M., Chilingar, G.V., Gorfunkel, M., Robertson, J., Seismic techniques of enhanced oil recovery: Experimental and field results. *Energy Sources*, 22, 877–890, 2002.

21. Kouznetsov, O.L. and Simkin, E.M., *Transformation and Interaction of Geophysical Fields in the Lithosphere*, p. 271, Oxford and IBH Publ, 1994.
22. Langnes, G.L., Robertson, J.O., Chilingar, G.V., *Secondary Recovery and Carbonate Reservoirs*, p. 304, Elsevier, New York, NY, 1972.
23. McAuliffe, C., Solubility in water of paraffin, cycloparaffin, olefin, acetylene, cycloolefin, and aromatic hydrocarbons. *J. Phys. Chem.*, 70, 1267–1275, 1966.
24. Perry Jr., E.A. and Hower, J., Late stage dehydration in deeply-buried pelitic sediments. *Am. Assoc. Pet. Geol. Bull.*, 56, 2013–2021, 1972.
25. Powers, M.C., Fluid-release mechanisms in compacting marine mudrocks and their importance in oil exploration. *Am. Assoc. Pet. Geol. Bull.*, 51, 1240–1254, 1967.
26. Pride, S., Flekkøy, E., Aursjø, O., Seismic stimulation for enhanced oil recovery. *Geophysics*, 73, 23–35, 2008.
27. Rieke III, H.H. and Chilingarian, G.V., *Compaction of Argillaceous Sediments*, p. 424, Elsevier, Amsterdam, 1974.
28. Sadeghi, K.M., Sadeghi, M.-A., Jang, L.K., Chilingarian, G.V., Yen, T.F., A New Bitumen Recovery Technology and Its Potential Application to Remediation of Oil Spills. *J. Pet. Sci. Eng.*, 8, 2, 105–117, 1992.
29. Serruya, C., Picard, L., Chilingarian, G.V., Possible role of electric currents and potentials during diagenesis (Electrodiagenesis). *Sediment. Petrol.*, 37, 695–698, 1967.
30. Simkin, E.M., Hydrodynamic description of oil displacement by water process within the elastic vibration field, in: *Proc. Int. Conf. Flow through Porous Media: Fundamentals and Reservoir Engineering Applications*, pp. 171–175, 1992.
31. Simkin, E.M. and Surguchev, M.L., Advanced vibroseismic technique for waterflooded reservoir stimulation, mechanism and test results. *Sixth European Symp. on Improved Oil Recovery*, 1991.
32. Sinnokrot, A.A. and Chilingar, G.V., Effect of polarity of oil and presence on carbonate particleson relative permeability of rocks, in: *The Compass of Sigma Gamma Epsilon*, vol. 38, pp. 115–120, 1961.
33. Surguchev, M.L., Kouznetsov, O.L., Simkin, E.M., *Hydrodynamic, Acoustic, and Thermal Cyclical Processing of Oil Layers*, p. 320, Nedra Publ., Moscow, 1975.
34. Surguchev, M.L., Zheltov, U.V., Simkin, E.M., *Physicochemical Microprocesses in Oil-Gas-Bearing Formations*, p. 215, Nedra Publ., Moscow, 1984.
35. C.H. Titus, J.K. Wittle, C.W. Bell, Apparatus for passing electric currents through an underground formation. US Patent 4495990 29 Jan, 1985.
36. Vahitov, G.G. and Simkin, E.M., *Utilization of Physical Fields for Enhancement of Oil Recovery*, p. 231, Nedra Publ., Moscow, 1985.
37. Welte, D.H., Petroleum exploration and organic geochemistry. *J. Geochem. Explor.*, 1, 117–136, 1972.
38. Wittle, J.K., Hill, D.G., Chilingar, G.V., Direct current stimulation for heavy oil production. Paper 2008-374, in: *Second World Heavy Oil Congress*, Edmonton, March 10-12, 2008.

39. Wittle, J.K., Hill, D.G., Chilingarian, G.V., Direct electric current oil recovery (EEOR) - A new approach to enhancing oil production. *Energ. Source Part A*, 33, 805–822, 2011.
40. Zaks, S.L., Effect of rock and bound water on value of pressures at which oil/gas system is transformed into one-phase gas condition. *Dokl. Akad. Nauk USSR*, 86, 1017–1020, 1952.
41. Zaks, S.L., Toward question of migration and accumulation of petroleum. *Dokl. Akad. Nauk USSR*, 105, 332–334, 1955.
42. Zhuze, T.P. and Yushkevich, G.N., Solubility of oil and its heavy fractions in compressed gases. *Trudy Inst. Nefti Akad. Nauk*, 13, 626–274, 1959.
43. V.P. Dyblenko, O.L. Kuznetsov, Ya.V. Eremenko, R.Ya. Sharifullin *et al.*, International application PCT/RU No. 2010120080. Method of productive reservoir treatment and downhole equipment for its implementation. Publ. in B.I. 2011. No. 33.139. Patent No. 2258803 RF, 2005.
44. V.P. Dyblenko, R.Ya. Sharifullin, A.P. Lysenkov *et al.*, Patent No. 2261990 RF, 2005.
45. V.P. Dyblenko, I.A. Tufanov, A.V. Ovsyukov *et al.*, Patent No. 2228437 RF, 2004.
46. V.P. Dyblenko, A.P. Lysenkov, Yu.S. Ashchepkov *et al.*, Patent No. 2366806 RF, 2009.
47. E.Yu. Marchukov, V.G. Tarasenko, V.I. Zhdanov, V.G. Bazarov, V.P. Dyblenko *et al.*, Patent No. 2087756 RF, 1997.
48. V.P. Dyblenko, E.Yu. Marchukov, V.I. Zhdanov, I.A. Tufanov, Patent No. 2144440 RF, 2000.
49. V.P. Dyblenko, A.P. Lysenkov, Yu.S. Ashchepkov *et al.*, Patent No. 2366806 RF, 2009.
50. V.P. Dyblenko and I.A. Tufanov, . Patent No. 2258803 RF, 2005.
51. V.P. Dyblenko, I.A. Tufanov, A.V. Ovsyukov *et al.*, Patent No. 2228437 RF, 2004.
52. V.P. Dyblenko and I.A. Tufanov, Patent No. 2258803 RF, 2005.
53. V.P. Dyblenko and V.S. Evchenko, Patent No. 2349741 RF, 2009.
54. V.P. Dyblenko, O.L. Kuznetsov, R.S. Khisamov *et al.*, Patent No. 2247828 RF, 2005.
55. V.P. Dyblenko, O.L. Kuznetsov, R.S. Khisamov *et al.*, Patent No. 2268996 RF, 2006.
56. V.P. Dyblenko, R.Ya. Sharifullin, I.A. Tufanov *et al.*, Patent No. 2291954 RF, 2007.
57. Donaldson, E.C., Chilingar, G.V., Yen, T.F., *Enhanced Oil Recovery, I: Fundamental and Analyses*, Developments in Petroleum Science 17A. Elsevier, New York, 357 pp, 1985.
58. Donaldson, E.C., Chilingar, G.V., Yen, T.F., *Enhanced Oil Recovery, II: Processes and Operations*, Developments in Petroleum Science 17B. Elsevier, New York, 604 pp, 1989.

59. Chilingar, G.V., Katz, S., Fedin, L., Fedin, A., Khilyuk, L., Liu, M., Primary Versus Secondary Migration of Oil. *J. Sustain. Energy Eng.*, 2, 4, 323, April 2015.
60. Sadeghi, K.M., Sadeghi, M.-A., Chilingarian, G.V., Yen, T.F., Developing a New Method For Bitumen Recovery from Bituminous Sands Using Ultrasound and Sodium Silicate. *Geologiya Nefti i Gaza/(Geology Oil Gas)*, 8, 5357, 1988.
61. Sadeghi, K.M., Sadeghi, M.-A., Yen, T.F., A Novel Extraction of Tar Sands by Sonication with the Aid of In-Situ Surfactants. *Energy & Fuel*, 4, 604608, 1990.
62. Sadeghi, K.M., Lin, J.R., Yen, T.F., Sonochemical Treatment of Fuel Components. *Energy Sources*, 16, 3, 439–449, 1994.
63. Sadeghi, M.-A., Sadeghi, K.M., Kuo, J.-F., Jang, L.-K., Yen, T.F., Sonication Method and Reagent for Treatment of Carbonaceous Materials. U.S. Patent 4,891,131, January 2, 1990 (38 Claims).
64. Chilingar, G.V. and Haroun, M., *Electrokinetics for Petroleum and Environmental Engineers*, p. 240, Scrivener Publishing (Wiley, New Jersey, 2014.
65. Wilkinson, D. and Willemsen, J.F., Invasion percolation: A new form of percolation theory. *J. Phys. A.: Math. Gen.*, 16, 3365, 1983.
66. Roberston, J.O. and Chilingar, G.V., *Environmental Aspect of Oil and Gas Production*, p. 394, Scrivner Publication - Wiley, 2017.
67. Bablyan, G.A., *Questions of Mechanism of Oil Recovery*, p. 254, Aznefteizdat, Baku, 1956.
68. Kuznetsov, O.L., Dyblenko, V.P., Chilingar, G.V., Sadeghi, K.M., Marchukov, E.Y., Sharifullin, R. Y., Chirkin, I.A., Simkin, E.M., *Oscillations and Waves in Fluid-Saturated Geological Environments: Phenomena, Processes, Patterns. Wave Paradigm.* Izhevsk: Institute of Computer Sciences, Moscow (In Russian), 420 pp, 2021.

2

Wave Spreading Patterns in the Porous Media

2.1 Spread of Vibration in Reservoir

Wave spreading in the media with fading is best studied for the conditions of low amplitude flat waves within a boundless and uniform porous medium saturated with a viscous fluid. Usually, two directions are identified in these studies.

The first direction involves construction of idealized models of porous medium. In such models, the solid phase is viewed as a system of variously packed grains with fluid-filled voids. It is assumed that the effect of a relative motion of the fluid and rock matrix on the wave spreading is negligible. This assumption is substantiated for low frequencies although exact frequency criteria of its applicability are absent. It is believed that this approach is applicable in seismology and seismic exploration as it allows for an approximate computation of major parameters of the elastic vibration field for idealized porous medium models with a certain grain packing at the assigned thermodynamic conditions.

Another study direction is based on mechanics of a continuous medium and on thermodynamics of irreversible processes. This direction was first described by Frenkel [24] and subsequently expanded by Bio and Rahmatullin [18], Nikolayevsky [15], Nigmatullin [14], and others. This direction presents substantially broader opportunities. In describing a porous medium by a set of thermodynamic variables ("observed" mechanical and concealed parameters), it is possible to determine various dissipative functions, to evaluate the system behavior in time and the relaxation effects. The elastic constants in the equations of wave spreading, according to the conformity principle, may be replaced by operators, and this way, various processes of absorption and dispersion may be accounted for. For instance, it may be processes associated with surface effects, dissipative phenomena directly in the solid phase or liquid, etc. Within a framework

George V. Chilingar, Kazem Majid Sadeghi, and Oleg Leonidovich Kuznetsov. Acoustic and Vibrational Enhanced Oil Recovery, (19–54) © 2022 Scrivener Publishing LLC

of this approach, temperature effects may also be considered, effects of porous medium compressibility changes at changing of frequency and other relaxation processes resulting in fading of the elastic waves.

This theory gives the fading coefficient values underestimated by two or three orders of magnitude at low frequencies compared with the values measured in real media. However, this discrepancy may be imaginary. It may be explained by that the preconditions of the theory relatively boundless and uniform nature of the media do not match the observation conditions at low frequencies when, due to the need of using large measurement bases, absorbing properties of a rock massif with a characteristic dimension no smaller the wave length is evaluated. Thus, the studied medium may not be viewed as uniform relative to its physical properties, and the standard mechanism of viscous friction becomes insufficient (at least in the low frequency area) for the description of elastic waves' fading patterns in saturated porous media.

To confirm this, an attempt was undertaken, remaining within the framework of this theory, to evaluate the effect of accidental nonuniformities in the medium by introducing a transformation energy exchange mechanism between different wave types [12]. The predicted absorption coefficients and their correlation with frequency well agreed with the experimental data. A more general approach was based on a well-developed procedure of averaging differential equations with rapidly oscillating coefficients [7]. The obtained results enabled a substantial expansion of applicability boundaries of the fading transformational mechanism for a sufficiently broad spectrum of nonuniformity values existing in the real media. This allowed explaining most substantial discrepancy between the theoretical conclusions and experimental data, which indicate the permanent measured experimental fading decrement within a wide frequency range. Also explained was experimentally observed low increase in wave spread velocity with increasing frequency. Frenkel-Bio-Nikolayevsky equations described a linear approximation of wave spreading. With an increase of the source vibration amplitude, the appearance in the medium of nonlinear effects resulting in the formation of stable wave fronts, increase of vibration amplitude far from the source and other phenomena is possible.

Another factor affecting the wave spreading is various boundaries and rock stratification in the productive reservoirs. Experimental observations showed that in a porous medium saturated with oil, gas, or water, reflection coefficients from various fluids' contact surfaces are comparable with reflection coefficients by geological boundaries separating different lithologies [13]. When boundaries are available with clearly expressed reflecting properties, interactions appear and energy exchange between different wave types.

At certain conditions, the medium stratification results in resonance, wave phenomena [27]. The influence of separation boundaries on wave spreading in the saturated porous media was studied by several authors. The analytical description of the reflection and refraction processes at the boundary of such media is complex. That is why only cases of flat wave normal incidence on the separation boundary of two media were studied or cases of an arbitrary incidence angle on the free surface (contact with vacuum). The problem of a flat wave at-angle incidence on the boundary corresponding with the contact of different fluids (oil, gas, and water) in a porous medium was numerically solved by Berson [1]. Insufficient knowledge of reflection and refraction processes for this case was due to a great number of parameters in the Frenkel-Bio-Nikolayevsky equations, the parameters characteristic for any specific situation. A special problem was selection of the source data for possible calculations as it was necessary, within several calculation situations, to identify specific features of wave transformation processes at the boundaries.

Within the framework of this theory, flat harmonic vertically and horizontally polarized waves within isotropic saturated layer limited by absolutely rigid semi-space were also studied [7]. Analytic formats of the fading coefficients were derived. Analytic formats for phase and group velocities of normal waves within the layer were derived. The presence of critical normal waves was established, similar to the appropriate concept of the wave theory in purely elastic wave guides near which the monotonous behavior of group velocities and fading coefficients disrupted. Thus, approximate analytical approach for the consideration of interaction between a layer (bed) with enclosing rocks and major parameters of flat monochromatic waves in the layer was used. Also, the escape of some vibration energy into the overlying and underlying rocks was considered.

Bio determined the following boundaries of a low-frequency area where theoretic of results using both directions have been in a good agreement:

$$f < 0.1\eta m/2\pi k\rho f$$

where η *and* ρf are, respectively, fluid viscosity and density; and k and m are rock permeability and porosity, respectively.

Sound dispersion in a micro-nonuniform medium is usually tied with the relaxation processes. These processes are leading to value leveling of some thermodynamic parameter ξ which depends on the wave pressure (or tension) in the enclosing medium and in inclusions. This leveling usually occurs through processes of heat conduction and diffusion through the surface of inclusions and is described by a relaxation time τ_r. For the frequencies $\omega >> 1\ \tau_r$, the dispersion law may look like follows [29]:

$$K \propto \frac{1}{C_\infty}\omega\text{-}\lambda i\sqrt{i\omega} \tag{2.1}$$

where K is the frequency limit of the sound velocity; λ is the constant depending on the difference between the sound velocity limit values at high and low frequency and on τ_r; and K is the complex wave vector.

A similar dispersion law is typical also for nonuniform porous media filled with a viscous liquid due to the sound wave dispersion on the surface of nonuniformities and their conversion to rapidly fading viscous waves [28].

Let us assume that prior to the time moment $t = 0$, there were no wave disturbances in the medium. Then, a flat wave spreading from the side of positive x-axis values, according to the dispersion law Equation (2.1), is described by the following causal first-order wave equation:

$$\partial_x u - \frac{1}{C_\infty}\partial_t u - \frac{\lambda}{\sqrt{\pi}}\int_0^t dt' \frac{1}{\sqrt{t'}}\partial_t u(t-t') = 0 \tag{2.2}$$

The last term in the left part of Equation (2.2) may be shown as $\varphi_{1/2}(t) * \partial_t u(t)$, where the asterisk indicates packing of two functions and $\varphi_{1/2}(t) = \frac{\lambda}{\sqrt{\pi}} t^{-1/2}$.

The core of the packing operator Equation (2.2) is a particular case of a more common Abel core:

$$\varphi_a(t) = \lambda \frac{t^{-\alpha}}{\Gamma(1-\alpha)} \tag{2.3}$$

where Γ is the gamma function.

The core of Equation (2.3) kind may also be used for the description of more complex media. A singular function $\varphi_\alpha(t)$ at $t = 0$ is integratable if $0 < \alpha < 1$.

Next, the authors exam the complete second order equation for flat single-dimension waves extracted from Equation (2.2) and generalized according to Equation (2.3) for mathematical modeling of wave spreading within micro-nonuniform media. The causal single-dimensional second-order wave equation is

$$\partial_{tt}u - \partial_{xx}u + (2\varphi_\alpha + \varphi_\alpha * \varphi_\alpha) * \partial_{tt}u = 0 \qquad (2.4)$$

2D and 3D scalar wave equations of the medium model under review is easy to write down for an isotropic case by way of replacing the operator δ_{xx} with the dimension-appropriate Laplace operator. These equations obviously describe the wave spreading in some viscous-elastic medium. The right part of Equation (2.4) is different from zero at the availability of diffuse sources. The fundamental solution of Equation (2.4) must satisfy initial conditions

$$u(0,x) = 0, \quad \partial_t u(0,x) = \delta(x) \qquad (2.5)$$

where $\delta(x)$ is Dirac delta function.

This fundamental solution enables the presentation of a general solution Equation (2.4) in the form of modified Duhamel integral, and in and of itself, it describes a wave impulse excited by an instantaneous point source. As Equation (2.4) is asymptotic at $\omega >> 1/\tau_r$, this model is applicable only near the wave impulse front during the time period or at a distance ($C\infty = 1$) smaller than τ_r (which at accepted dimensionless units equals one).

The wave impulse front that emerged at the moment $t = 0$ and at the point $x = 0$ reaches the points $+x$ and $-x$ at the moment $t = |x|$. Solution of the problem Equations (2.4) and (2.5) represents the wave after this moment if to record $t = |x| + \tau$ and $0 < \tau < \tau_{r=1}$:

$$|x|^{\frac{1}{\alpha}-1}\int_0^{\tau/|x|^{\frac{1}{\alpha}}} \varphi_\alpha(\zeta) * f_\alpha(\zeta)d\zeta = \frac{|x|^{\frac{1}{\alpha}-1}}{\alpha}\int_0^{\tau/|x|^{\frac{1}{\alpha}}} \zeta f_\alpha(\zeta)d\zeta \qquad (2.6)$$

where the function $f_\alpha(\zeta)$ may be represented through an inverse Laplace transform:

$$f_\alpha(\zeta) = L^{-1}_{s\to\varsigma}\left[\frac{1}{2}\exp(-\lambda \cdot s^\alpha)\right] \qquad (2.7)$$

Its expansion in a series is

$$f_\alpha(\zeta) = H(\zeta)\sum_{k=1}^{\infty} \frac{(-1)^{k-1}\lambda^k \sin(\pi k\alpha)}{k!} \cdot \frac{\Gamma(k\alpha+1)}{\tau^{k\alpha+1}} \qquad (2.8)$$

where $H(\zeta)$ is Heaviside step function. In a special case at $\alpha = ½$, the series (2.8) converges to

$$f_{1/2}(\zeta)=H(\zeta)\frac{\lambda}{4\sqrt{\pi}}-\zeta^{3/4}\exp(-\lambda^2/4\zeta) \tag{2.9}$$

Due to nonuniform convergence of the series Equation (2.8), may also be found asymptotic formula for $f_\alpha(\zeta)$. It is determined using the saddle point method:

$$f_\alpha(\zeta)\approx\frac{(\lambda\alpha)^{1/2(1-\alpha)}}{2\sqrt{2\pi(1-\alpha)}\zeta^{(2-\alpha)/2(1-\alpha)}}\exp\left[-\frac{1-\alpha}{\alpha}\left(\frac{\lambda\alpha}{\zeta^\alpha}\right)^{1/(1-\alpha)}\right] \tag{2.10}$$

This correlation becomes equal to the exact expression Equation (2.9) at $\alpha = 1/2$.

The function $f_\alpha(\zeta)$ defined by the equality Equation (2.7) converges to $\delta(\zeta)$ at $\lambda \to 0$. There is the equivalent representation of the last term in expression Equation (2.6) for cores of Equation (2.3) type. The packing $\varphi_\alpha(\zeta)^* f_\alpha(\zeta)$ may be computed explicitly for this case and the left part of Equation (2.6) may then be written in a form more convenient for computation as it no longer includes the second integral. The function t = ($|x|$ + $\tau|x|$) is smooth and is faster tending to zero with the approach to the front corresponding to $\tau = 0$ than nay power of τ, remaining infinitely differentiable which is obvious from (2.10).

The trivariate Green function for the medium model under review represents a solution of the spherically symmetric Cauchy problem [30]:

$$\partial_{tt}G_\alpha - \Delta G_\alpha + (2\varphi_\alpha + \varphi_\alpha{*}\varphi_\alpha){*}\partial_{tt}G_\alpha = 0, \tag{2.11}$$

$$G_\alpha(0, \tau) = 0, \partial_t G_\alpha(0, \tau) = \delta(x)\delta(y)\delta(z), \tag{2.12}$$

where $\tau = \sqrt{x^2 + y^2 + z^2}$.

The solution of Equations (2.11) and (2.12) has the following format:

$$G_\alpha(t,r)=-\frac{1}{2\pi\tau}\partial_\tau u_\alpha(t,r) \tag{2.13}$$

where it is assumed that

$$u_\alpha(t,r)=H(t-r)\int_0^{(t-r)/r^{1/\alpha}} f_\alpha(\zeta)\left(1+\frac{r^{\frac{1}{\alpha}-1}}{\alpha}\zeta\right)d\zeta. \tag{2.14}$$

The Green's function (2.13) behind the front ($\tau << \alpha r$) as presented in space-time has the following format:

$$G_\alpha(r+\tau,r)=\frac{H(\tau)}{2\pi\alpha}\cdot\frac{(r+\alpha r)}{r^{\frac{1}{\alpha}+2}}\cdot f_\sigma\left(\frac{\tau}{r^{1/\alpha}}\right)+O(\tau^2). \quad (2.15)$$

According to the theory of generalized functions, $\frac{1}{r^{1/\alpha}}\cdot f_\sigma\left(\frac{\tau}{r^{1/\alpha}}\right)$ tends to $\delta(\tau)/2$ at $\lambda \to 0$. The Green function $G_\alpha(t,r)$ shown in Equation (2.15) converges in this case to a simple Green function of the classical wave differential equation, i.e., to $G_\alpha(t,r)=\frac{1}{4\pi r}\delta(t-r)$. The solution at $\alpha = 1/2$ may be presented in a simple explicit format:

$$G_{1/2}(\mathrm{t,r})=\frac{1}{4\pi r}[F_0(t-r,\lambda r)+2F_1(t-r,\lambda r)+\lambda^2 F_2(t-r_1,\lambda r)] \quad (2.16)$$

where

$$F_0(\tau,\xi)=\frac{\xi}{2\sqrt{\pi}}\tau^{-3/2}\exp(-\xi^2/4\tau), \quad (2.17)$$

$$F_1(\tau,\xi)=\frac{1}{\sqrt{\pi}}\tau^{-1/2}\exp(-\xi^2/4\tau), \quad (2.18)$$

$$F_2(\tau,\xi)=H(\tau)\,\mathrm{erfc}\,(\xi/2\sqrt{\tau}) \quad (2.19)$$

As two independent polarization types are available in a solid isotropic medium, the total displacement vector $\mathbf{u}(r,t)$ should be presented in the form of in-plane $\mathbf{u}_l$ and lateral $\mathbf{u}_r$ displacements:

$$\mathbf{u} = \mathbf{u}_l + \mathbf{u}_r. \quad (2.20)$$

Each of these displacements may be directed by integrodifferential equation of Equation (2.11) type but with different multiplier before the terms including time derivatives due to the differences in the high frequency sound velocity limit for these two wave types. The integral terms creep cores also differ (displacement creep function and triaxial creep function). However, their time correlation is identical if identical processes lead to local relaxation. For this reason, the methods designed for scalar equations may be applied with every scalar component of every polarization displacement vector simply by scaling the derived solutions.

2.2 Effect on the Wave Spread in the Oil Accumulations by the Geologic-Geophysical Conditions

As an indicator of efficiency at a certain frequency may serve encompassment radius within which are maintained certain interrelations between the threshold values of vibration parameters—vibratory displacements ξ and vibratory accelerations $\ddot{\xi}$. These parameters are determined from the density of vibratory energy flow E at a given point of the medium and through vibration frequency f as follows:

$$\xi = \frac{1}{2\pi f}(E/\rho C)^{\frac{1}{2}}, \ddot{\xi} = 2\pi f (E/\rho C)^{\frac{1}{2}},$$

where ρC is the wave resistance or acoustic impedance of the medium.

Numerical modeling was conducted by Sherifulling *et al.* [26] in order to determine the vibrations' space-energy distribution. This enabled the computation of the energy picture of the wave distribution field for the assigned vibration frequency and of the vibratory accelerations and offsets field in the reservoir accounting for petro-physical properties of the top and base of the reservoir. Modeling was conducted using a method of the wave spreading statistical testing in the plane dissecting the reservoir with the plane-parallel boundaries [20]. The source of harmonic waves was distributed along the circumference of a well with the radius R_c with the center in the origin. The top and base of a reservoir having thickness H was positioned parallel to the X axis. At calculating by the statistical testing method, the spreading of the low-frequency harmonic waves is modeled using acoustic "quanta" flying out from the source in a random direction. Acoustic energy is assigned to each "quantum", the energy equal to surficial density of a cylindric pulsator with the pressure amplitude P_0:

$$E_0 = \frac{P_0^2}{2\rho C}(2\pi R_c)^{-1}.$$

This quantum "is traveling" in the reservoir along a ray trajectory defined by the indicatrix of reflection from the top and base. The quantum's energy is continuously declining to the local patterns of absorption in the reservoir rocks, wave divergence and angular parameters of reflection coefficients on the porous boundaries. When the quantum is flowing out of the studied portion of the reservoir, a transition is implemented to playing out the subsequent history of the acoustic quantum.

For the evaluation of acoustic properties of the saturated porous medium and description of two-dimensional processes of wave transfiguration on the boundaries, one can use the approach proposed by Sharifullin based on Bio's linear theory. The offsets of solid and liquid phases (appropriately, **u** and **V**) may be determined from Bio's equations:

$$\lambda^2\nabla^2\mathbf{u}+\text{grad}\,[(\lambda_1+\lambda_2)e+Q\varepsilon]=\frac{\partial^2}{\partial t^2}(\rho_{11}\mathbf{u}+\rho_{12}\mathbf{V})+b\frac{\partial}{\partial t}(\mathbf{u}-\mathbf{V}),$$

$$\text{grad}\,(Qe+R\varepsilon)=\frac{\partial^2}{\partial t^2}(\rho_{12}\mathbf{u}+\rho_{22}\mathbf{V})+b\frac{\partial}{\partial t}(\mathbf{u}-\mathbf{V}),$$

$$e=\text{div}\,\mathbf{u}\quad \varepsilon=\text{div}\mathbf{V}.$$

Here, λ_1, λ_2, Q, and R are Bio's elastic moduli for the medium; $b = m^2\mu/k$ is Bio's resistance coefficient; μ is the fluid's dynamic viscosity; k the permeability of porous medium; m is porosity; ρ_{11}, ρ_{12}, and ρ_{22} are Bio's density parameters expressed by rock matrix density ρ_1 and fluid density ρ_2 as follows:

$$\rho_{11} = (1 - m)\rho_1 - \rho_{12},\ \rho_{22} = m\rho_2 - \rho_{12},$$

where ρ_{12} is the so-called "reduced density".

The Bio's moduli may be expressed directly by the measured parameters: rock matrix triaxial compression modulus K, shear modulus G for a "dry" porous medium, and compressibility factors β_1 and β_2.

On assuming that in plane OXZ on the separation boundary of two saturated porous media (the boundary coinciding with the OX axis), first or second kind compressional wave or shear wave incidence angle to OZ axis is α. The OZ axis is directed into the second medium. In the general case, six types of waves are generated on the boundary, three are reflected and three are refracted.

This approach was implemented for computing phase velocities, free-space attenuation factors, angular wave transformation factors, and energy reflection parameters at reservoir's boundaries. These parameters are needed for a "raffle" of the acoustic "quanta" trajectories.

The same way was modeled a great number of the acoustic quanta trajectories flowing from the source in all possible directions. A field intensity was computed in 2.5 to 5 thousand of the reservoir points using the energy flux vector:

$$\mathbf{E} = \rho_{12}\omega^2 A^2/C_m,$$

where A is amplitude of wave potential in the medium and C_m is the phase velocity of *m*-wave.

The exit angle of each "quantum" trajectory ray of 0° to 180° was "raffled" using pseudorandom number sensor for the numbers uniformly distributed in the zero to one interval. The field near every of 5,000 observation points was averaged on the surface of a cell with the side $H/50$, where H is the reservoir thickness. Thus, phase displacement at the quanta superposition in a given point was taken into account by that separate cells have been provided for accumulating two orthogonal components of the wave vector at every point of the reservoir. Modeling was being performed in the mode of continuous information accumulation with releasing results over a certain number of histories—"acoustic quanta exits" from the source. The result at each release was normalized per a number of hits in every reservoir cell. For calculating the hit numbers, a special two-dimensional massif was included in the memory.

Computations have been conducted of the spatial-energy distribution of the well vibration source field in the oil-saturated productive reservoirs restricted by a low permeability water-saturated top and base. At constructing of the reservoir and enclosing rocks' design model have been used real field-geologic property data of the reservoirs, oils, reservoir water and enclosing rocks. The reservoir thickness varied in the range of 5 to 30 m. As source parameters have been taken average parameters for each type of boundary of porosity, permeability, solid phase and saturating fluid density, rock triaxial compression, and shear moduli. The top and base rocks have been assumed practically impermeable and saturated with reservoir water having density of $1.188 \cdot 10^3$ kg/m^3 and viscosity of 0.05 cP or gas with density of 70 kg/m^3 and viscosity of 0.015 cP. In Figure 2.1, relative fields of vibration intensity for various frequencies in an oil-saturated reservoir, 20-m thick, are shown. The obtained spatial distributions of vibrational energy flow density from the vibration source in the reservoir are wave pictures with interference maxima and minima illustrating a substantial influence by the top and base depending on frequency.

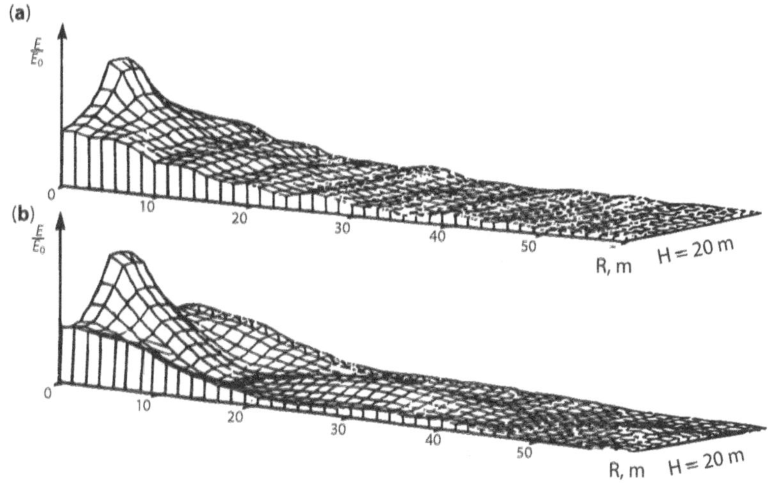

Figure 2.1 Relative intensity of vibration fields E/E_0 vs. reservoir thickness H, distance from vibration source R and frequency: (a) 12-Hz frequency and (b) 120-Hz frequency.

Figure 2.2 is relative change in the intensity of elastic vibrations vs. the distance to the vibration source for various frequencies. The curves are derived by averaging of field massifs over the reservoir thickness. At frequency over 1 kHz, monotonous and substantial decline in vibration intensity with an increase in distance to the source is observed. At frequency over 10 kHz, the influence of boundaries on the field in the reservoir is negligible (curve 5). At low frequency, there are certain frequencies at which a sharp energy decline starting at some distance from the source is replaced by a smooth alternation of maxima and minima.

This phenomenon indicates that at those frequencies in the reservoir vibrational mode are generated expanding without a noticeable energy radiation from the reservoir into enclosing nonproductive rocks. In these rocks, the vibration intensity decline with distance is caused only by cylindrical divergence and spatial absorption, which is not too great at low frequency. The most clearly expressed normal mode for a 20-m-thick reservoir is recorded near the frequency of 120 Hz. An increase in energy flow density in the reservoir at substantial distance from the vibration source in this frequency area in comparison with the field intensity even for very low frequencies is perhaps indicating a possibility of existence in the oil reservoirs of the resonance activation regimes; these are characterized by a substantial decline in the energy fraction dispersed in the nonproductive rocks enclosing the reservoir and an increase of the energy fraction penetrating within the reservoir.

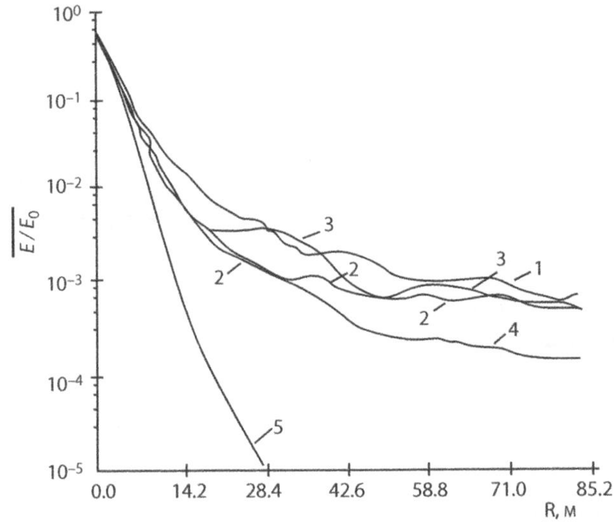

Figure 2.2 Distribution of relative vibration intensity $\overline{E/E_0}$ averaged over the reservoir thickness vs. the distance to the vibration source R and frequency: 1, 12 Hz; 2, 50 Hz; 3, 120 Hz; 4, 1,000 Hz; and 5, 10,000 Hz.

Phenomena similar to the described ones are also observed at different reservoir thickness values. For example, in a 5-m-thick reservoir a "strong" normal vibration mode near the frequency of 480 Hz is present. The derived vibration intensity fields for assigned frequencies enable computation of the fields of vibrational displacements and vibrational accelerations.

2.3 Wave Spreading From the Vibrating Surface of the Reservoir Matrix Into the Saturated Medium

Currently, the problem of wave propagation from the vibrating surface of the reservoir matrix into the various media has not yet been satisfactorily solved.

The motion of a viscous incompressible liquid filling half-space over the flat surface performing extension vibrations has been the earliest considered by Stokes.

A corresponding solution of the linear problem is easily generalized for a case of periodic vibrations [11].

We will review first a case of a horizontal fracture in the reservoir when it is performing straight-linear incremental harmonic vibrations in the same plane:

$$\eta = V_0 \cos \omega t, \tag{2.21}$$

where η is the vibration velocity of the horizontal surface, V_0 is its amplitude, and ω is the vibration frequency.

For the velocity of uncompressible liquid over a vibrating surface $V = V(x, t)$ from the Navier-Stokes equation:

$$\frac{\partial V}{\partial t} = \nu \frac{\partial^2 V}{\partial x^2}, \tag{2.22}$$

where ν is kinematic viscosity factor. Equation (2.22) is similar to the diffusion and heat-conductivity equations. Boundary conditions of the problem are

$$V(0,t) = \dot{\eta} = V_0 \cos \omega t, \qquad \left.\frac{\partial v}{\partial x}\right|_{x=h} = 0 \tag{2.23}$$

where h is the facture width.

The first equality is the condition of a liquid sticking to the surface. The stationary solution of Equation (2.22) satisfactory to the conditions Equation (2.23) is [2, 3]:

$$\mathrm{V} = \mathrm{u(x)} \sin \omega \mathrm{t} + \mathrm{v(x)} \cos \omega \mathrm{t} \equiv V_0 \sqrt{R^2(x) + S^2(x)} \cos[\omega \mathrm{t} + \varphi(x)], \tag{2.24}$$

where

$$U(x) = \mathrm{V_0 S}(x), \ \mathrm{v}(x) = \mathrm{V_0 R}(x);$$

$$\mathrm{S}(x) = \mathrm{NK_1}(\beta x) - 2\mathrm{K_2}(\beta x) - 2\mathrm{MK_3}(\beta x); \ \beta = \sqrt{\omega / 2\nu},$$

$$M = \frac{1}{2} \cdot \frac{sin2\beta h - sh2\beta h}{sh^2 \beta h + cos^2 \beta h}, N = \frac{1}{2} \cdot \frac{sin2\beta h + sh2\beta h}{sh^2 \beta h + cos^2 \beta h};$$

$$K_0(z) = \mathrm{ch}\, z \cos z, \qquad K_1(z) = \frac{1}{2}(\mathrm{ch}\, z \sin z + \mathrm{sh}\, z \cos z),$$

$$K_2(z) = \frac{1}{2} \mathrm{sh}\, z \cos z, \qquad K_3(z) = \frac{1}{4}(\mathrm{ch}\, z \sin z + \mathrm{sh}\, z \cos z);$$

$$\sin\varphi(x) = S(x)/\sqrt{R^2(x)+S^2(x)},$$
$$\cos\varphi(x) = R(x)/\sqrt{R^2(x)+S^2(x)}. \qquad (2.25)$$

Stokes solved this problem for a case of h → ∞:

$$R(x) \rightarrow e^{-\beta x}\cos\beta x,\; S(x) \rightarrow e^{-\beta x}\sin\beta x,$$

$$V = V_0 e^{-\beta x}\cos(\omega t - \beta x). \qquad (2.26)$$

It follows from Equation (2.26) that the facture's surface involves the liquid in a vibratory motion with the same frequency ω and with the amplitude rapidly (exponentially) declining with distancing from the surface. The thickness δ of the liquid's layer involved in vibration by the fracture surface due to liquid's viscosity may be described by a distance at which the velocity amplitude is equal to 5% of its value on the fracture's surface. This s a value determined, according to Equation (2.26), as

$$\delta \approx 3/\beta = 3\sqrt{2\nu/\omega} = 3\delta_0 \qquad (2.27)$$

Here, the value $\delta_0 = \sqrt{2\nu/\omega}$ may be called the "vibration penetration depth".

At viscosity factor ν = (1.007–1.519)·10^{-2} cm^2/s (which corresponds to water temperature change from 20° to 50°) and the vibrations frequency 2.5 to 5 Hz, the penetration depth is about 1 mm, which, is quite commensurate with the fracture width.

Figure 2.3 includes distribution diagrams of nondimensional vibratory velocity amplitudes of liquid $(V/V_0) = \sqrt{R^2+S^2}$ vs. the width drawn corresponding with Equations (2.24) and (2.26). At the fracture width $h < \delta$, the velocity distribution with fracture height may significantly differ from the free surface ($h \rightarrow \infty$). If, however, $h > \delta$ (i.e., $\beta h > 3$), then this difference is small. In such a case, Equations (2.24) and (2.26) provide close results. For this reason, liquid velocity in a δ-wide fracture may be determined with an accuracy sufficient for practical calculations from a simple Stokes formula Equation (2.26) considering the liquid outside of this layer immobile. If, however, $h < \delta$, ($\beta h < 0.5$), the calculations should be conducted using a more complex formula Equation (2.24). Liquid layers for which $h < \frac{1}{6}\delta$ practically move together with the fracture surface (as a solid body).

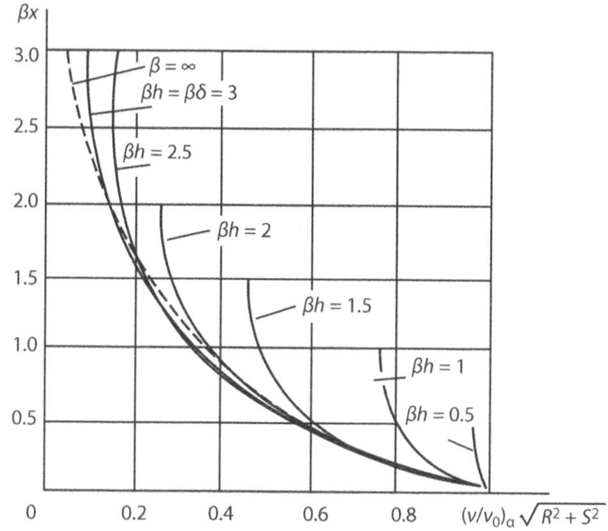

Figure 2.3 Vibration velocity distribution in a fracture (solid curves) and over a free surface (dashed line).

In a finite thickness layer above the harmonically vibrating fracture surface, for which $\beta h \geq c \approx 3.69$ (where c is the root of equation $\mathrm{sh}^2 c + \mathrm{ch}^2 c = 400$), and the penetration depth determined as previously is somewhat greater than the value $\delta = 3/\beta$, but does not exceed the value c/β which it assumes at $\beta h = c$. In layers where $\beta h = c$, the 20-fold decline in the velocity amplitude is not reached.

Due to linearity of the problem, the above results are easily generalized for cases of rectilinear periodic arbitrary vibrations and periodic vibrations in two mutually perpendicular directions on the fracture surface.

At periodic rectilinear motion of the fracture surface (represented by a sum of harmonics with the frequency $k\omega$, where k 1, 2, ...), the penetration depth for some m^{th} harmonic, according to Equation (2.27), will be $\sqrt{m}$ times lower than for the first one. For this reason, at sufficient distance from the fracture surface, liquid vibration is determined by the first harmonic; in other words, it approaches harmonic character regardless of the pattern of periodic vibrations (of course, on condition that h $\geq \delta$ and the harmonic amplitudes do not grow with the increase of their numerical order).

If the fracture surface is making random periodic vibrations in two mutually perpendicular directions in two mutually perpendicular directions lying on its surface, the liquid velocity in each of these directions are determined independent from one another. But, the aforementioned qualitative patterns remain valid. In particular, if periods of mutually perpendicular vibrations coincide, then the liquid particles' trajectories at a sufficient distance from the fracture surface approach ellipses, and the size of these ellipses is exponentially declining. In a case when periods are substantially different, the liquid motion at distancing from the fracture surface approaches rectilinear harmonic vibrations coinciding in the direction and frequency with longer period vibrations. In a manner of speaking, the liquid layer has properties of a high frequency filter.

The quoted results are comparable with the solution of a problem of fading the agitations caused by a spherical fractured surface in the process of its harmonic vibrations along some direction in an incompressible liquid.

For instance, we will review for a case of Reynolds' small numbers a spherical surface within which 95% of the total energy loss in a viscous incompressible liquid is dispersed (Figure 2.4a). Assume that R_0 is radius of vibration penetration and r_0 is radius of the spherical surface. Then, the value $\delta = R_0 - r_0$ may be considered the appropriate depth of vibrations' penetration into the liquid. Then, correlation $\delta = R_0 - r_0$ may be treated as the appropriate depth of vibrations' penetration into the liquid. For the correlation δ / r_0 vs. the parameter $\beta = r_0 / \delta_0 = r_0 \sqrt{\omega / 2v}$, we obtain a solid line diagram in Figure 2.4b. At that, $\delta_0 = \sqrt{\omega / 2v}$ is the same value with the dimension of length as in Equation (2.27). The same diagram within the considered limits of β variation may be approximated as follows:

$$\frac{\delta}{r_0} = \frac{3}{\beta} = \frac{3\delta_0}{r_0} = \frac{3\sqrt{2v/\omega}}{r_0} \tag{2.28}$$

(dashed line in Figure 2.3). Upon reducing by r_0, this equality is exactly equal to (2.27).

Thus, there is practically total coincidence of the results for a case of the flat and spherical fracture surfaces' vibrations. The study result is also close if the agitation area from the spherical surface is defined not by the energy dissipation, but by leveling of the pressure field or by fading of the velocity field as it has been done for a case of a flat fracture. The agitation zone in the case of a spherical surface's incremental advance in a viscous liquid under the problem's conditions is by an order of magnitude greater in size than in the case of its vibration. This conclusion is in the qualitative agreement with correlations of Equations (2.27) and (2.28), which indicate an expansion of the agitation zone with the decline in the vibration frequency.

In conclusion, one may note that under the assumption of incompressibility of liquids, a case of lateral fracture surface vibrations is trivial as the liquid vibrates in this direction together with the fracture surface. If compressibility is present, then the process is described by a wave equation which includes an addend accounting for the dissipation of the vibration energy.

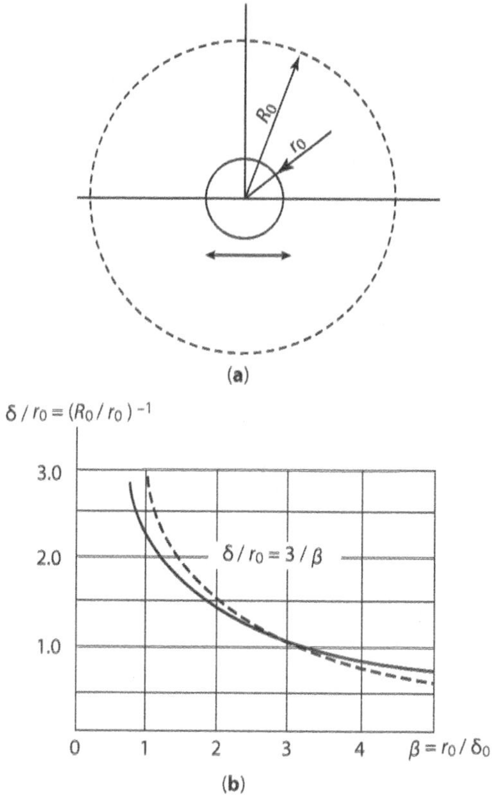

Figure 2.4 Vibration in a liquid volume at progressive vibrations of the fracture's spherical surface.

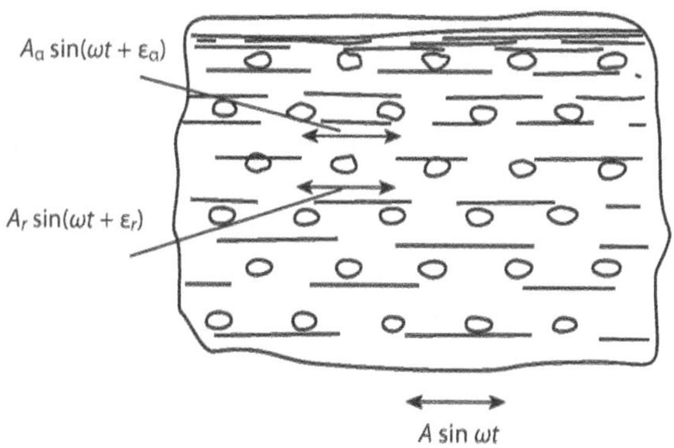

Figure 2.5 Schematic advance of a mixture of a wetting and nonwetting liquids in a unit pore volume in a field of elastic vibrations. A_r and A_a are the relative and absolute displacement amplitudes of the nonwetting liquid; ε_r and ε_a are the relative and absolutes dislocations.

In the case of pore volume filled up with a wetting liquid with floating droplets of nonwetting liquid, the vibration penetration depth may be estimated treating the mixture equivalent to some liquid with viscosity greater than that of the wetting liquid with droplets of a nonwetting liquid. In order to estimate the absolute and relative amplitudes of displacements, force, and energy expense needed for providing vibrations in a reservoir, we will review a unit pore volume restricted by the rock matrix. Let us assume that a selected volume is filled up with a wetting liquid with floating droplets of a non-moisturizing liquid. To this volume are applied harmonic vibrations with a frequency ω with the maximum displacement amplitude *A* (Figure 2.5). Walls of the pore volume and the wetting liquid are vibrating in phase, coherently with the displacement amplitude:

$$x = \mathrm{A} \sin \omega t,$$

where *t* is the vibration time.

The simplest case is with nonwetting liquid droplets being balls of equal diameter $d = 2r$ suspended in the wetting liquid and the volume concentration *C* within the study volume of no greater than 5%. It such a case, the distances between the droplets are greater than 2 or 3 their diameters and their mutual influence may be disregarded. On assuming further that the Reynolds' acoustic numbers are much smaller than a unit:

$$Re_{\mathrm{ac}} = A_r \frac{ad}{\nu} \ll 1;$$

here, ν is the kinematic viscosity of the wetting liquid.

A solution for the nonwetting liquid displacement amplitude relative to the surrounding wetting liquid is [6]:

$$x_r = \mathrm{A}_r \sin(\omega t + \epsilon_r),$$

and the absolute amplitude of the displacement is

$$x_a = \mathrm{A}_a \sin(\omega t + \epsilon_a).$$

The force *F* needed for the occurrence of harmonic vibrations of such unit volume is:

$$\mathrm{F} = \rho \ddot{x} + K \dot{x},$$

where

$\rho = \rho_s - \rho'$ is the effective density of liquids' mixture at vibration;
$\rho_s = \rho_e(1 - C) + C\rho_n = \rho_c[1 + C(\Delta - 1)]$ is the static mixture density;

$$\rho' = \rho_c C(\Delta + 9/4\alpha^{-1} + 1/2)(A_r/A)^2; \quad \alpha = r\sqrt{\omega/2\nu};$$

$K = 9/4\rho_c C\omega(1+\alpha)\alpha^{-2} A_r^2/A^2$ is the effective vibration damping factor;
$\Delta = \rho_n/\rho_c$ is the density ratio of the nonwetting to the wetting liquids.

Due to a greater mobility of the nonwetting liquid relative to wetting one, the effective density of liquid mixture at vibration lower than the density in a dormant state by a positive value ρ′. Power that needs to be expended for the vibration support of a unit pore volume is $N = KA^2/2$. For the most practically important cases $\alpha = r\sqrt{\frac{\omega}{2\nu}} \gg 1$. The equations for the displacement amplitudes are:

$$A_r/A = [\Delta - 1]/(\Delta + 1/2) = 2(\rho_n - \rho_c)/(2\rho_n + \rho_c);$$

$$A_a/A = 3/(2\Delta + 1) = 3\rho_c/(2\rho_n + \rho_c); \varepsilon_r = -\pi, \varepsilon_a = 0.$$

These equations show that non-wetting liquid's relative displacement maximum amplitudes A_r are greater, the greater the difference between its density and the density of wetting phase, the non-wetting liquid's vibration is leading in phase the vibration of the wetting liquid. At absolute motion, of the non-wetting liquid, the absolute amplitudes of displacement exceed those of wetting one.

As the equivalent viscosity at certain conditions may exceed by one order of magnitude (and even greater) the wetting liquid's viscosity, the vibration penetration depth for a liquids' mixture may be significantly greater. Various equations have been proposed for calculating effective dynamic viscosity factor. For example, at 30% content of nonwetting phase, the effective dynamic viscosity factor is almost three times greater than the similar viscosity factor for a wetting liquid. Appropriately, the mixture's kinematic viscosity is 1.4 times greater than for the wetting phase. Thus, the penetration depth is 1.4 times greater than in wetting phase.

Some qualitative patterns (regularities) in vibration's penetration into a unconsolidated reservoir for the simplest case of circular vibrations follow from the solution of motions by a material particle touching a flat surface. If the surface vibration acceleration

$$A\omega^2 < fg,$$

where ω is the frequency, A is the radius of circular vibration trajectory, f is the slipping friction factor, and g is the gravity acceleration, then the particle is moving together with the vibrating surface.

At $A\omega^2 > fg$, the particle acceleration W, velocity U, and trajectory radius r at steady absolute motion are defined by the following equations:

$$W = fg,\ U = fg\omega,\ r = fg\omega^2, \tag{2.29}$$

i.e., the trajectory velocity and radius rapidly decline with growing frequency.

At sufficiently large surface vibration frequency, the particle remains practically immobile in space. This pattern is manifested to some extent also at vibratory motions of unconsolidated reservoir layers. There is some range of the surface vibration accelerations, within which vibration penetrates the reservoir in a certain most efficiently.

A basic assumption is explaining the motion of one relative to another of the elementary layers composing the reservoir is that the dry friction factor f between the layers is considered a function of the normal overburden pressure, i.e., the weight of the overlying layer. This is supported by a series of direct and indirect experiments by Yamshchikov [27]. Designating the overlying layer weight per unit area as G, the relationship between the dry friction force F ("resistance force to a relative layers' displacement") and G and f values is as follows:

$$F = Gf(G) \tag{2.30}$$

where $f(G)$, according to our assumption, is some increasing G function. Then, at a growth of vibrating surface vibration acceleration $A\omega^2$, the motion looks as follows.

At $A\omega^2 = (A\omega^2)_1 < gfl$, where fl is the value of friction factor at the upper boundary of the layer, all layer's particles move together with the vibrating surface as one solid body. After the acceleration exceeds, the $A\omega^2$ value, relative sliding begins of the upper layer's part on the lower one. It continues moving together with the vibrating surface. At further increase of the acceleration, the relative motion expands over the underlying layers and at

$$A\omega^2 = (\omega^2 A)_2 = \sqrt{F_{m\left(\frac{d^2F}{dG^2}\right)} + \left(\frac{dF}{dG}\right)^2} > (A\omega^2)_1$$

reaches the lower boundary of the reservoir (subscript $_m$ belongs with values related to this boundary). At $A\omega^2 > (A\omega^2)_2$, the entire reservoir is sliding relative the vibrating surface (sliding variously at different levels). The particles' absolute velocity at this layer is decreasing from the lower boundary to the upper one. Velocity moduli at the points with certain G values are found as

$$U = \frac{g}{\omega}\sqrt{F\frac{d^2F}{dG^2} + \left(\frac{dF}{dG}\right)^2} \tag{2.31}$$

If the particles' material is uniform in the reservoir, then the value u,

$$0 \leq u = \frac{G}{G_m} = \frac{x}{h} \leq 1,$$

(where x is absolute coordinate, G_m is the total weight of the larger per vibrating surface unit area, and h is the reservoir thickness), is a relative coordinate counted measured vertically down from the reservoir upper boundary. If the correlation of friction factor f with G is linear, then Equation (2.31) becomes

$$U(u) = \frac{g\,f_0}{\omega}\sqrt{1 + 6u\xi(1 + u\xi)} \tag{2.32}$$

where $\xi = f_m - f_0/f_0 > 0$ is the relative difference of friction factors at the lower and upper reservoir boundaries. At these boundaries

$$U(0) = U_0 = \frac{gf_0}{\omega}, \quad U(1) = U_m = \frac{gf_0}{\omega}\sqrt{1 + 6u\xi(1 + u\xi)} \tag{2.33}$$

The former equation is equivalent to Equation for a particle velocity if the friction factor f is replaced with its f_0 value for the upper boundary of the reservoir. The other patterns of vibration's penetration for the subject model of an unconsolidated reservoir are also similar to those established for the particle. Maximum vibration acceleration of the reservoir particles

$$W_{max} = U_m \omega = gf_0 \sqrt{1+6u\xi(1+u\xi)} = (\omega^2 A)_2$$

is reached at its lower boundary at the same value of the vibration acceleration of the surface $A\omega^2 = (A\omega^2)_2$. At the further increase of $A\omega^2$ increase, the reservoir particles accelerations remain unchanged, whereas the velocity and the trajectory radius at a constant vibration amplitude decline with the increase in frequency. Thus, with increasing surface acceleration vibrations $A\omega^2 < (A\omega^2)_1 = gf_0$, the vibrations completely penetrate the entire reservoir thickness, and at $A\omega^2 > (A\omega^2)_1$, they do only partially. The "penetrating acceleration" with its increasing value $A\omega^2$ stabilize at the gf_0 level for the upper reservoir's boundary, and "the velocity and amplitude penetration" declines with increasing frequency, in proportion to ω and ω^2.

The maximum amplitude of relative migrations close to the amplitude of absolute surface vibrations occurs at a sufficiently large value of acceleration $A\omega^2$, when particles of an unconsolidated reservoir remain practically immobile in space.

A case of rectilinear extensional vibration of the surface displays the same patterns of vibration penetration as the considered case of circular vibrations. Solution of the problem here is more complex than in the case of circular vibration, especially for the second model, because the particle motion occurs with two long or instantaneous stops in each vibration period [5].

A different situation occurs with the vibration penetration process in individual reservoir layers in a case of transverse surface vibration of one of the layers. Experimental data indicate exponential nature of particles' vibration amplitude decline with certain distance from the vibrating surface of one of the layers. Let us denote v a pulsating component of the reservoir particle velocity and introduce a function

$$U = U_0^{(1+R)/2R},$$

where R is the velocity recovery factor at a strike, and V_0 is some characteristic velocity value. This function will satisfy equation of heat conductivity type equation, i.e., will play the role of quasi-temperature [9]. Figure 2.6 shows diagrams of the function U_x/U_0 at various values of the recovery factor R. Fading of the velocity of reservoir particles has exponential nature, which agrees with the above mentioned experimental results.

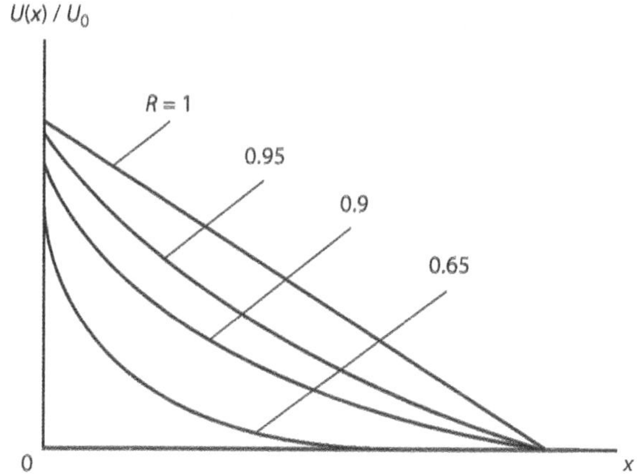

Figure 2.6 Velocity distribution in unconsolidated reservoir at transverse vibrations of its confining surface.

Certain specificity is typical of the vibration penetration dynamics in the cement grouts. The cement grouts are a multi-component medium composed of large and small filling aggregate, cohesive material, and water especially selected by the composition. In the process of mixture preparation, air unavoidably is getting into it. The amount of air in the mixture is relatively small (and is changing in the preparation process). Physicomechanical properties of the mixture and of the solidified cement substantially depend on its contents. As result, there is a rather complex rheological property of the cement mixture. Most important among them are resistance of a dry friction type and also the viscous friction and elasticity (the elasticity is mostly defined by the presence of air component). The vibration causes transformation of the mixture rheological parameters and properties, in particular, its pseudo-fluidization, which provides for the efficient compaction necessary for obtaining a cement with high strength parameters. For this reason, the cement mixture is modeled in theoretical studies as a continuous medium with elastic, plastic, and viscous properties.

2.4 Excitation of Vibration in Oil Reservoirs

Vibration excitation in oil reservoirs is best studied in cases of power load application directly to the land surface. It is implemented for a directional vibroseismic action on shallow oil accumulations and for a solution of seismic exploration tasks. Of greatest interest are issues of coordinating the sources with the medium from a point of view of the best transformation of the source power in the power of irradiated waves. Reviewed have been the models of spherical sources and sources distributed over the surface of elastic half-space [25]. A solution of the problem of creating efficient method of forming highly directional radiation suggests that the use of acoustic antenna theory constructed for ideal uniform acoustic medium in a case of stratified media results in significant errors. Taking into account vertical nonuniformity of the medium plays an important role for obtaining acceptable practical results. Distribution studies of energy supply arriving in multilayer elastic half-space from a harmonic surface power loading established that, in the multilayer half-space, may exist the frequency range wherein emerge reflected waves with the opposite direction of phase and group velocities.

An analysis of energy flow lines, which may be strongly involuted up to generating the areas where energy is circulating on closed trajectories, indicated the following. In the vicinity of resonance frequencies, the power flow of these internal circular flows may substantially exceed the flow coming from the source. This indicates a possibility of a vibration source energy accumulation with an unlimited amplitude increase in certain areas of semi-infinite laminated space. In these studies, the actual rocks have been modeled by an elastic medium. The solutions of dynamic problems for elastoporous media are known where, applying the Frenkel-Bio equations was conducted an analysis of the wave field excitation by a load varying in time according to the harmonic law. At that, the excited field energy distribution by the wave type has been estimated.

Similar tasks related to the half-space force excitement had a practical value for selecting efficient vibroseismic action regimes on shallow oil reservoirs using powerful surficial vibration sources developing force up to 10^6H.

The energy transfer processes from well to a reservoir are associated with substantial difficulties. These difficulties are associated with well geometry and paucity of a well as an irradiator compared with the excited wave length. If a source is creating vibration in the well fluid, then the field generated in the surrounding reservoir is equivalent to the field of an infinitely long cylinder on the surface of which one of possible is excited vibration modes. In this case, a solution of the wave equation is

$$\Phi_{\mathrm{m}} = \left[AH_m^{(1)}(k_r r) + BH_m^{(2)}(k_r r)\right]\cos(k_z z)\cos(m\varphi)e^{i\omega t},$$

where $H_m^{(1)}(k_r r)$ and $H_m^{(2)}(k_r r)$ are Hankel functions of the first and second kind, $m = k_y R_c$; R_c is the well radius; k_r, k_y, and k_z are wave numbers.

The first term in this equation corresponds to a wave converging to the cylinder axis, and the second term corresponds to the wave expanding from the cylinder. For the expanding running waves, the k_r number must be a real value, i.e., the condition $k^2 - k_z^2 > 0$ must be observed. If $k_z > k_r$, then k_r is a purely imaginary value, and in such a case, the field sharply exponentially declining with distancing from the cylinder surface. No irradiation occurs in such a case. The length of a flexural wave in the axial direction turns out smaller than a sound wave in the encircling medium regardless of another wave number $k_y = m/R_e$, which corresponds to the circular system of nodal lines. If we introduce the angle $\theta = \arcsin(\xi/k)$, $k = \omega R_e/c$ as the incidence angle or reflection angle of cylindrical wave, normal or modulated by front, then at incidence angles smaller than $\theta_1 = \arcsin(c/c_p)$, the energy enters the reservoir, and at the angle larger than θ_1, it returns in the well for forming head-waves in the liquid. Within the angle range between a critical value θ_1 to the second critical value $\theta_2 = \arcsin(c/c_s)$, the energy flow into the medium is caused only by irradiation of the shear waves. At high frequencies, such that $k^2 \gg k_z^2$ we have $\cos\theta \approx 1$, and the waves irradiated by the cylinder spread perpendicular to its surface. With the k decline, the θ angle is increasing, and at $k = k_z$, the wave is spreading only tangentially to the cylinder's surface. As soon as the axial flexural wave length becomes shorter than the sound wave length in the enclosing medium, appropriately k becomes smaller than k_z, and the cylinder-well is not radiating in the enclosing medium at all. Irradiation in a case $k^2 > k_z^2$ is studied in great detail, the results are practically applied in geophysical acoustic studies in wells and at high-frequency thermoacoustic action on the reservoir's bottomhole zone.

Energy interrelations at wave irradiation from wells by an axisymmetric source have been studied in [10]. Connections have been identified between field energy parameters and well's acoustic impedance. Also studied has been the nature of frequency and amplitude distribution on the irradiation surface onto the energy transmission into the massif. At the fixed spatial frequency ξ of a source u_0 harmonic amplitude offset distribution, equations have been derived for per-unit impedance of the well-loading massif. Radial resonance frequencies of a ring liquid layer between the emitter body and the casing and antiresonance frequencies also have been determined. The energy release into massif at the resonance frequency turned out to be most efficient.

Based on a solution of the dispersion equation for Rayleigh waves in the reservoir, a case has been reviewed when the irradiation was totally disappeared. The well acoustic impedance in this case becomes equal to zero and the well becomes absolutely "soft". The total energy flow through the well's walls at that becomes equal to zero, although in a near-surface layer quite significant densities are created of the kinetic and potential energy. At $\theta > \theta_R$, the acoustic energy may enter the reservoir from the well only in a case of normal fading of waves in the liquid. This moment corresponds with a case of an infinitely long cylinder surface-wave mode of which is excited by a low-frequency action within the well $(k^2 > k_z^2)$. Therefore, at the wave radiation from a well in the reservoir, there are critical frequencies, below which the irradiation in the conventional sense is absent.

Wave excitement form well considering reservoir elastoporosity was studied by Rakhmanov *et al.* (1985) [19]. The boundary conditions at the surface of a liquid-filled cylindrical well were assigned in the form of a tension in the solid phase and pressure in the liquid generated by moving at a constant speed load charge as follows:

$$P_0 \exp[iK(z - Cr)].$$

Restricted solutions of the Bio equations satisfying the irradiation conditions allows one to derive equations for calculating offsets and tensions in the solid phase as well as the pressure in a fluid-porous medium.

White [23] reviewed the absorption mechanism of normal waver spreading in the well fluid caused by fluid pulsation flow through the permeable walls of an uncased well. His results were appropriate for low frequency, when the tube-wave length is large compared with the well diameter. If to evaluate the contribution of such fading by a normal wave in its radiation at subcritical frequencies, then with the frequency decline, approximately from 100 Hz and to very low frequencies, the fraction of vibration energy irradiated into the massif monotonously grows to some value determined by the media permeability. Regardless of permeability, at frequency above 100 Hz, this phenomenon make practically no contribution to the irradiation of vibration energy. These results apply to uncased wells and did not take into account the mechanism of wave energy reradiation through a well fluid.

In order to increase the efficiency of energy introduction into the reservoir at low frequency, it is necessary to find a possibility to increase substantially the irradiation from a well into the reservoir by changing loading parameters and optimally utilize action of resonance properties of the well and reservoir systems. The existence of resonance regimes of vibration present in a well, which are associated with parameters of the enclosing medium and emerging in a pre-critical frequency area, is experimentally confirmed. For instance, if to run in a fluid-filled well a receiver of sound vibrations and to measure the energy spectrum of the noise, then in a fluid-saturated reservoir the resonance frequency may be identified.

The resonance excitation in a well may be achieved at the regime of high-frequency radial resonance of the fluid layer as well as a lengthwise resonance of the fluid column at low frequency. The radial resonance is determined by the well radius and emerges at high frequency on the order of dozen kilohertz and higher. Thus, coordination of the operation regime of the vibration source and the reservoir achieved, so that practically the entire power of the vibration source is transferred in the reservoir. By varying the frequency and load distribution on the surface of the vibration source, it is possible to control field energy structure in the reservoir. However, practical utilization of the radial resonance is substantially complicated for the following reasons. The frequency even of the first radial resonances at the existing well radiuses are too high both for favorable manifestation of elastic vibration action mechanisms on the reservoir and for the frequency coordination with the reservoir excitation resonance regimes. Besides, high-frequency elastic waves experience a strong fading in the reservoir. As the motions of porous medium at such resonances are radial, the generation of high-pressure amplitudes in wells is restricted by the acceptable radial displacement of the casing and cement column.

At lengthwise resonance, the vibration frequency of the fluid column in a well is defined mostly by the distance between reflecting surfaces in the well. As the lower reflecting surface is usually taken the dib hole and as the upper, the fluid and gas contact next to wellhead. At long distances between these boundaries, the vibration resonance frequency may reach the value on the order of 1 Hz and lower. Nevertheless, efficient utilization of lengthwise resonances at low frequencies is associated with certain difficulties. For creating the resonance lengthwise vibration of the entire fluid column in a well, it is necessary to know the fluid level in the well and the phase velocity of a tube wave propagation in it. It is practically impossible to support a constant fluid level in well in most technological operations. On taking the reflection properties from the lower boundary (the dib hole), they are clearly low because the wave resistance in the cement and in the surrounding reservoir are only slightly different from the wave resistance of a fluid. That is a reason why, due to a significant length of the fluid column, plenty of energy is expended on wave fading in the fluid itself and uselessly irradiated into the nonproductive zones above the reservoir.

That is why various submersible resonance devices are used, in which a set frequency vibration source is rigidly tied with acoustic reflecting filters [17]. These devices include a generator and a resonator enclosed in the body. At that, the fluid column filling the cavity of a resonator chamber forms a vibration contour in which a standing wave is generated. The resonance excitation of vibrations is achieved if the reflector length L and generator operating frequency v_0 are tied as follows:

$$L = c/4\,v_0,$$

where c is the wave propagation phase velocity.

Using of similar devices with fixed parameters in practice is not sufficiently efficient because the elastic wave phase velocity in a well fluid depends on the reservoir elastic properties and immersion depth and besides may change from one well to the next [4]. If the wave running in a productive interval of a well with perforation holes in the casing, then its phase velocity also depends on the reservoir porosity, permeability, compressibility, and well fluid viscosity [23]. All this results in a substantial instability of the existing devices resonance regime because these devices have been designed according to simplified rules without considering the aforementioned factors. Besides, these devices cannot be too long, and this in restricting the frequency regime of vibration excitation. As a result, resonance lengthwise vibrations of the fluid within the device are created, and at irradiation into the well, the vibration energy is passed along the column of the well fluid beyond the productive interval limits and is irradiated into overlying and underlying unproductive rocks. A possibility of achieving a well fluid effective lengthwise resonance was studied using "sliding" reflecting filters installed within the productive interval.

For instance, a well's resonance excitation regime may be achieved using hollow downhole reflector-filters filled with gas and fit on the production tubing. Such design does not require a hermetic contact with casing walls. Before running into a well, positions of the reflecting filters along the well length and relative the vibration generator may be changed. Distances between the reflectors are selected so that most energy from the generator was concentrated in the productive interval. The field excited in the enclosing rocks is close in its character to a finite length pulsating cylinder.

At the operation of the downhole vibration source within the limits of productive interval, a vibrating energy may be additionally irradiated which creates the emergence of pulsating fluid flows in the well's perforation holes. As the perforation channels are narrow, the fluid motion velocity in then is high compared with the fluid flows within the well proper. Within some volume small compared with the wave length is alternatively created the excess or shortage of a given medium's matter; its surface is permeable for a given medium and is a sovereign type radiator. That is why every perforation channel of radius r_k and length l_k at low frequency may be considered a point monopole exciting a spherically symmetric wave in the enclosing rocks. This wave is mostly defined not by geometric size of the perforation channel but by the size of generator's created outflowing fluid flow.

The vibration energy irradiated by an individual perforation channel, in considering of its geometrical parameters, may be presented as follows:

$$N_k = \frac{1}{8\pi} \rho c K^2 S_k^2 V^2, \tag{2.34}$$

where V is the velocity amplitude of fluid motion in a perforation channel, $S_k = \pi r_k^2$ is the channel's cross-sectioned area, and ρ is the fluid density.

The irradiation efficiency of each individual perforation channel is low mostly due to a low resistance to the irradiation at low frequency. However, if a system is available of closely spaced monopole irradiators and a condition $L \le \frac{\lambda}{6}$ is satisfied where L is the distance between irradiators and λ is the wave length in the medium, then these irradiators interact between themselves. At that, every individual irradiator is working in the pressure field of all other irradiators, which is equivalent to increased resistance to irradiation. As a result, active power increases and passive power, decreases.

At synphase operation of a group of similar irradiators, the total irradiation power is equal to the power of every one of them multiplied by squared number of the irradiators [16]. If for a perforation interval H_p the condition $H_p \le \frac{\lambda}{6}$ is satisfied, then accrual of the estimated irradiation power in the well should be expected to be as follows:

$$N = (H_p n)^2 N_k, \tag{2.35}$$

here, n is the perforation density (number of holes per unit length of the perforation interval).

For studying additional energy of individual channel, it is necessary to provide a sufficiently efficient transition mechanism of pulsating pressure in wells into vibration velocity fluid motion in channels. This process is accompanied by a high extent of tube wave fading in the well fluid caused by a low absorption on the casing walls and in the fluid as well as by the elevated acoustic irradiation of the perforation channels. Most of the casing wave energy is absorbed through acoustic irradiation in channels in the reservoir perforation interval.

A similar mechanism is operating in a well with a fluid and perforation channels as an oscillator with lumped parameters. A low basic frequency of vibration in such as oscillator is reached either by the participation in vibration of two media with drastically different properties (for instance, gas bubbles, and liquid) or by the realization of mechanism present in a Helmholtz resonator (where at a small actual mass of the vibrating medium it is possible to create a large amount of effective mass).

When a vibration source is operating in a well, fluid flows are generated in perforation holes. Due to a narrow channels' diameter, the motion flow in them is high compared with the fluid's velocity in the well. The kinetic energy

$$E_k = \frac{1}{2}\rho l_k S_k V^2 \tag{2.36}$$

is concentrated in perforation channels despite that the actual fluid mass in the well is much greater than the fluid mass in the channels. Whereas the elastic energy is concentrated within the well. Therefore, as the potential and kinetic energy are localized in different media (channel medium and in the well medium), the well may be considered an analog of Helmholtz resonator. On coordinating the generator operating frequency with eigen frequency of such resonator effective transmission of the generator energy in to kinetic energy of a fluid in the perforation channels occurs. The perforation channels represent monopole sources having purely active resistance at resonance frequency. The resonance vibration frequency of such an oscillator is [8]:

$$\omega = \left(\frac{S_k}{\rho \beta l_k \Omega} \right)^{1/2}, \tag{2.37}$$

where S_k and l_k are, respectively, the area and length of a channel, respectively; ρ and β are the density and compressibility of a vibrating fluid, respectively; and Ω is the volume of fluid amount possessing potential elastic energy.

One can now estimate the amount of fluid volume possessing a store of the potential elastic energy.

For this case, $\Omega = H_p \pi R_c^2$. The total area of channel openings in the well length H_p is equal to $S = \pi r_k^2 H_p n$. Using Equation (2.37), the resonance frequency for a well with perforation channels is equal to

$$\omega = \left(\frac{r_k^2 n}{\rho \beta l_k R_c^2} \right)^{1/2}. \tag{2.38}$$

At the resonance frequency, irradiated on average the reservoir power is

$$N = \frac{(nH_p)^2}{8\pi} \rho c k^2 S_k^2 V^2, \tag{2.39}$$

and this is the reason why the effective absorption of the generator energy which is expended for irradiation through the perforation channels is determined by the density of perforation holes. As Equation (2.38) indicates, this value also defines the generator's resonance frequency. Considering Equations (2.36) and (2.39), we will evaluate fading factor of the casing wave (caused by the acoustic irradiation through perforation channels) as

$$\alpha = \frac{1}{2} \frac{N}{E_k} = \frac{(nH_p)^2}{8\pi} \frac{\omega k S_k}{l_k} \tag{2.40}$$

Switching to the casing wave fading factor over the well length H_p, one obtains

$$\delta = \frac{\alpha H_p}{c} = \ln \delta(E),$$

where $\delta(E) = E_k/E_0$ is a relative loss of the generator energy to the irradiation over the well length; here, E_0 is kinetic energy on the generator wall. Taking Equations (2.40) and (2.38) into account one obtains an equation connecting n, H_p, and $\delta(E)$ values:

$$H_p N = \frac{l_k R_c C}{r_k^2}\left(\frac{8\ln\delta(E)\rho\beta}{nH_p}\right)^{1/2}. \tag{2.41}$$

Of special interest is the experimental verification of these equations through modeling field conditions. The experimental testing unit was a horizontal model of the reservoir with a well and perforation channels.

The reservoir model had a cylindrical core holder filled up with a cemented porous medium. The core holder, through a steel disk with the perforation hole, was joining a special stand for testing hydraulic vibrators. The space between the steel disk and the porous medium was filled up with a cement stone imitating the well cement shell. The aforementioned special stand imitated a well with hydraulic generator and was represented by a segment of casing with flanges. It was connected with a high-pressure and high-performance water-saturated reservoir. Water run-off through the hydraulic generator, a static pressure of the stand and the model was controlled by valves and the pressure was shown on a manometer. The hydraulic generator operating parameters have been determined on a gauge; elastic vibration parameters within the porous medium at the model input and at the model output have been observed using gauges and probes. The gauges have been connected with a normalizer and then the signals have been supplied to the remembering oscillograph and spectrum analyzer. The porous medium has been placed in a core holder that is 0.31-m long and 0.14 m in diameter. The porous medium has been a mixture of quartz sand and has epoxy resin. The perforation channel was 0.003 m in diameter and 0.01-m long. Between the stand and the model has been established a pivot crane intended for switching the model off the stand and measuring background noise and if necessary, for controlling the pressure fluctuations level.

The experimental methodology consisted of measuring fluctuation of vibration parameters at the input, output and within the model at a stable operation of the hydraulic generator. The measurement results have enabled evaluation of the vibration weakening amplitude at different points of the model and at different frequencies. Correlation diagrams of relative pressure fluctuation levels (registered at various points in the model) vs. frequency of the hydraulic generator were obtained. The experimental results have indicated that at transmission of vibrations into the reservoir their weakening occurred increasing with the frequency growth. At the same time, behavior of the curves has been notably different from the anticipated linear correlation of vibrations' weakening with frequency observed for a saturated porous medium under regular conditions at low frequencies. This unexpected result has been recorded in the maximum deviation area. The recorded pressure fluctuation level in a perforation channel was lower than the relative pressure pulsation level at distant points of the porous model.

The obtained result confirmed the theoretical analysis. The behavior of experimental diagrams may be explained as follows. Within the studied frequency range defined by the parameters of the used experimental equipment the conditions have been determined with optimum conversion of the generator pressure fluctuations into the kinetic energy of liquid within the perforation channel. An elevated efficiency has been observed of the perforation channel monopole irradiation due to an increase of the fluid volumetric flow rate. A relative level of pressure fluctuations in the reservoir has turned out higher than expected; pressure fluctuations in the perforation channel have converted in fluid rate fluctuations. At that, the relative level of pressure fluctuations in the channel has been declining.

In conclusion, an estimate of vibrations' resonance frequency according to experimental conditions. Parameters used in Equation (2.37) are as follows:

- Cross-section area of a perforation channel, $S_k = 7.065 \times 10^{-6}$ m^2;
- The perforation channel length, $l_k = 0.01$ m;
- Fluid density, $\rho = 1{,}000$ kg/m^3;
- Fluid compressibility $\beta = 0.455 \times 10^{-9}$ Pa^{-1}.

Substituting these data, one obtains $\omega = 134$ Hz. Comparison of this frequency with the one measured in the experiments are in good agreement. Results obtained indicate a real possibility of increasing efficiency by introducing vibratory energy from a well into the reservoir. The reason is that the real field values of parameters τ_k and l_k in the well with perforation density of 10 to 30 holes/meter result in the excitation of resonance frequencies within a 50- to 300-Hz range.

References and Bibliography

1. Berzam, V.A. Interaction of seismic waves with separation boundaries in porous saturated media. *Issues of Nonlinear Geophysics. Collection VNIIYaGG*, p. 66–74, Moscow, 1981.
2. Blekhman, I.I. *Vibration mechanics*. Moscow, Nauka, 400 p., 1994.
3. Blekhman L.I., Kizelwalter B.V. Distribution of flow velocities at the decks of vibration separators for gravity benefication. *Ore benefication*, 4, p. 53–54, 1981.

4. Valiullin A.V. *et al.* Some specifics in productive reservoir vibro-treatment technology. *Oilfield Engineering*, 12, p. 13–16, Moscow, 1973.
5. Gortinsky V.V. *et al. Separation Processes at Grain Processing Plants.* Moscow, Kolos, 1973, 180 p.
6. Granat N.L. Steady-state vibrations of vessels with two-phase mixture. Izv. AN USSR, *Mechanics and Machine-Building*, 5, p. 70–75, 1964.
7. Egorov A.G. *et al.* Consolidation and Acoustic Waves in Saturated Porous Media. Kazan, Kazan University, 1990, 102 p.
8. Isakovich M.A. *General acoustics.* Moscow, Nauka, 1976, 527 p.
9. Kremer E.V. *et al., A single-dimensional dynamic continual model of a loose medium.* DAN USSR, 309, 4, 1989.
10. Krutin V.N. *Energy interrelations at elastic waves irradiation from wells.* Coll. VNIIYaGG, p. 76–88, Moscow, 1982.
11. Landau L.D., Lifshitz V.M. *Mechanics of continuous media.* Moscow, Gostekhizdat, 1954, 350 h.
12. Lopatnikov S.A. *et al. Elastic wave fading in saturated porous random medium.* DAN USSR, 292, 3, h. 574–576, 1986.
13. Mirchink M.F. *et al. Estimate of a possibility to apply seismic exploration for direct exploration of oil accumulations.* Moscow, AN USSR, 1962, 123 h.
14. Nigmatullin R.I. *Fundamentals of heterogenous media mechanics.* Moscow, Nauka, 1978, 350 p.
15. Nikolayevsky V.N. *et al. Mechanics of saturated porous media.* Moscow, Nedra, 1976, 335 p.
16. Peshkovsky S.L. *et al.* Effect of ultrasound vibrations on the flow of viscous-elastic fluid. *Mechanics of Polymers*, 6, h. 12–14, 1971.
17. Zay V.V. *et al. Submersible resonance hydraulic-shock machine.* A.s. USSR, No. 259754, 1970.
18. Rakhmatullin Kh. A. *Waves in dual component media.* Tashkjent, FAN, 1974, 266 p.
19. Rakhmanov T.G. *et al.* Waves in the elastoporous medium saturated with a liquid with cylindrical void. *Izv. AN UzSSR, Technical Series*, 2,k p. 46–59, 1985.
20. Selin E.A. *Application of statistical testing for calculating acoustic fields.* (Monte-Carlo methods' application). Novosibirsk, SO USSR, 36 p., 1971.
21. Skuchik E. *Fundamentals of acoustics.* Moscow, Mir, 527 p., 1976.
22. Stepanov G.Yu. *Why Dean apparat is impossible.* Priroda, 7, p. 40–42, 1963.
23. White J. *Exitation and spreading of seismic waves.* Moscow, Nedra, 261 p., 1986.
24. Frenkel, Ya.I. On a theory of seismic and seismo-acoustic phenomena in a moist soil. *Izv. AN USSR, Geography and Geophysics*, 8, 4, p. 134–150, 1944
25. Chichinin S.I. *Vibration irradiation by seismic waves.* Moscow, Nedra, 224 p., 1984.
26. Sherifulling R.Ya. *et al.* Estimate of resonance excitation factors for wells and layers. *Ufa, VNIINefteotdacha*, 16 p., 1989.

27. Yamshchikov V.S. Wave processes in a rock massif. Nedra, 271 p., 1984.
28. Gurevich V. *et al.* Velocity and attenuation of elastic waves in finely-layered porous rocks. *Geoph. Journ. Intern.*, 121, p. 933–947, 1995.
29. Rok V.E. Near-front spatial profile of wave impulses in micro-inhomogeneous media. *7th EAOG Conf., Glasgow, Extended abstr.*, 1, B008, 1995.
30. Rok V.E. Time-domain representation of waves in media with frequency power law of dispersion. *58th EAGE Conf. and Tech. Exhibition*, Amsterdam, Extended abstracts, C008, 1996.
31. Stokes G.G. On the effect of internal friction of fluids on the motion of pendulums. *Trans. Cambr. Phys.*, IX, No. 8, p. 100–102, 1901.

3

Directional Displacement of a Dispersed Phase

Directional displacement of the dispersed phase is a central issue in explaining the vibration effect on variations of water saturation, phase permeabilities, and decolmatation of the near-bottomhole zones of wells.

One can present the motions emerging on vibration as a sum of two components: (1) a "slow, little changing in one vibration period, and (2) "fast" ("vibrational"). In most cases, of the greatest interest is in the slow motion.

Let us consider an observer who does not notice either rapid (as a rule, small) motions or the rapid forces, i.e., is the observer who either used special glasses through which rapid motions of the system are not visible or watched the motions at a stroboscopic illumination with the flash frequency equal to the vibration frequency [5]. Such an observer (**V**), as opposed to a regular observer (**O**), who is "seeing everything", will notice only a slow component of the motion. It is quite natural that he, not wishing to contradict the laws of mechanics, will have to explain all phenomena he observes that together with regular slow forces some additional slow forces or moments began to act.

3.1 Simplest Models of the Vibrational Directional Displacement

In terms of differential equations, the situation looks as follows (we will quote deliberations for a system with one degree of freedom as generalization for a case with an arbitrary number of degrees of freedom is not a conceptual difficulty). May the system's motion be described by the following equation:

$$m\ddot{x} = F(\dot{x}, x, t) + \Phi(\dot{x}, x, t, \omega t),$$

George V. Chilingar, Kazem Majid Sadeghi, and Oleg Leonidovich Kuznetsov. Acoustic and Vibrational Enhanced Oil Recovery, (55–88) © 2022 Scrivener Publishing LLC

where m is the mass, x is the coordinate, F is the "slow" and Φ is the rapid force, and ω is the vibration frequency. The accent means differentiation with respect to the regular, "slow" time t. The force Φ as opposed to F depends not only on t but also on the "rapid" time ωt, which is proportional to the "large parameter" ω, in the simplest case it is a periodic function ωt with period 2π. May the motion be presented as follows:

$$x = X(t) + \psi(t, \omega t),$$

where X is the slow component, ψ is the rapid (not necessarily small compared to X) component.

Then, it will appear to the observer **V**, who does not notice ψ rapid motions and rapid force Φ, that the slow motion is described by the following equation

$$m\ddot{X} = F(\dot{X}, X, t) + V(\dot{X}, X, t),$$

where V is vibration force.

Thus, one arriving at a situation to a great extent similar to a known theorem from the relative motion mechanics under which an observer tied with a rapidly moving coordinate system must add the inertia forces to all regular forces acting upon the system. In our case, the observer **V**, who does not notice either rapid forces nor rapid motions, must add vibration forces to all the regular forces.

The last equation is called the basic equation of vibration mechanics. It is a result of averaging the first, source differential equation performed in a certain way. The vibration force reflects the "accumulated" effects of action by vibrations on a nonlinear system. For instance, Figure 3.1 illustrates water flow in a segment of the cylindrical channel of some length L. In the left-hand upper part of Figure 3.1, a laminar flow, which occurs when the Reynolds number does not exceed a certain critical value. Then, figure's lower portion shows turbulent flow. Such a regime is described by the quite complex pulsations of the liquid emerging "automatically," for instance, on application of an external vibration. As a result, for assuring the same water discharge rate, a much greater pressure gradient is needed than at a laminar flow. Observer **O** can explain this fact by that part of the head which is lost at turbulent pulsations in the vibration field. As for observer **V**, he, not noticing water pulsation, will see the averaged motion as previously laminar but will explain the need of a greater pressure gradient by a sharp increase in efficient water viscosity factor on vibration. For such an observer, water "turned" into syrup.

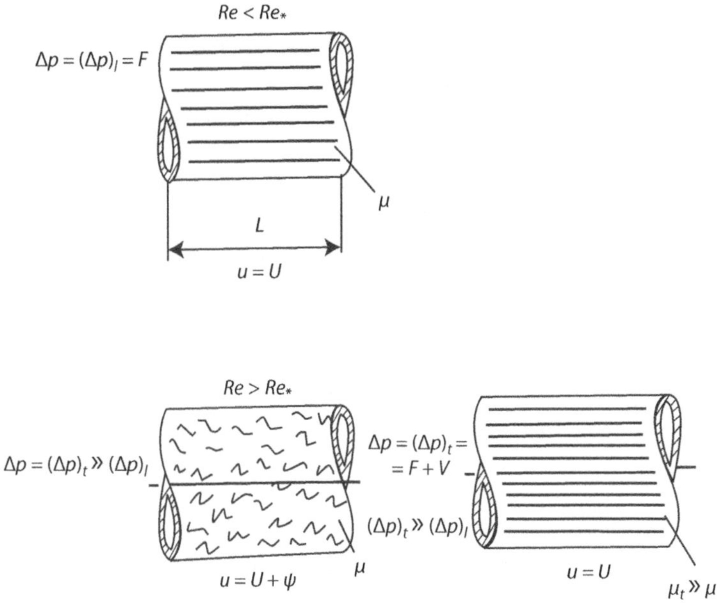

Figure 3.1 Water flow in a cylindrical channel of some length L in the vibration field. Re* is the Reynolds critical number; Δp_l and Δp_t are pressure gradients, respectively, at laminar and vibration flows; u is the total flow velocity; U and ψ are, respectively, slow and rapid components of the velocity; F and V are, respectively, slow and rapid components of the pressure gradient; μ is the water viscosity at laminar flow; and μ_t is the effective water viscosity at flow in the vibration field.

As a result of on average nondirectional (vibratory) actions, "on average directional" changes (for instance, motions) occur. It this case, it is a matter of directional vibratory displacement in the vibration field. Most problems associated with directional vibratory displacement boils down to the solutions of nonlinear differential equations, with periodical by rapid time right parts for which the rates of generalized coordinate changes can be presented as follows:

$$\dot{x} = \dot{X}(t) + \dot{\psi}(t, \omega t), \tag{3.1}$$

where $\dot{X}(t)$ and $\dot{\psi}(t, \omega t)$ are, respectively, slowly changing component with

$$\psi(t', \omega t) = 0. \tag{3.2}$$

The appearance of sought for solutions (3.1) makes it natural and expedient to use in solving displacement theory problems of the concept and methods of vibratory mechanics [1].

Next, the authors review displacements on a vibrating flat surface of sticking nondeformable small particles of a dispersed phase (oil, water, and gas bubbles) within a resistive medium.

Let us examine first to the motion of an arbitrary-shape particle m on a horizontal rough surface under action of a length-wise harmonic coercive force $\Phi_0 \sin\omega t$ (Figure 3.2a). If adhesion factors f_-, f_+ at sliding the particle back and forth on the flat surface are identical and equal to f, it is clear that, at $\Phi_0 < mgf$ the particle remains immobile and, at $\Phi_0 > mgf$ (where g is the gravity acceleration), it will be performing symmetrical vibrations relative to some average position. If we assume, however, that, for instance, $f_- > f_+$ then the symmetry will be distorted and, as the characteristic of the adhesion force F shows, at $\Phi_0 > mgf_+$, the particle will be displacing in the positive direction (in case $f_- < f_+$, at $\Phi_0 > mgf_+$, the particle will be displaced in the negative direction, i.e., same as earlier, in the direction of smaller resistance).

Within sometime interval in the first half-period $0 < \omega t < \pi$, when $\Phi_0 \sin\omega t > mgf_+$, the particle will be displaced to the right. Within the second half of the period $\pi < \omega t < 2\pi$ when $\Phi_0 \sin\omega t < 0$, the particle will remain in place (if $\Phi_0 < mgf_-$) or will move to the left at a smaller distance than to the right as resistance to the movement leftward is lower than the resistance

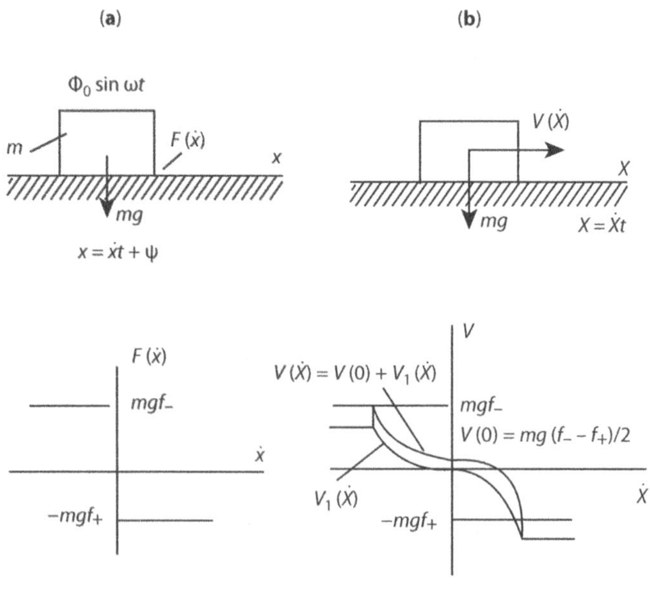

Figure 3.2 Movement of droplet of the horizontal non-uniform surface. (a) Picture for observer of O. (b) Picture for observer of V.

to the movement rightward. Therefore, in each period of the force change, some displacement of the particle to the right will be occurring.

The **V** observer, who does not notice the rapid force $\Phi_0 \sin \omega t$ and the adhesion force, which also "rapidly" changes, will ascribe a forward motion of the particle, to the appearance of some vibration force *V*, acting in the positive direction of the *X*-axis (Figure 3.2b). This force, in particular, may even cause the particle moving up the flat surface, i.e., against the gravity force if the flat surface is tipped at some not too large an angle.

Let us now review the problem based on a solution of the particle motion equation:

$$m\ddot{x} = F(\ddot{x}) + \Phi_0 \sin \omega t, \qquad (3.3)$$

where m is the mass of the particle and

$$\begin{aligned} F(\dot{x}) &= -mgf_+ \quad \text{at } \dot{x} > 0 \\ F(\dot{x}) &= mgf_- \quad \text{at } \dot{x} < 0 \end{aligned} \qquad (3.4)$$

$$-mgf_+ < F(\dot{x}) < mgf_- \text{ at } \dot{x} = 0$$

Here, $F(\dot{x})$ is the adhesion force corresponding to Figure 3.2a.

Equation (3.3) may be solved exactly

$$m\ddot{x} = V, \qquad (3.5)$$

$$m\ddot{\psi} = F(\dot{X} + \dot{\psi}) - \langle F(\dot{X} + \dot{\psi}) \rangle + \Phi_0 \sin \omega t \qquad (3.6)$$

where

$$V = (F(\dot{X} + \dot{\psi})) \qquad (3.7)$$

Here, *V* is the vibratory force which is in this case average for the period the adhesion force is acting on the particle.

If we assume that the amplitude of coercive force Φ_0 is much greater than the maximum adhesion force $F_+ = mgf_+$ and $F_- = mgf-$, the adhesion force may be disregarded for the solution of rapid motion Equation (3.6). Periodical solution of this equation is

$$\varphi = -A \sin \omega t \qquad (3.8)$$

where

$$A = \Phi_0/m\omega^2 \tag{3.9}$$

An approximate expression of the vibratory force is

$$V = V(\dot{X}) = \begin{cases} -mgf_+ \text{ at } X \geq A\omega \\ \dfrac{mg}{\pi}\left[(f_+ + f\_)\arccos\dfrac{\dot{x}}{A\omega} - f_+\pi\right] \text{at } |\dot{X}| \leq A\omega, \\ mgf\_ \text{ at } X^{\cdot} \leq \text{-} A\omega \end{cases} \tag{3.10}$$

then vibratory mechanics basic equation may be written as follows:

$$m\ddot{X} = V\dot{X} \tag{3.11}$$

The particle's steady motion velocity on the plane surface OX is found from an equation $V(\dot{X}_*) = 0$, from which

$$\dot{X}_* = A\omega\cos\frac{\pi f_+}{f_+ + f\_} = \frac{\Phi_0}{m\omega}\cos\frac{\pi f_+}{f_+ + f\_} \tag{3.12}$$

As $V(\dot{X}_*) < 0$, the motion at this velocity is steady.

When $f_+ = f\_$, $X_* = 0$, that is the particle "on the average" remains immobile (is at the state of quasi-equilibrium). Also, with the directional displacement velocity $\dot{X} = 0$ corresponds the vibratory force:

$$V(0) = \frac{1}{2}mg(f\_ - f_+) \tag{3.13}$$

All derived equations remain valid also in the case when the coercive force $\Phi_0\sin\omega t$ works on the particle and the flat surface performs assigned harmonic vibrations with amplitude A and frequency ω. As shown in Figure 3.2b, a vibratory smoothing has occurred of the rupture characteristics adhesion forces. The vibratory force $V(\dot{X})$ may be represented as the sum of two addenda:

$$V(\dot{X}) = V(0) + V_1(\dot{X}) \tag{3.14}$$

where $V(0)$ is determined from Equation (3.13). It is reasonable to call the $V(0)$ force the moving vibratory force and the force $V_1(\dot{X})$ the vibroconverting resistance force. According to Equation (3.14) $V(0) = 0$, the latter force has the nature of viscous friction. This means that the resistance force to the particle movement has an adhesion nature is needed for the particle motion if some finite force is needed. If, however, the particle motion is caused by an indefinitely small force, then one would consider a viscous friction resistance force.

Therefore, as a result of the vibration occurs not only imaginary (i.e., taking place only for the observer **V**) conversion of adhesion forces into viscous forces, but also the formation of a moving vibratory force. A review of this simple system demonstrates a possibility of the directed vibratory displacement.

3.2 Physical Mechanisms and Major Types of Asymmetry Causing Vibratory Displacement

Directional displacements of the dispersed phase emerge only in a case of the system asymmetry. The case reviewed above is only one case of asymmetry emerging at vibration. Figure 3.3a is a schematic representation of four possible types of asymmetry and vibratory displacement of the dispersed phase.

Type *I* asymmetry is usually called "strong". It may be caused either by the action of a constant force T (variant *1*) or by the inclination of a flat surface to the horizontal (variant *2*), which is in effect no different from variant *1*, or by a different value of the module of resistance force F_+ to rightwise motion and to leftwise motion F_- (variant *3*). The arrows along the x-axis, which have equal length in both directions, tentatively show that the vibrations of the surface points are totally symmetric, i.e., occur, for instance, under a purely harmonic law. In the case of type *I* asymmetry, vibration trajectories are also symmetrical relative to the vibrating flat surface and are straight line segments either parallel or perpendicular to this flat surface.

Type *II* asymmetry is called "kinematic". It is caused by one or another asymmetry of the surface points vibration trajectory or of the motion law along this trajectory. As an example, harmonic vibrations are shown along a straight-line trajectory inclined to the surface at some angle. Type *III* asymmetry is called "gradient" asymmetry. In this case, the vibratory displacement is caused by a substantial dependence of parameters defining the particle motion, on the x coordinate. These may be, for instance, the amplitude ψ and phase ϵ, frequency ω and the phase of the force vibratory action, adhesion forces F, vibration trajectories and laws of the surface points, and surface inclination to horizon angle. Forces T and type *IV* asymmetry are sometimes called "wave asymmetry," a particle here is

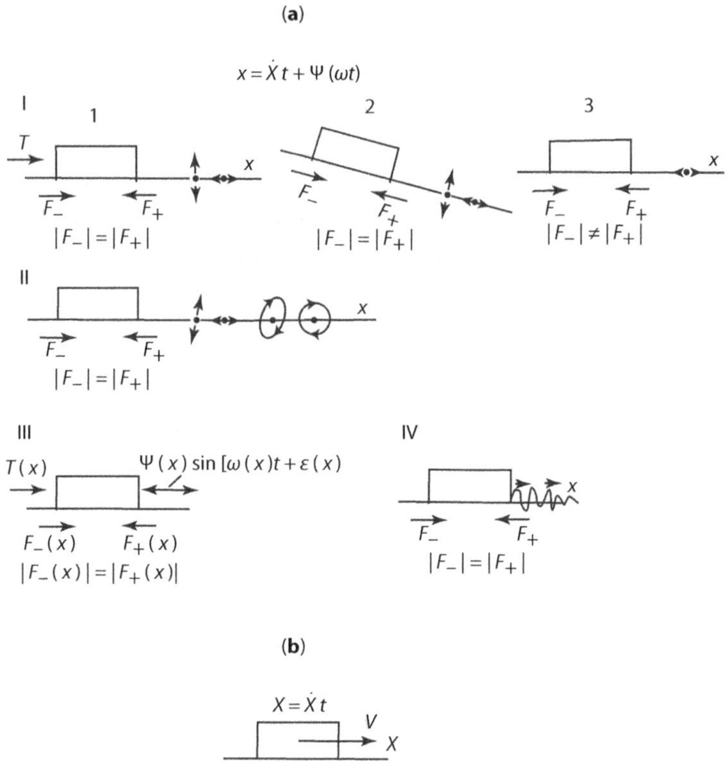

Figure 3.3 Asymmetry types causing vibratory displacement. (a) As observed by observer O. (b) As observed by observer V.

displacing in the spreading direction of the running compression or shear regular wave or individual impulse character waves. A combination of several, e.g., two asymmetry types is possible. The dispersed phase displacement in all these cases may be interpreted within the framework of vibration mechanics: it appears to the observer **V** that it occurs as affected by the vibratory force *V* (Figure 3.3b).

The formation mechanism of vibratory force for the scheme 3 belonging to asymmetry type *I was* already reviewed. As for the scheme 1, in the case of lengthwise vibration it boils down to the scheme *3* assuming that $F_- = F + T$ and $F_+ = F - T$. Scheme *2* is a variant of scheme *1*, for which $T = mg \sin\alpha$, where mg is the weight of the particle and α is the angle of the flat surface to horizon. At shear vibration of the flat surface in the case of schemes *1* and *2*, the particle during one vibration half-period is either torn away from the flat surface or the force pressing it decreases. In either case, force *T* may move the particle along the axis despite the fact that in the absence of vibrations it would have remained immobile. Type *II* asymmetry is best studied. Here, in the case of straight-linear harmonic vibrations at some angle β to the flat surface (different from zero and from 90°), the vibratory displacement emergence mechanism is as follows. During a half-period, when the inertia force in the relative motion is directed to the right and up; it either tears the particle away from the surface or weakens pressure on the surface decreasing the normal reaction and, therefore, the adhesion force. During another half-period when the inertia force is directed the left and down, this force, on the contrary, additionally presses the particle to the surface, thereby increasing the adhesion force. This result in a prevailing particle motion to the right. A prevailing possibility is similarly created for the particle displacement to the right during a certain vibration half-period in the other three cases.

The vibratory displacement mechanism of the type *III* asymmetry does not require any special explanation.

In the case of type IV asymmetry, the running wave as if it is pushing or "carrying" the particle in the direction of its propagation. In the former case, it is asynchronous and, in the latter case, synchronous particle displacement by the running wave.

Next, a model of the directional displacement will be reviewed, which is a generalization of the aforementioned scheme. This is a single-dimensional motion of the particle m in a viscous medium. The medium possesses resistance characterized by the force $F(\dot{x})$ dependent on the relative velocity of the particle in the medium (Figure 3.4). On assuming that is some constant force T acting on the particle and rapidly changing force $\Phi(\Phi = -m\ddot{\xi}\omega t)$ where $\xi(\omega t)$, is some 2π-periodic function of the rapid time

$$(\dot{\xi}) = 0. \tag{3.15}$$

Thus, particle motion equation is

$$m\ddot{x} = Tm\ddot{\xi}(\omega t) + F(\dot{x}). \tag{3.16}$$

In order to solve Equation (3.16), one has to switch from this equation to the appropriate equation from the vibratory mechanics using the method of direct motion separation. Considering the force T as slow and the forces $m\ddot{\xi}$ and $F\_$ rapid, one can derive the following basic equation of the vibratory mechanics:

$$m\ddot{X} = T + V(\dot{X}), \qquad V(\dot{X}) = V(0) + V_1(\dot{X}) \tag{3.17}$$

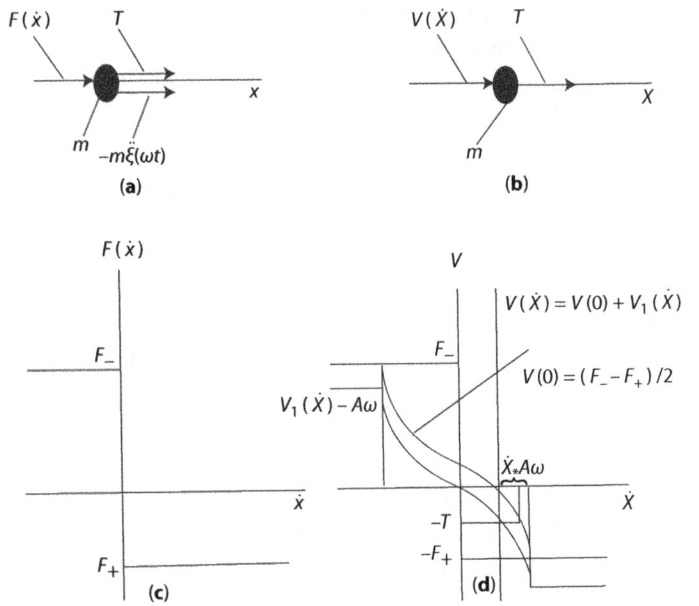

Figure 3.4 Particle motion within the vibrating medium with resistance. (a) Picture for observer O. (b) Picture for observer picture for observer V. (c) Character of the adhesion force. (d) Vibratory force in the vibration field.

As a first approximation, the following correlation of the resistance force to velocity may be accepted:

$$F\dot{x} = -k_{\mp}|\dot{x}|^{n-1}\dot{x}, \tag{3.18}$$

where the value of factor $k_+ > 0$ corresponds to $\dot{x} > 0$; the value of factor $k_- > 0$ corresponds to $\dot{x} < 0$, $n \geq 0$; whereas the value $n > 0$ corresponds to the viscous and $n = 0$, the adhesive action. One can assume that n takes integer-valued values.

First one can consider, the case of a viscous friction ($n \geq 0$):

$$\begin{aligned} &V(\dot{X}) = -\langle k_{\mp}\,|\dot{X} - q\dot{\xi}\,|^{n-1}\,(\dot{X} - q\dot{\xi})\rangle, \\ &V(0) = -q^n\langle k_{\mp}\,|\dot{\xi}|^{n-1}\dot{\xi}\rangle, \\ &V_1(\dot{X}) = V(\dot{X})\text{-}V(0), \quad V_1(0) = 0 \end{aligned} \tag{3.19}$$

At linear resistance, when $n = 1$ and $k_+ = k$, so that $F(\dot{x}) = -k(\dot{x})$, we have $V(0) = 0$ and $V_1(\dot{X}) = -k\dot{X}$, that is, as was to be expected, no moving vibratory force emerges and the slow motion does not depend at all on the vibration nature. At $n = 3$ and $k_+ = k_- = k$ (cubic resistance),

$$V(0) = -kq^3\langle\dot{\xi}^3\rangle, \quad V_1(\dot{X}) = -k(X^3 + 3q^2\dot{X})\langle\dot{\xi}^2\rangle, \tag{3.20}$$

i.e., the vibration results in the emergence of a moving vibratory force and an additional linear resistance $3q^3k\dot{\xi}^2$. For instance, at biharmonic vibration

$$\xi = A \sin \omega t + B \sin(2\omega t + \delta) \tag{3.21}$$

one obtains

$$\langle\dot{\xi}^2\rangle = \frac{1}{2}\omega^2(A^2 + 4B^2); \quad \langle\xi^3\rangle = \frac{3}{2}A^2B\omega^3 \cos\delta$$

consequently.

$$V(0) = \frac{3}{2}kq^3A^2B\cos\delta,$$
$$V_1(\dot{X}) = -k\left[\dot{X}^3 + \frac{3}{2}q^2(A^2 + 4B^2)\omega^2X\right]. \tag{3.22}$$

A case is of interest under conditions of viscous friction when the constant force is absent ($T = 0$). Then, as $V_1(0) = 0$, it occurs that a necessary condition of the vibratory displacement is the distinction from zero of the moving vibratory force $V(0)$. For this, in turn, it is necessary, according to Equation (3.19), that at $k_+ = k_-$, the vibratory action $m\ddot{\xi}(\omega t)$ would be "asymmetric" in the sense that the equality

$$\ddot{\xi}(\omega t + \pi) = -\ddot{\xi}(\omega t) \tag{3.23}$$

would not be complied with, and, therefore, according to Equation (3.15), also the equality

$$\dot{\xi}(\omega t + \pi) = -\dot{\xi}(\omega t) \tag{3.24}$$

In other words, it is required for the curves $\ddot{\xi}(\omega t)$ and $\dot{\xi}(\omega t)$ at some half-period not to repeat, with the opposite sign, its course in the following half. Symmetric are, in particular, the harmonic vibrations. At $k_+ = k_-$, i.e., at equal resistance forces to the particle motions in the positive and negative directions, the vibratory displacement is also possible under a symmetrical vibratory action.

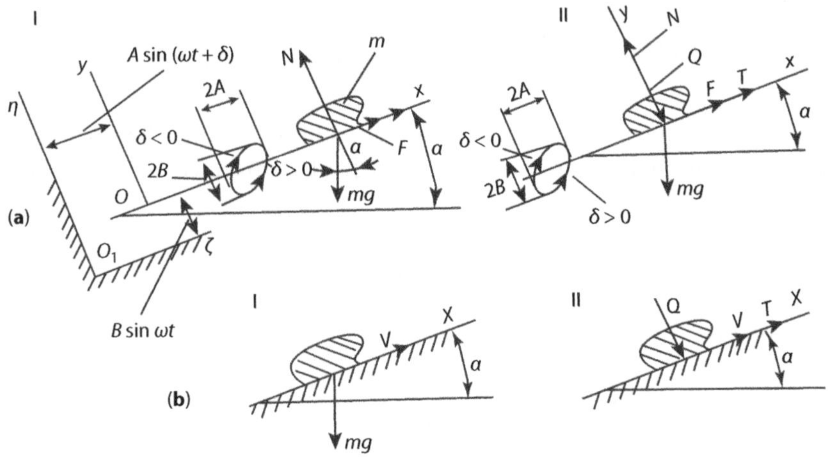

Figure 3.5 Particle motion on an inclined flat surface in a harmonic vibration field in two perpendicular directions. (a) As viewed by observer O (I. initial position; II. action of lengthwise and crosswise constant forces). (b) As viewed by observer V.

Now, one can consider motions of a particle with the mass m on the rough flat surface inclined to a horizontal position at some angle α and performing harmonic progressive advance in two mutually perpendicular directions (Figure 3.5a):

$$\xi = A \sin (\omega t + \delta), \quad \eta = B \sin \omega t. \tag{3.25}$$

Here, ξ and η are displacements of the flat surface points by immobile coordinate axes; at that, the $O_1\xi$ axis is parallel and the $O_1\eta$ axis is perpendicular to the Ox surface—A and B are the amplitudes of, respectively, longwise and crosswise components of the vibrations: δ is the phase shift between these components; at that, the values $\delta < 0$ go with clockwise elliptic points' motions and the values $\delta > 0$ go with counterclockwise elliptic points' motions; ω is the vibration frequency. Vibration trajectories of the flat surface points are ellipses. In xOy coordinates rigidly tied with the vibrating flat surface, the particle motion equations are

$$m\ddot{x} = mA\omega^2 \sin(\omega t + \delta) - mg\sin\alpha + F(\dot{x}) \tag{3.26}$$

$$m\ddot{y} = mB\omega^2 \sin \omega t - mg\cos\alpha + N, \tag{3.27}$$

where $F(x)$ and N are, respectively, adhesion force and a normal reaction of the flat surface.

At $y \equiv 0$, i.e., during the time interval when particle is on the flat surface, it follows from Equation (3.27):

$$N = mg\cos\alpha - mB\omega^2 \sin\omega t, \tag{3.28}$$

and the adhesion force is determined from

$$F(\dot{x}) = \begin{cases} fN\,\mathrm{sgn}\,\dot{x} & \text{at } \dot{x} \neq 0, \\ -f_1 N < F(\dot{x}) < f_1 N & \text{at } \dot{x} = 0, \end{cases}$$

where f and f_1 are the friction factors, respectively, of sliding and rest.

It is necessary to add to the above equations the equalities defining the tie between the lengthwise and crosswise projections of the particle's velocity before and after collision with the flat surface:

$$\dot{y}_+ = \dot{y}_- = -R, \qquad \dot{x}_+ = (1-\lambda)\dot{x}_- \tag{3.29}$$

where $\dot{x}_-$ and $\dot{y}_-$ are projections of the particle velocity before the collision; $\dot{x}_+$ and $\dot{y}_+$ are projections of the particle velocity after the collision; R is the recovery factor; and λ is the instantaneous friction factor.

If the nonequality for parameter ω is complied with:

$$\omega = B\omega^2/g\cos\alpha < 1, \tag{3.30}$$

then, according to Equation (3.28), the normal reaction is $N > 0$ and the particle, once having gotten on the flat surface, remains on it. The intro duced parameter ω plays a significant role and is called the overcharge parameter. Thus, the problem solution boils down to the integration of a quite complex differential equations' system, the differential equations whose substantial nonlinearity of which is caused by the adhesion force.

Of the greatest practical interest are the solutions corresponding with steady regimes of the vibratory displacement, i.e., the solutions similar to

$$\dot{x} = \dot{X} + \dot{\psi}(\omega t), \qquad \dot{y} = \dot{y}(\omega t), \tag{3.31}$$

where $\dot{X}$ is a constant (displacement velocity) and $\dot{\psi}$ and $\dot{y}$ are periodical functions of ωt with the $2\pi n$ period ($n = 1, 2, \ldots$), thus

$$(\dot{\psi}) = 0, \qquad (\dot{y}) = 0. \tag{3.32}$$

The steady regime is understood such motion as is described by Equation (3.31) with a certain set and sequence of stages, i.e., time intervals where by the motion is described as "smooth" (in this case, linear) differential equations (sliding forward, back, flight above the flat surface, relative quiet). Of special interest for applications is $\dot{X}$ velocity determination under steady regimes.

Analytically accurate solutions of this problem are known for the particle motion without tossing. Such regime occurs, for instance, at compliance with the condition of Equation (3.29). As for the regimes with tossing, due to their exceptional multiplicity, they are not completely studied. It was established that in certain domains of the space parameters, especially at the factors' values close to one, chaotic or very complex long-periodic motions occur. With every certain parameter of a steady particle motion regime corresponds to a certain existence and stability area in the parameter space of the system and a certain analytical expression for the displacement velocity $\dot{X}$. Thus, for a motion with tossing, in certain area of the parameter space, such expression may exist and be steady. The displacement velocity $\dot{X}$ at a steady regime, being an integral parameter of the motion, as a rule shows certain "stability" relative to the nature of the regime and continuously changes with the changing system parameters. Such "stability" occurs even under quite complex regimes. In the case of a regime with intense tossing with parameter ω values satisfying the condition

$$3.5\frac{1+R^2}{(1+R)^2} \geq \omega > 2\frac{1+R^2}{(1+R)^2} \tag{3.33}$$

this circumstance enabled to come up with the following approximate equation for the steady displacement velocity:

$$\dot{X} \approx A_\omega \left[\frac{\varphi(\omega,R)}{\omega} \frac{1-R}{1+R}\cos\delta - \sqrt{1-\left(\frac{\varphi(\omega,R)}{\omega}\right)^2 \left(\frac{1-R}{1+R}\right)^2}\sin\delta \right]$$

$$-\frac{\varphi(\omega,R)g}{\omega}\frac{(2-\lambda)}{\lambda}\sin\alpha \tag{3.34}$$

where

$$\varphi(\omega,R) = \frac{1}{2}\left[\frac{\omega(1+R)}{1-R} + \frac{\sqrt{\omega^2(1+R)^4 - 4(1+R^2)^2}}{(1-R^2)} \right] \tag{3.35}$$

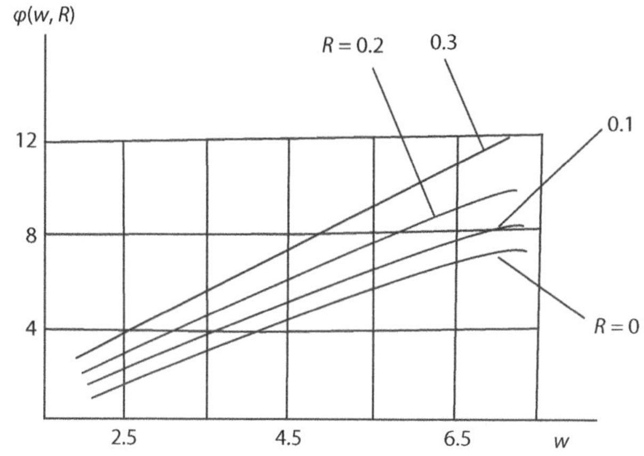

Figure 3.6 φ parameter vs. ω and R.

The diagrams of this function are displayed in Figure 3.6.

At an increase of tossing intensity, when

$$\omega > 3.5\frac{1+R^2}{(1+R)^2}, \tag{3.36}$$

the correlation of Equation (3.34) is replaced by a simpler one

$$\dot{X} \approx A\omega\cos\delta - \frac{1+R}{1-R}\frac{(2-\lambda)}{\lambda}B\omega\,\mathrm{tg}\alpha \tag{3.37}$$

3.3 Directed Motion of the Dispersed Phase in Vibrating Pore Channels

A substantial feature of vibrations is, in that, the phase displacement at which occurs either as a result (1) vibration of the pore channel walls or (2) volume vibration of the dispersed phase particles themselves or (3) their joint vibration. This circumstance predetermines a number of important technical and production benefits of the vibration. Most commonly, there are displacements of the dispersed phase particles, trajectories of which are inclined at some acute angle relative to the pore channels' direction, which corresponds to the case 1 in Figure 3.3.

In this case, the vibration mechanics' basic equation for a dispersed phase single-dimensional displacement of an individual particle may be presented as follows:

$$m\ddot{X} = mg\sin\alpha + T + V(\dot{X}), \tag{3.38}$$

where m is the particle mass, α is the inclination angle of a channel wall to horizontal, T is some constant or slow lengthwise force, $V(\dot{X})$ is a vibratory force which, in a case of pore channel vibrations, substantially depends on the overcharge parameter

$$w = \frac{m|\ddot{\eta}|_{max}}{mg\cos\alpha + Q}. \tag{3.39}$$

Here, Q is a constant or slow force pushing the particle to a pore channel wall, and $|\ddot{\eta}|_{max}$ is the maximum value of the surface acceleration crosswise component.

The T and Q forces are introduced for generality: they may represent, for instance, the action of hydrodynamic pressure, of inertia forces, etc.

Equation (3.38) corresponds to a picture as seen by observer **V** as shown in Figure 3.5b. The displacement velocity at a steady regime is determined from the following equation:

$$mg\sin\alpha = T + V(\dot{X}) \tag{3.40}$$

The stability condition of such regimes is inequality $\dot{V}(\dot{X}) < 0$. Correlations for V may be obtained when the exact or approximate expression for the displacement velocity is known.

Let us take, for instance, the motion of a dispersed phase particle within a cylindrical pore channel. Crosswise and lengthwise vibrations are delivered to the pore channel wall. The crosswise vibrations frequency is half of the lengthwise vibrations frequency ω [1, 2]. The concept of such an action is as follows: Under a proper vibration phasing, a lengthwise inertia force affecting the displacement particle m in relative movement is directed up exactly during those time intervals when it is least of all pressed to the channel walls by the crosswise inertia force's action. During those time intervals when the lengthwise inertia force is directed down, the particle is strongly pushed to the channel walls. The result is the emergence of vibration force directed up. This force is surmounting the particle weight and assuring its rise in the pore channel.

A study covering such system may be performed based on constructing the exact solution of the particle motion's differential equation. The following approximate solution also may be obtained.

Provided there is no slit between the particle m and the channel wall, a differential equation of the particle relative motion in the channel is as follows:

$$\text{m}x = -mg - fm|(\dot{\eta}(\omega t)|\text{sgn})(\dot{x} - m\ddot{\xi})(2\omega t), \tag{3.41}$$

where x is the particle coordinate up along the pore channel's axis, m is the particle mass, f is the sliding friction factor, $\ddot{\xi}(2\omega t)$ and $\eta = \ddot{\eta}(\omega t)$ are periodical functions of own arguments (with the 2π period) describing the law respectively of lengthwise and crosswise channel vibrations.

Let one consider the solution of Equation (3.41) kind:

$$\dot{x}(t) + \dot{X}(t) + \dot{\psi}(t, 2\omega t), \tag{3.42}$$

where $\dot{X}(t)$ is the velocity's slow component (displacement velocity), $\dot{\psi}(t, 2\omega t)$ is the rapid, 2π-periodical component, satisfying for the $2\omega t$ argument the condition:

$$(\dot{\psi}(t, 2\omega t)) = 0. \tag{3.43}$$

The vibration mechanic equations are

$$m\ddot{X} = -mg + V, \tag{3.44}$$

$$m\ddot{\psi} = fm(|\ddot{\eta}(\omega t)|\text{sgn}(\dot{X} + \dot{\psi})) - m\ddot{\xi}(2\omega t), \tag{3.45}$$

where

$$V = -fm\langle fm(|\dot{\eta}(\omega t)|\text{sgn}(\dot{X} + \dot{\psi})\rangle \tag{3.46}$$

On assuming that the friction force is small compared to the lengthwise component of the inertia force $m\ddot{\xi}$, a 2π-periodical solution of the rapid motions' Equation (3.45) for $2\omega t$ is

$$\dot{\psi} = -\dot{\xi} \tag{3.47}$$

and Equation (3.46) for the vibratory force will look as follows:

$$V = V(\dot{X}) = -fm\langle|\ddot{\eta}(\omega t)|\text{sgn}[\dot{X} - \dot{\xi}(2\omega t)]\rangle \tag{3.48}$$

In the case of the pore vibration harmonic channel,

$$\xi = A_x \sin 2(\omega t - \delta), \quad \eta = A_y \sin \omega t \tag{3.49}$$

introducing dimensionless vibratory force V_r and dimensionless velocity $\dot{X}_r$:

$$V_r = V / \left(\frac{2}{\pi} fmA_y\omega^2 \right), \quad \dot{X}_r = \dot{X}/2A_x\omega. \tag{3.50}$$

After averaging Equation (3.48), one derives for

$$V_r = V_r(\dot{X}_r,\delta) = \begin{cases} 1 & \text{at } \dot{X}_r \leqslant -1, \\ -\cos\delta\sqrt{2(1+\dot{X}_r)}+1 & \text{at} -1 \leqslant \dot{X}_r \leqslant \cos 2\delta, \\ \sin\delta\sqrt{2(1-\dot{X}_r)}-1 & \text{at } \cos 2\delta \leqslant \dot{X}_r \leqslant 1, \end{cases}$$

$$(0 \leq \delta \leq \pi/2). \tag{3.51}$$

It is easy to derive expressions for any δ from this equation taking into account the following interrelation:

$$V_r(-\dot{X}_r,\delta \pm \pi/2) = -V_r((\dot{X})_r,\delta),$$

which follows from Equations (3.48) and (3.49). Correlations $V_r(\dot{X}_r,\delta)$ and $V_r(0,\delta)$ are displayed in Figure 3.7c and d. The maximum positive value of V_r occurs at $\delta=\pi/2$. At $0 < \delta < \pi/4$ and $3\pi/4 < \delta < \pi$, the vibratory force is negative; therefore, a displacement up is not possible.

The basic equation of the vibratory mechanics Equation (3.44), in consideration of Equation (3.50) as follows:

$$q\ddot{X}_r = -q_r + V_r(\dot{X}_r,\delta), \tag{3.52}$$

where

$$q = \frac{A_x\pi}{A_y f\omega}, \quad q_r = \frac{g\pi}{2A_y f\omega^2}. \tag{3.53}$$

Velocity values under the stable motion regime are determined from the following equation:

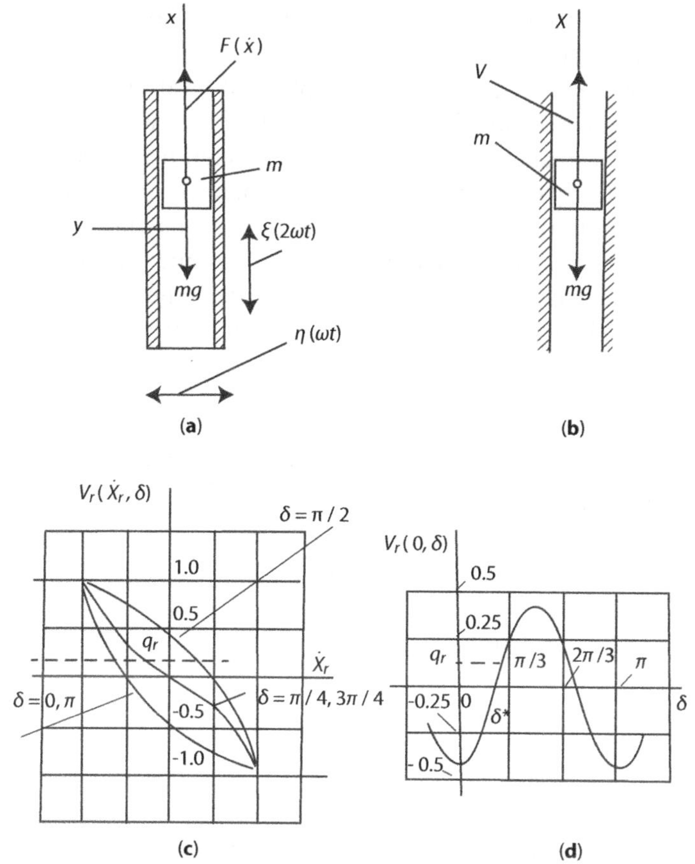

Figure 3.7 Particle motion in a straight-linear cylindrical pore channel. (a) Picture for the observer O. (b) Picture for the observer V. (c) Diagram of a vibratory force $V_r(\dot{X}_r, \delta)$ at $\xi = A_x \sin(2\omega t) - \delta$; $\eta = A_y \sin \omega t$. (d) $V_r(0, \delta)$ for the same case.

$$V_r(\dot{X}_r, \delta) = q_r, \tag{3.54}$$

and this regime is stable because according to Equation (3.51), $\dot{V}_r(\dot{X}_r) < 0$. The solution of Equation (3.54) is determined as the abscissa of Figure 3.7c correlation curves intersection point with the strait line depicting equality Equation (3.54). The figure shows that at $\delta = \pi/2$, the velocity is positive if the crosswise vibration is sufficiently intensive ($q_r < 0.4$). Varying parameters δ and $A_y\omega^2$, it is possible to control the velocity $\dot{X}$ of the particle rise within relatively broad limits. In particular, it is possible to turn $\dot{X}$ to zero, i.e., "hanging" of the particle in a pore channel. The value of a parameter

$\delta = \delta^*$ for such, case is easy to determine from the known q_r value using a correlation diagram in Figure 3.7.

Let us now review the motion of a spherical oil droplet in an infinite length vertical pore channel filled up and wetted with water [8]. The droplet's diameter is smaller than the channel diameter. If the droplet initially has been in a steady state, its buoyancy in the channel occurs nonuniformly. Initially, the motion is accelerated; however, later the buoyancy is occurring practically at a constant velocity. Let us assume that water remains immobile. One may first review the simplest case of the droplet's buoyancy at a constant velocity. In this case, the force of gravity is acting on it:

$$F_1 = \rho_1 v g,$$

where ρ_1 and v are the droplet's density and volume. The buoyancy (Archimedes) force is equal to

$$F_2 = \rho v g,$$

where ρ is the water density.

The viscous resistance force is

$$F_3 = \varphi \rho d^2 V^2.$$

where φ is the viscous resistance factor (a function of the Reynolds number), d is the equivalent diameter of a spherical (equivoluminar) droplet, and V is the droplet velocity. At small Reynolds numbers (Re < 1), the correlation of resistance factor vs. Reynolds number is $\varphi = 3\pi/\mathrm{Re}$. For a spherical droplet, the Stokes correlation is valid:

$$F_3 = 3\pi\mu d V,$$

where μ is the dynamic viscosity of a surrounding water.

The adhesive interaction force between the droplet and the pore channel walls F_{adh} is perpendicular to the walls. Its value is determined based on double layers' interaction condition [4]. Clearance size between the droplet and the channel wall changes depending on the duration of the droplet's contact with the surface. For instance, at the hydrocarbon's fluid droplet contact with the walls of a water-filled channel is observed thinning with time of the water film up to a total break. The thinning velocity is higher for a smaller channel diameter, and, therefore, a film layer is thinner [7]. The film layer's thinning must be accompanied by increasing force of the adhesive interaction. For instance, in the interaction of quartz particles, 3 mkm in diameter, with a quartz surface in a solution of 0.001 to KCl in water, the adhesive force in 10 minutes of contact increases by the factor of 300 and then stabilizes at the maximal level [9]. The film thinning on contact of a droplet with the surface is due to the fact that with the passage of time, affected by the adhesive force pushing the droplet to the surface, the water layer separating the surfaces is squeezed out. At the droplet displacement relative to the channel surface, the force of adhesive interaction with the surface causes the friction force F_4 which may be written according to Amonton's law as follows:

$$F_4 = 0 \qquad (V = 0)$$

$$F_4 = kF_{adh} \qquad (|V| > 0),$$

where k is the friction factor.

Based on the reviewed forces, the droplet's equilibrium condition in a pore channel is

$$vg(\rho\text{-}\rho_1) = \varphi\rho d^2 V^2 + kF_{adh}.$$

From this, one can to determine the droplet's final buoyancy velocity in a pore channel

$$V = \sqrt{\frac{vg(\rho - \rho_1) - kF_{adh})}{\varphi\rho d^2}}.$$

For a spherical droplet surfacing at a low final velocity (Re < 1), one obtains

$$V = \frac{d^2 g(\rho - \rho_1)}{18\mu} - \frac{kF_{adh}}{6\pi\mu r} \tag{3.55}$$

As shown in this equation, increased droplet volume and the water film thickness on the pore channel wall cause a higher droplet buoyancy velocity. The same result is caused by the temperature increase as it decreases water viscosity and the adhesion force.

The oil droplet gravity displacement may be noticeably intensified in a vibration field [3]. In order to prove this, one will bring to the pore channel walls lengthwise vibration with a set velocity $\varphi(t)$. We will assume that water in the pore channel is incompressible and that its every particle has the same velocity as the channel walls. In this case, a similar acceleration of all liquid particles in the pore channel corresponds to the similar pressure gradient in the acceleration's direction in the entire flow. It is equal to

$$\frac{dp}{dx} = -\rho \frac{d\varphi(t)}{dt}.$$

As a result of this pressure drop, an oil droplet in water receives "lifting force" F_{lift} in the direction of acceleration:

$$F_{lift} = v\rho_1 \frac{d\varphi(t)}{dt}.$$

The physical meaning of this force becomes apparent by the following reasoning. If one replaces the surrounding water for an oil droplet, force equal to the product of the liquid mass and its acceleration on the droplet will be acting. As the droplet's density is different from the density of the surrounding fluid, then at equal value of the force acting both on water and on the droplet, the droplet gets acceleration $\frac{dV}{dt}$ different from the acceleration of the enclosing medium. Therefore, droplet motion's acceleration relative to the enclosing medium is equal to $\frac{dV}{dt} - \frac{d\varphi(t)}{dt}$.

This accelerated motion of the oil droplet corresponds to the enclosing water resistance caused by the transfer of the droplet acceleration to its particles. Water resistance force F_c directed opposite to the droplet motion is equal to

$$F_c = \rho_1 \xi v \left[\frac{dV}{dt} - \frac{d\varphi(t)}{dt} \right]$$

where ξ is the factor of the joined mass depending on the droplet shape. For a spherical droplet, $\xi = 0.5$. Therefore, in this case

$$F_c = 0.5 \rho_1 v \left[\frac{dV}{dt} - \frac{d\varphi(t)}{dt} \right].$$

A resulting force acting on the droplet in the channel must be equal to the product of the droplet mass $\rho_1 v$ and its acceleration $\frac{dV}{dt}$. Taking into account forces reviewed above, the droplet motion equation at lengthwise vibrations of the capillary is as follows:

$$\frac{dV}{dt}=\frac{1}{(\rho_1/\rho+\xi)}\left\{(1+\xi)\frac{d\varphi(t)}{dt}+g\left(1-\frac{\rho_1}{\rho}\right)\right.$$
$$\left.-\frac{1}{v}\left[\frac{kF_{adh}}{\rho}+\varphi^2 d^2(V-\varphi(t))^2\right]\right\} \tag{3.56}$$

Two important conclusions are reached from this equation:

1. The lower droplet's density, the higher its acceleration. Therefore, at vibration of the enclosing fluid, later droplets get ahead of heavier ones.
2. The smaller droplet volume, the smaller its acceleration. Therefore, on vibration, larger droplets volume get ahead of smaller ones.

Disregarding resistance force of the enclosing water, the following condition of the droplet motion follows from (3.53):

$$\frac{(1+\xi)\dot{\varphi}(t)}{g}+\left(1-\frac{\rho_1}{\rho}\right)\geq\frac{6kF_{adh}}{\pi d^3\rho g}.$$

In a field of sinusoid vibrations, $\varphi(t) = \varphi_0 \sin \omega t$, where φ_0 is the amplitude of vibration velocity and ω is the cyclic frequency. Thus, estimate of the maximum vibration amplitude is

$$\left[\frac{\omega^2 A}{g}+\left(1-\frac{\rho_1}{\rho}\right)\right]\geq\frac{6kF_{adh}}{\pi d^3\rho g},$$

where A is the maximum vibration amplitude.

Equation (3.56) with variable factors is integrated in quadratures only in exceptional cases. In the case of a slow droplet motion relative the pore channel and spherical shape of the droplet, one obtains a non-homogenous linear equation. The general solution is

$$V-\varphi(t)=Z\exp(-pt)\int_0^t\dot{\varphi}(t)\exp(pt)dt+\frac{R}{p}[1-\exp(-pt)];$$

where $R=\frac{g}{a}\left(1-\frac{\rho_1}{\rho}\right)-\frac{6kF_{a,\pi,r}}{\pi a d^3\rho}$, $a=\left(\frac{\rho_1}{\rho}+\xi\right)$, $p=\frac{18\mu}{ad^2\rho}$,

$$Z=\left(\frac{1+\xi}{a}-1\right). \tag{3.57}$$

For a sinusoidal vibration after integrating:

$$V-\varphi(t)=\frac{Z\omega\varphi_0}{p^2+\omega^2}[p\cos(\omega t)+\omega\sin(\omega t)$$

$$-p\exp(-pt)]+\frac{R}{p}[1-\exp(-pt)], \tag{3.58}$$

The vibration velocity amplitude, appropriately, is

$$V_0-\varphi_0=\frac{Z\omega\varphi_0}{p^2+\omega^2}\left[\omega-p\exp\left(-\frac{p\pi}{2\omega}\right)\right]+\frac{R}{p}\left[1-\exp\left(-\frac{p\pi}{2\omega}\right)\right].$$

If the frequency does not exceed 1–3 KHz, the droplet velocity amplitude in the pore channel is

$$V_0-\varphi_0=\frac{Z\omega^2\varphi_0}{p^2}+\frac{R}{p}.$$

Equations (3.56) to (3.58) remain valid, also when considering initial period of rising droplet and immobile liquid in pore cannel, from the quiescence's immobile liquid: thus,

$$V=\frac{R}{p}(1-\exp(-pt)).$$

It is easy to show that at $t\to\infty$ this equation determines the end velocity of droplet buoying in a fluid-filled capillary channel and agree with Equation (3.55).

There is an emergence of a "directional on average" saturating fluid motion as affected by the vibration of the porous medium with various configuration of pore channels. Such a vibration, as a rule, results in the emergence of vibratory lifting force or thrust force $V(\dot{X})$ and causes, in a general case, some directional, on average, flow through the porous medium (Figure 3.8).

Figure 3.9 is a schematic of creating a directional fluid flow by way of vibrating (as a rule, symmetrical) of the porous medium with narrowing holes or slit channels [10]. Due to a substantial system's asymmetry (the resistance to fluid motion in the direction toward of holes' narrowing and slits is lower than at motion toward widening), a pulsating fluid flow emerges directed on average toward the narrowing. The observer **V** will tie the emergence of this

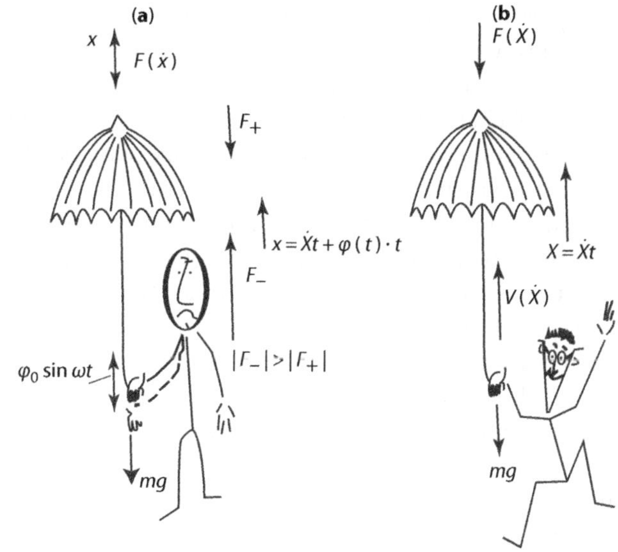

Figure 3.8 Schematic of emergence of vibratory lifting force. (a) As seen by observer O. (b) As seen by observer V.

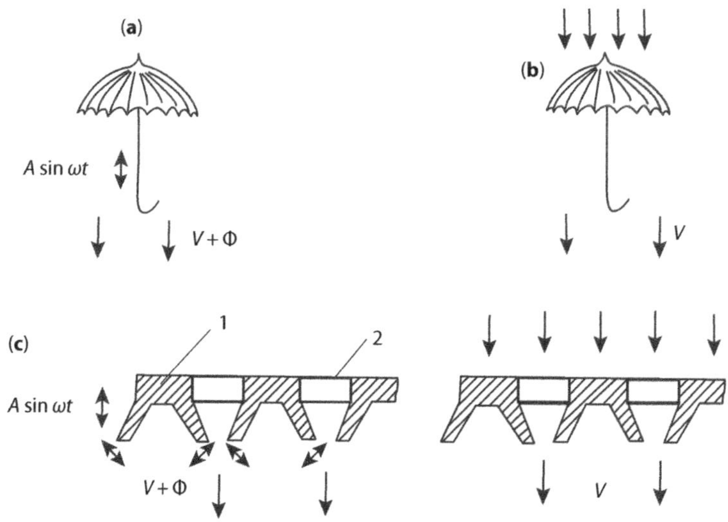

Figure 3.9 Effect of the vibrating pore channels' configuration on the directional fluid motion. (a) As seen by observer O. (b) As seen by observer V. (c) schematic directional fluid motion through porous channels: 1. Perforated slab; 2. Narrowing holes or slit channels.

flow's constant portion with the appearance of some pressure gradient. This gradient, algebraically summing up with the available gradient, will either strengthen or weaken the flow. The latter may even cause the flow stoppage (vibratory closing of channels) or emergence of a reverse flow. Some believe that exactly this phenomenon is the cause of stoppage of fuel supply from the airplanes' fuel tanks, which in some cases resulted in emergency situations. If narrowing channels look like deformable petals (rigidity of which is selected based on the condition of closeness to resonance with the frequency increase of excited vibrations), then the directional flow is strengthening. The petals play the role of special valves which at the matrix vibration permit more fluid in the direction of channel narrowing than in the direction of channel widening. As a result, directional on average flow is formed, which may be more intensive than with underformed channel walls.

For a quantitative evaluation of the emerging directional flow, one can examine vibrations of a porous layer placed in the immobile fluid (Figure 3.10). The velocity $\dot{X}$ of the layer's motion relative to the fluid will be connected with the fluid Q throughput of the layer by the following equation:

$$Q = -S\dot{x} \tag{3.59}$$

where S is the layer's area. On assuming that a harmonic coercive force $\Phi_0 \sin\omega t$ and some constant force P are acting on the layer with which the

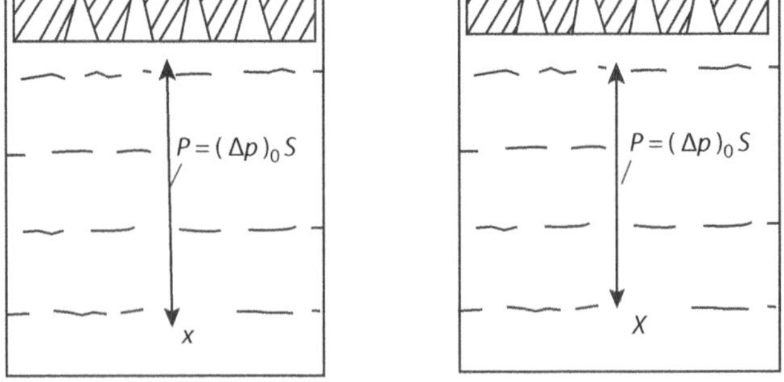

Figure 3.10 Schematic directional fluid motion in vibrating pore channels. (a) As seen by observer O. (b) As seen by observer V.

corresponding to the static pressure gradient $(\Delta p)_0$ along the layer's thickness, so that

$$P = -(\Delta p)_0 S. \tag{3.60}$$

One may consider that the resistance to the layer motion correlates with the squared velocity $\dot{x}$ of the layers motion relative to the fluid, where up on the resistance factor k_+ at the layer's motion in the positive direction of the x-axis is smaller than the resistance factor k_- at the layer's motion in negative direction of the x-axis. This case corresponds to the direction of the holes' narrowing or slits in the layer. Thus, the resistance force is defined by the following equation:

$$F(\dot{x}) = -k_{\pm}\dot{x}^2 \operatorname{sgn}\dot{x} = -k_{\pm}\dot{x}|\dot{x}| \quad (k_- > k_+) \tag{3.61}$$

This correlation is in agreement with a known hydraulics' equation for fluid effluence through a hole:

$$Q = \mu S_0 \sqrt{2\Delta p/\rho}, \tag{3.62}$$

where S_0 is the hole area in its narrow portion, ρ is the fluid density, Δp is the pressure gradient, and μ is the nondimensional factor of the fluid flow rate ($\mu < 1$). The fluid effluence through a hole in the direction of hole narrowing is greater than at the fluid effluence through a hole in the direction of hole widening ($\mu_- > \mu_+$).

Equation (3.62) becomes Equation (3.61) if one assumes $Q = -S\dot{x}$, $\Delta p = -F(\dot{x})/S \operatorname{sgn}\dot{x}$ at $\mu = \mu_{\pm}$
and define

$$k_- = \frac{\rho S^3}{2\mu_+^2 S_0^2}, \qquad k_+ = \frac{\rho S^3}{2\mu_-^2 S_0^2} \qquad (\mu_+ < \mu_-,\ k_+ < k_-) \tag{3.63}$$

The factors μ_+, μ_-, k_+, and k_- may be significantly different from their static values and be dependent both on the frequency and amplitude of the layer vibration.

Under the indicated assumptions, the layer motion equation is

$$m\ddot{x} = P + \Phi_0 \sin\omega t + F(\dot{x}), \tag{3.64}$$

where m is the layer's mass. The forces P and $F(\dot{x})$ are determined from Equations (3.60) and (3.61).

The basic equation of the vibratory mechanics will thus be

$$m\ddot{X} = P + V(\dot{X}), \tag{3.65}$$

where the vibratory force $V(\dot{X})$ is determined from

$$V(\dot{X}) = \frac{1}{4}A\omega^2(k_- - k_+) - \frac{3}{2\pi}A\omega(k_+ + k_-)\dot{X}. \tag{3.66}$$

where $A = \Phi_0/m\omega^2$. The following equation for the determination of an average fluid Q throughput through the layer is as follows:

$$Q = -S(\dot{x}) = -S\dot{X} = \frac{\pi}{3}\left[\frac{2(\Delta p)_0 S^2}{A\omega(k_- + k_+)} - \frac{1}{2}A\omega S\left(\frac{k_- - k_+}{k_- + k_+}\right)\right]. \tag{3.67}$$

At $\Delta p = 0$, i.e., in case of absence of the static pressure gradient

$$Q_* = \frac{-\pi}{5}A\omega S\left(\frac{k_- - k_+}{k_- + k_+}\right) \tag{3.68}$$

The minus sign in front of this formula indicates that the flow direction is toward narrowing of the holes in the layer (a reminder: $k_- > k_+$). From this equation follows one simple expression for a stable fluid flow velocity, which corresponds to the vibratory closing effect of the pore channels in the layer:

$$(A\omega) = 2\sqrt{(\Delta p)_0 \frac{S}{k_- - k_+}} \tag{3.69}$$

This equation may be used for experimental determination of the resistance factors difference $(k_- - k_+)$.

3.4 Directional Motion of the Vibrating Dispersed Phase in Pore Channels

Authors investigated conditions of a dispersed phase particle directional motion emerging due to its periodical volume vibration. A particle interaction with the external medium requires a change in the average velocity of its center of inertia motion. Otherwise, a known position of the mechanics of the impossibility of such a change only at the expense of the internal forces would be violated.

Such motion is one manifestation of the directional displacement—getting directional on average motions at the expense of nondirectional actions. Energy needed for the motion may be extracted either from an internal source relative to the particle or may be coming from outside. A basic element of such directional displacement is sufficiently large oil or water bubble found, for instance, in a narrowing pore channel. The channel shape is such that the medium resistance against the particle motion up along the axis is much smaller than resistance against the particle motion down (in this sense, the particle is asymmetric and imaged in Figure 3.8 as an umbrella). As a result of the particle volume axial vibrations (excited by one or another method) emerges the vibrational lifting force $V(\dot{X})$ which under certain conditions may exceed the particle weight: then the particle would be rising. A concept of the vibratory force emergence in this case is similar to the one shown in Figure 3.3 (force asymmetry).

On exploring a directional particle displacement schematically shown in Figures 3.4a and b. We will assume that the medium resistance is different at the particle motion up and down and the coercive force $\Phi(\omega t)$ is harmonic. Then, the particle motion equation may be presented as follows:

$$m\ddot{x} = -\mathrm{mg} + \Phi_0 \sin\omega t + F(\dot{x}) \tag{3.70}$$

Equation (3.67) is a particular case when

$$T = -mg, \quad -\mathrm{m}\,\ddot{\xi}(\omega t) = \Phi_0 \sin\omega t, \quad (n=2). \tag{3.71}$$

The vibratory mechanics' basic equation in this case will be

$$m\ddot{X} = \mathrm{mg} + V(0) + V_1(\dot{X}) \tag{3.72}$$

where, according to Equation (3.20),

$$\begin{aligned} &V(\dot{X}) = -\langle k_{\pm}\lceil \dot{X} - \dot{\xi} \rceil (\dot{X} - \dot{\xi})\rangle, \\ &V(0) = \langle k_{\pm}\lceil \dot{\xi} \rceil (\dot{\xi})\rangle, \qquad V_1(\dot{X}) = V(\dot{X}) - V(0) \end{aligned} \tag{3.73}$$

$$\dot{\xi} = \frac{\Phi_0}{m\omega}\cos\omega t. \tag{3.74}$$

Introducing a denotation:

$$A = \Phi_0 / m\omega^2, \tag{3.75}$$

Equation (3.72) can be presented as follows:

$$\begin{aligned} V(\dot{X}) &= -\langle k_{\pm} \lceil \dot{X} - A\omega \cos \omega t \rceil (\dot{X} - A\omega \cos \omega t) \rangle, \\ V(0) &= (A\omega)^2 \langle k_{\pm} \lceil \dot{\xi} \cos \omega t \rceil \cos \omega t \rangle. \end{aligned} \tag{3.76}$$

These approximate equations are valid when $\Phi_0 \gg F(\dot{x})$. Calculating Equation (3.76) one gets at $t = \arccos\ \dot{X}/A\omega$

$$\begin{aligned} &V(\dot{X}) = V(0) + V_1(\dot{X}) = \\ &= \frac{1}{2\pi}\left[-k_+ \int_{t_1}^{2\pi - t_1} \left(\dot{X} - A\omega\cos\omega\right)^2 dt + k_- \int_{2\pi - t_1}^{2\pi + t_1} \left(\dot{X} - A\omega\cos\omega\right)^2 dt\right] = \frac{1}{2\pi}\left\{ \quad -k_+\left[\quad (2\dot{X}^2 + \right.\right. \\ &A^2\omega^2) \cdot \left(\pi - arccos\frac{\dot{X}}{A\omega}\right) + 3A\omega X\sqrt{1 - \frac{\dot{X}^2}{A^2\omega^2}}\,\Bigg] + k_-\Bigg[\quad (2\dot{X}^2 + A^2\omega^2)\arccos X/_{A\omega} - \\ &3A\omega\dot{X}\sqrt{1 - \frac{\dot{X}^2}{A^2\omega^2}}\,\Bigg]\Bigg\} = \tfrac{1}{2}\pi\left\{(k_+ k_-)\left[(2\dot{X}^2 + A^2\omega^2)\arccos\dot{X}/_{A\omega} - 3A\omega\left(\dot{X}\sqrt{1 - \frac{\dot{X}^2}{A^2\omega^2}}\right)\right] - \right. \\ &\left.\pi k_+(2\dot{X}^2 + A^2\omega^2) \qquad \right\}, \text{at } |\dot{X}| < A^2\omega^2; \\ &V(\dot{X}) = -k_{\pm}\left(\dot{X}^2 + \frac{A^2\omega^2}{2}\right)\text{sgn}\dot{X} \text{ at } |\dot{X}| > A\omega;\ V(0) = \frac{1}{4}(A\omega)^2(k_- - k_+) = \frac{1}{4}\Phi_0/m\omega^2(k_- - k_+) \end{aligned} \tag{3.77}$$

It follows from Equations (3.71) and (3.76) that for the particle going up this condition must be complied with:

$$\frac{1}{4}\Phi_0/m\omega^2(k_- - k_+) > mg. \tag{3.78}$$

i.e., resistance to the particle moving down must exceed the resistance to the particle moving up.

Experiments with directional motion of vibrating hydrocarbon fluid droplets colored with a fast pigment within capillaries of molybdenum glass (modeling cylindrical pore channels) qualitatively support the above equations [6]. Capillary diameters have been 0.8 to 1.5 mm. The studies have been conducted in a closed system: the pore channel where the droplet displacement had been occurring has been connected by its ends with the other fluid-filled channels. This way has been modeled a porous medium in which an elementary pore channel has been identified. A droplet of hydrocarbon fluid (kerosene) in the channel was moving. It has been done by the lower end of the capillary has been placed in the fluid and on the upper end of the capillary has been slipped on a polyvinylchloride tube the other end of which has been placed in the same vessel. Using a waterjet pump, a capillary has been filled up with water or appropriate solution as well as the capillary tube. Then, the tube has been connected to the same vessel which the capillary has been connected to. A 100% water saturation of the system was achieved. The kerosene droplet was introduced by a medical syringe in the lower end of the capillary made in the shape of a funnel. After the introduction of the droplet and measuring the velocity of its displacement, the pore channel has been flushed through sequentially with benzene, potassium dichromate, and distilled water and then dried up. The length of the capillary, where the droplet displacement duration and velocity have been measured, was 400 mm. The length of the introduced kerosene droplets varied between 3 mm and 20 mm.

The effect of the droplet volume vibrations within the channel has been modeled by a way of lowering interfacial tension at the boundary between the droplet and enclosing medium. As the enclosing medium, solutions of alkali-mineralized water with concentration of 0.1 and 2% NaOH in a 10% water solution of NaCl have been used. Interfacial tensions of the medium at the droplet boundary have been, respectively, 6 mkH/m and 0.015 mkH/m. Experimental results (Figure 3.11) indicates that the kerosene droplet has been displacing in the channel of a constant diameter at a velocity directly proportional to the droplet length and inversely proportional to the interfacial tension value at the droplet boundary. At the same droplet's length, a decline of the interfacial tension caused by changes in its volume has resulted in a substantial increase of its raising velocity within the pore channel. Based on these experimental results, the displacement model of vibrating hydrocarbon fluid droplet within a cylindrical porous channel with a hydrophilic surface was as follows. The droplet displacement was occurring as a single body with stable boundaries. A fluid flow within the channel between the droplet's side surface and the internal wall of the pore channel is absent. The volume variations of the droplet result not only in lowering of the interfacial tension but also in lowering of the saturating medium is viscosity at the boundary with the droplet.

1 and 2. Droplet diameter is 1.5 mm; 3 and 4. Droplet diameter is 1.0 mm; 2 and 4. Interfacial tension is 6 mk H/m; 1 and 3. Interfacial tension

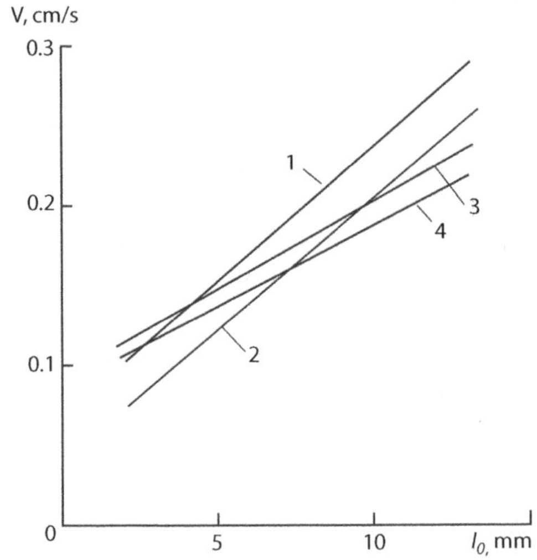

Figure 3.11 Directional displacement of a vibrating kerosene droplet within a capiullary.

is 0.015 mk H/m. V is the steady droplet displacement within the capillary. l_0 is the length of the droplet.

Let us present the droplet displacement equation within a cylindrical channel inclined at the angle α to horizontal. Along the x-axis, on the droplet are acting:

- Pressure from the fluid

$$p_1 = p_0 - \rho gh \sin\alpha;\ p_2 = p_0\ \rho g(h + l_0) \sin\alpha,$$

where p_0 is the atmospheric pressure, ρ is the fluid density, h is the depth of droplet position, and l_0 is the droplet's length;

- The droplet weight:

$$P_{\mathrm{k}} = \rho_1 gv \sin\alpha$$

where ρ_1 and v are droplet's density and volume:

- The friction force acting on the droplet's side surface:

$$F_{fr} = \pi\mu d_k\, l\dot{x} / l_0,$$

where l is the distance to the droplet's upper end, $\dot{x}$ is the droplet's displacement velocity, and d_k is the droplet's diameter.

- the force of dynamic pressure (flow resistance):
- $F_d = \pi k_0 \mu d_k \dot{x}$,

where k_0 is the factor accounting for the irregularity of the velocity field at the frontal surface of the droplet;

- The water solution resistance emerging at the droplet rise within an L-long capillary

$$F_c = 8\pi\mu L\dot{x}.$$

Taking the above described forces into account, the motion equation is

$$(\rho - \rho_1)v\ddot{x} = V_k g\Delta\rho \sin\alpha - (\pi\mu l d_k / l_0 + \pi k_0 \mu d_k + 8\pi\mu L)\dot{x}$$

with the initial conditions $x(0) - \dot{x}(0) = 0$. The solution of this equation for the droplet's displacement steady velocity is

$$V = \frac{g\Delta\rho l d_K^2 \sin\alpha}{4\mu(l d_K / l_0 + 8L)};$$

It was assumed here that $L \gg d_k$ and $k_0 \approx 1$.

For the steady-state velocity of the droplet displacement in the capillary within which the velocity distribution is controlled by law the one obtains

$$V = \frac{g l d_k^2 \sin\alpha}{8\mu L}.$$

The viscosity decline of the saturating medium at the boundary with the droplet, caused by the droplet's volume vibration, results in its increased speed of motion in the capillary.

References

1. Blkekhman, I.I. and Dzhanelidze, G.Yu., *Vibration displacement*, vol. 32, p. 250, Nauka, Moscow, 1964.

2. Brumberg, P.M., On unseparated motion of a solid body within a vibrating pipe. *Izv. AN USSR, Ser. Mechanics solid body*, 5, 135–139, 1970.
3. Ganzha, V.L., Simkin, A.M., Zhuravsky, G.I., Heat- and mass-transfer in multiphase media. *Minsk, Sci. Technol.*, 280, 1990.
4. Kapitsa, P.L., Dynamic stability of a pendulum at vibrating suspension point. *ZhETF*, 21, 5, 260–261, 1951.
5. Zimon, A.D., *Adhesion of the dust and powders*, p. 300, Chemistry, Moscow, 1976.
6. Kusakov, M.M. and Mikenitskaya, L.I., *On the thickness of bonded water thin layers*, p. 70, Publ. AN USSR, Moscow, 1955.
7. Kuznetsov, O.L. and Simkin, E.M., *Transformation and interaction of geophysical fields in lithosphere*, p. 267, Nedra, Moscow, 1990.
8. Simkin, E.M., Mathematical model of an oil droplet gravity buoyance in a capillary at its lengthwise vibration. *Coll. Proc. VNIINeft*, 54–58, 1986.
9. Surguchev, M.L., Zhltov, Yu.V., Simkin, E.M., *Physicochemical processes in oil-gas-bearing reservoirs*, p. 300, Nedra, Moscow, 1984.
10. Usakovsky, V.M., *Inertia pumps*, p. 180, Machine building, Moscow, 1973.

4
Formation Damage Control and Cement Sheath Stability

Production of oil/gas from the oil wells in near-well bore zones occurs at elevated pressures and temperatures and requires special attention. The depth of zone of linear filtration law may reach 6 to 23 m from the wall of a well [10].

Thermodynamically, a near-well wall reservoir rock zone resides in a substantially unbalanced state, in the phase of active energy and mass exchange with the well and reservoir. Its properties are continuously changing in the process of field development. This area to a large extent determines the development success of a productive accumulation.

4.1 Status of the Reservoir

In the development process of oil and gas fields, a gradual deterioration of reservoir properties is observed. This is especially true in low-permeability and thickness-nonuniform reservoirs. The deterioration of natural permeability is occurring already during well drilling process when compressing stresses emerge in a ring zone near the borehole. Drilling mud filtrate is filtering into the reservoir and forming on the walls of a well (2 to 3 mm thick). Many studies [29] found a possibility of even deeper penetration of the drilling mud filtrate into productive reservoir through fractures some emerging due to high hydrostatic pressure exceeding formation fracturing pressure. Such a pressure may occur at restoration of the flushing fluid circulation or at round trip operations. This is a reason of the observed decline in the rock absolute permeability by a factor 1.8 to 42. In some cases, the absolute permeability declines to zero.

The drilling mud filtrate is capable of penetrating productive reservoirs for great distances reaching 0.2 to 3 m [30]. A notable deterioration of the natural reservoir properties occurs for the following reasons:

George V. Chilingar, Kazem Majid Sadeghi, and Oleg Leonidovich Kuznetsov. Acoustic and Vibrational Enhanced Oil Recovery, (89–122) © 2022 Scrivener Publishing LLC

- Swelling of the clays, when the filtrate water is different in salinity from the reservoir water.
- A decline in the relative permeability to oil at the appearance of water blocks [6].
- Emergence of capillary phenomena at water-oil contact.
- Formation of stable emulsions (mostly, of a "water in oil" type); they are poorly filtering not only due to a higher viscosity than a pure oil, but also due to the thixotropic properties.
- Blockage of reservoir pores by insoluble in water and in oil solid deposits forming chemical resulting form the interaction of drilling mud filtrate with reservoir fluids.

The drilling mud filtrate penetration of the reservoir for a distance of just a few centimeters decreases the reservoir sweep on waterflooding by 30% to 40%. Studies of oil wells indicated that, due to the aforementioned reasons in an overwhelming majority of wells, the operating perforated portion of the productive reservoir is not performing; the performing part is making up only 25% to 30%.

The reservoir is subjected to significant changes upon installing casing and cementing operations. There is a physicochemical and hydrodynamic interaction between the cementing slurry and reservoir fluids, depending on thermodynamic conditions and lithology. All this results in changes in the filtration parameters. Upon the completion of cementing, mechanical stress acting in the ring-space around the borehole begins unloading on the casing with the set cement. That causes changes in the porosity/permeability properties of the rocks.

In the process of exploitation, further deterioration of the reservoir properties occurs. Beside the already mentioned causes of reservoir property deterioration, there is the deposition of naphthenic and paraffin fractions. This creates additional resistance to filtration and decreases the permeability [2]. For a detailed treatment of drilling and drilling fluids, reader is referred to the book entitled "Drilling and Drilling Fluids" by Chilingarian and Vorabutr (1981) [42].

The permeability deterioration occurs also in injection wells. This process is going on especially intensively at water injection in the reservoir due to a gradual blocking of pores in the reservoir by suspended mechanical particles and oil products contained in water. The mechanical particles content in the injected water above a certain limits is a cause of a rapid and noticeable decrease in intake and, sometimes, of a total stoppage of flow [1]. The water injected in a reservoir is, as a rule, different in chemical composition and temperature from the interstitial water in the pores. This causes a disruption of the natural physicochemical balance in the reservoir and swelling of clays. Distortions of the physicochemical balance may cause the formation of insoluble deposits, deposition of salts and precipitation of paraffin crystals in the pore channels [21, 22].

A decline in permeability may be caused by adsorption of asphaltic-resinous substances and on the pore channel surfaces, which will decrease the effective cross-section of filtration channels [11]. An interaction of the injected water with the oil present in the oil formation, at a sufficiently high filtration rate and at the presence in the oil of natural emulsifiers stabilizers (asphaltenes and resins), may cause the formation of stable water-in-oil emulsions [4, 44, 45]. As a result, there is a drastic increase in the viscosity of an injected fluid.

Cleaning up the damaged zone, by improving its thermodynamic status, i.e., measures facilitating the recovery of permeability. As the field experience indicates, most perspective in this sense are such methods which do not result in the formation of new nonuniformities. This is a reason, for instance, why the reservoir hydraulic fracturing, while causing an increase in the current oil production, is increasing the reservoir nonuniformity and, therefore, may decrease the reservoir oil yield as a whole [34].

Among favorable methods of defeating formation damage, it may turn out to be the vibration. Low-frequency vibrations accelerate by two or three orders of magnitude the mechanical tension relaxation processes [15]. This may eliminate negative consequences of well drilling and production associated with undesirable tension of walls of a well and of the perforation channels and with mechanoactivation of the rock surface. Thus, natural balanced state of the reservoir rock is restored and the initial reservoir permeability is achieved. The experimental results indicate changes in absolute permeability of the saturated porous media under the high-amplitude low-frequency vibrations of liquid on the order of 0.3 MPa as produced by a hydraulic vibrator [20]. Relative permeability changes in artificial cores reach 30%. This is due to the formation of (1) new filtration channels in porous medium, (2) porosity changes, (3) fractures, and (4) media's grain repacking and orientation change. These processes strengthen when the clay content increases to 35%.

Other effects are directly associated with the effect of colmatants (sealants) and fluids upon their interaction with the reservoir matrix. The writers experimentally observed changes in the rheological behavior of non-Newtonian fluids, which possess viscous-elastic and viscous-plastic properties, in the vibration field. Studies of oil shearing viscosity affected by the elastic vibrations with intensity of 8 to 100 kW/m^2 and frequency between 20 Hz and 4.5 MHz showed that the shear viscosity after the vibrations declined by 20 to 30% and then recovered after some time (to the pre-cavity regime) or not completely recovered (after development of cavitational action regime) [31]. The maximum viscosity changes under the pre-cavity regime at low frequency were observed with an increase in the content oil of asphalt-resinous and paraffinic components. The viscosity recovery time after the vibration is 5 to 6 hours and longer.

Studies have been conducted of a rheological behavior of oils with a large resins (29%) and asphaltenes (11.3%) content as affected by high-frequency vibrations (18 kHz) using a rotational viscosimeter. Immediately after the treatment, the oil viscosity measured at low deformation speed declined by 28%. At increasing rotor revolution rate and, correspondingly, the deformation speed, viscosity recovered to its initial value. A regular correlation format of oil viscosity vs. deformation rate is recovered with time after the treatment; however, the viscosity, increases by 10% to 25% compared to the viscosity prior to the treatment. This result is due to the formation of new structure of oil after the treatment, which may serve as an example of changing thermodynamically unstable state by vibration.

Another result of this study (measured immediately after the treatment is unexpected correlation between the oil viscosity and deformation speed) may be explained by studying a periodical deformation of structured polymer fluids described by Briedis (1978) [9]. The high-frequency vibrations are acting mainly on the fluid's boundary layers with the solid phase. They do not render any noticeable effect on the non-Newtonian fluid volume structure. Contrary to that, low-frequency vibrations deform the fluid structure mostly within its entire volume. Taking these results into account, it is possible to assume that at measuring oil viscosity with a rotational viscosimeter at slow rotor revolution rate, the deformation occurs in the near-surface oil layer destroyed by the high-frequency action. The thickness of an oil layer subjected to the shear deformation increases with increasing of the deformation speed. Parameters are measured already in the oil volume layers with the undestroyed structure, and that shows viscosity increase to the initial value. This is supported by the fact that, in similar studies in a low-frequency range (20 to 70 Hz), the viscosity decline in structured oils immediately upon the vibration has been observed in the entire range of the deformation rates. At low rate, an increase of the vibration frequency notably affected the viscosity decline.

Multiple studies [33, 41] indicate degassing of the reservoir fluids affect by either low or high frequency vibrations. The degassing process continues up to establishment of a new balanced concentration value, which is always lower than the gas concentration balanced value without vibration. The vibration intensity and frequency define the rate of the gas concentration change and the timing of establishing new balanced concentration.

The gas release effect by the reservoir fluids in the vibration field depend on the specific conditions that may most diversely affect the reservoir properties.

Beside changes in the permeability and viscosity affecting the flow through the properties of fluids themselves, the following experimental results have been observed:

- An increase (up to 18-fold) in the relative water or model Newtonian fluid flow rate through sandstone cores. The vibration intensity has been a few hundred kW per 1 m^2 and frequency of 3 to 10 kHz [10];
- An increase (up to 10-fold) in the flow rate of polar and nonpolar fluids, dielectrics, and electrolytes. The vibration intensity has been 1.9 kW per 1 m^2 and frequency of 17 kHz [10];
- An increase (almost 100-fold) in the flow rate of distilled water and salt solutions through the natural cores upon vibration with frequency of 26.5 kHz [36].

The authors explain the obtained results by the destruction of surficial obliterated layers in the vibration field, which causes a decrease in the resistance to the fluid flow in the pore channels and an increase of their effective cross-section.

Another mechanism of a multifold flow increase in the vibration field has been suggested at the solution of a problem of directional motion by a viscous compressible fluid through a narrow capillary with the walls being deformed in the form of advancing waves [13]. The following has been established in the course of a numerical solution. The wave of the wall crosswise displacements is exciting an internal wave in the fluid. A substantial feature of this internal wave is that, even at small amplitudes of crosswise displacements of the capillary walls, the fluid's pressure and longwise speed amplitudes may reach values substantially exceeding the speed and pressure causing the acoustic flow. As applied to the flow in porous media, this phenomenon is an example showing that the speed and pressure pulsation on the order of the pore radius may cause unidirectional flow at a speed substantially exceeding the speed of normal filtration.

Of a maximum practical interest are study results obtained at sufficiently low vibration intensity (on the order of 1 kW/m^2) in the low frequency range (less than 1 kHz). The studies showed the vibration effect on the various liquids and gas flow process at the fluid absorption by a dry porous medium, at mutual displacement of immiscible fluids, at a multiphase flow, at gravity segregation and at other similar phenomena. Experimental studies by various authors indicated that under the vibration: (1) wettability hysteresis decreases, (2) the fluids penetrate deeper and faster into slits and capillaries, (3) capillary soaking processes are substantially intensified, (4) relative permeabilities to oil and water are substantially changing, and (5) the oil discharge from a porous medium is notably increasing [14, 16, 23].

The porous medium decolmatation studies have been conducted in the range of 5 Hz to 10 kHz with transparent slit-like models and with core material. In the experiments, the porous medium has been initially subjected to clogging either by mechanical admixtures (corrosion products or by artificial clogging) or by natural clogging down to a notable permeability decline. Then, at a constant throughflow for the filtered fluid, a cleanup of the porous medium has been conducted under the action of elastic vibration with various parameters. The cores with a natural clay content have also been used. Water and drilling mud acid (for the cores with a natural clay content) have been used as filtered (=flowing) fluids. The studies have shown that the vibration not only speeded up the cleanup process but actually "included" this process under conditions when without such action no suffusion of a colmatated porous medium had been occurring (for instance, at a slow flow speed). There is a frequency range (10 to 300 Hz) where maximum cleanup degree and speed is achieved at the assigned vibration intensity. At a combination of the vibration and mud acid treatment, the chemical reactions speed and a removal extent of the colloidized reaction products drastically increased. Permeability of cores with a natural clay content increased 2- to 10-fold [40].

Speaking of energy parameters at the vibration's action on the reservoir rock, it should be remarked that needed intensity values are defined not only by the nature of changes caused in the medium but also substantially depends on the initial thermodynamic status of the medium itself. The action results are defined by the interrelation between the elastic field energy and the energy necessary for the transfer of the system from a state of a stable thermodynamic equilibrium/balance to a new stationary state. If for the transfer of the system from a state of a stable thermodynamic equilibrium/balance to a new stationary state requires a relatively high energy of external action, then in a case of the system being in a metastable state, even a low intensity external action has been capable of creating a qualitatively new state. The action on the medium in consideration of a possible metastability of its characteristic parameters is energetically most beneficial.

A thermodynamic nonequilibrium of a reservoir suggests a possible manifestation of natural forces in the development process. It includes the emergence of saturations and nonequilibrium capillary state leaps at contacts of varisaturated zones. For instance, capillary pressures may at reaching the value of 0.03 to 0.05 MPa and may exist during extended time intervals. All these factors facilitate the efficiency of the vibration at sufficiently low vibration field intensity.

4.2 Vibration Effect on the Reservoir's Heat Properties

Our studies showed, for the first time, the vibration effect on the effective heat conductivity of saturated porous media. The studies have been conducted in the range of 6 to 20 kHz with poured sand and cemented synthetic samples as well as with natural cores. A description of the saturating fluids is included in Table 4.1.

Poured sand samples have been prepared using dry fine-grained quartz sand with rounded grains, 0.16 to 0.25 mm in diameter. Synthetic samples have been made out of a high-porosity ceramic material and limestones of Bodrak outers.

Natural cores have been selected from the siltstone and limestone interbeds of the Chikuduk Field.

The samples have been saturated with the indicated fluids under vacuum and placed in a calorimeter intended for the determination of samples' heat conductivity using a method of a first kind regular regime. The calorimeter was a copper cylinder with the internal diameter of 50 mm, height of 70 mm, and wall thickness of 1 mm. The installation for heat conductivity determination included the calorimeter, thermostat, thermocouple unit, and potentiometer for the temperature recording. On the removable lid of calorimeter electrical lead in of a vibration irradiator and a pocket for the thermocouple have been assembled. The thermocouple has been soldered

Table 4.1 Saturating fluid description at a study of heat conductivity in a vibration field.

Fluid	Field	Density (g/cm^3)	Viscosity (cP)
Oil	Pavlova Gora	0.949	540
Oil	Kirmakin	0.935	274
Kerosene Purified nonpolar		0.819	1.49
Model oil		0.877	138

in at an angle but in such a way that the thermocouple soldered joint was in the center of calorimeter.

The experimental methodology has been in placing the calorimeter (filled with a tested sample and the thermocouple) and heating up in a hot water to the temperature of 45°C to 50°C. Then, it has been placed in a thermostat where it has been cooling down with recording the temperature change in time. The cooling down period lasted 15 to 30 min.

Similar tests have been conducted with the same samples also with vibration applied. The impulse generator has been turned on at the moment of placing the calorimeter in thermostat; the wave picture (i.e., the moment of sending and shape of vibrations penetrating the sample) has been photographed on oscillograph screen [18].

The results obtained indicated that the effective heat conductivity was increasing both in loose and in cemented media. A largest increase has been observed in water-saturated samples, a smallest, in dry samples. If to determine the degree of cementation of the medium as the product of grain compressibility and matrix compression elasticity modulus, then the maximum effect has been observed in the least cemented samples. It has been recorded that the higher the sample porosity, the greater the relative increase in effective heat conductivity.

The influence of the saturated fluid viscosity η on the porous medium is heat conductivity has been characterized by a relative heat conductivity increase for the sand saturated with an oil and kerosene mix vs. individual mix components. As shown in Figure 4.1, the relative heat conductivity is increasing with increasing viscosity. The conducted studies indicated an increase in effective heat conductivity of rocks for various reservoirs in a vibration field. The extent of an increase depended on the viscosity, heat content and the type of saturating fluid.

The change in heat conductivity of a medium at spreading heat must result in temperature changes. Thus, temperature distribution has been studied in the heated porous medium heat conductivity of which has been changing as affected by the vibration. These studies have been performed using a lab installation represented by a triple-layer cylindrical reservoir model and enclosing rocks placed in a thin-wall steel housing. The model diameter has been 300 mm; thickness was 400 mm. In the model center, along the axis, a copper tube, 10 mm in diameter, perforated in the reservoir interval was placed. On the top and at the bottom of the reservoir have been layers modeling the enclosing rock. The thickness of the upper layer has been 150 mm and the lower layer has been 130 mm. The reservoir thickness has been 120 mm. The tube was insulated from the body by special rubber sealer preventing the spread of elastic vibration in the model body.

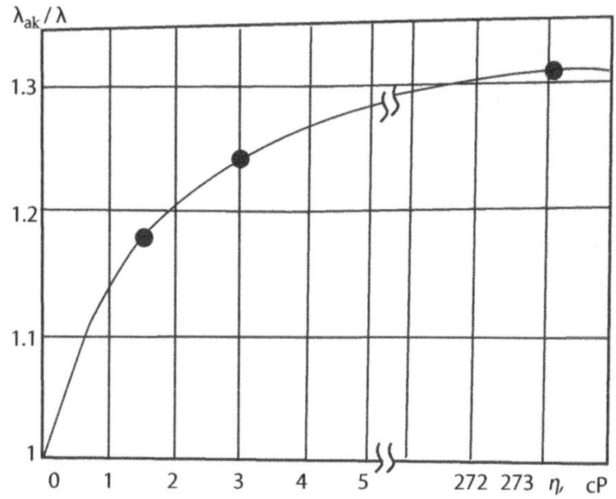

Figure 4.1 Effective heat conductivity for porous samples in a vibration field vs. viscosity η of the saturating medium. λ_{ak} and λ are sample effective heat conductivities in the vibration field and without the vibration.

The vibration irradiator and the heater have been placed in the tube so that the irradiation of vibratory energy has been uniform along the entire tube length. The heater has been placed in the reservoir interval. Fifteen thermocouples have been placed in the reservoir and enclosing rocks. The reservoir had been a carefully washed and dried up quartz sand of average rounding. The grain diameter has been 0.2 to 0.4 mm; porosity was 34%. The sand has been, completely (100%) saturated with an electric insulating oil having viscosity of 0.24 P. The enclosing rocks have been made from powdered bentonite clay with distilled water. Their porosity has been 5%, density has been 1.442 g/cm^3, and water saturation has been 15.6%.

The leading frequency in the irradiated vibration spectrum has been 21 kHz. A constant temperature has been maintained on the heater wall. For this purpose, the heater power was varied between 150 and 175 w. Five experimental series have been conducted at various vibration intensity. During the preliminary series has been temperature field was studied during the vibration, for maximum intensity. There has been no noticeable temperature increase (within the potentiometer error of 0.5°C). Thus, it was concluded that vibration within this range does not affect the thermocouple readings. In the second series, the temperature field has been measured at the reservoir heating without vibration. In the third, fourth, and fifth series, the reservoir temperature field has been measured during simultaneous heating and of vibration various intensity. The heater power and temperature in different points of the model have been continuously recorded.

Every experiment was continued until the establishment of quasi-stationary temperature regime. Effective heat conductivity of the media has been determined from the heater-created temperature field distribution. A computer experiment has been conducted for the determination of effective heat conductivity.

The following is the system of differential equations describing the experiment conditions:

$$c_p \frac{\partial T}{\partial t} = \frac{\partial}{\partial r}\left(\frac{\lambda_p}{r}\frac{\partial T}{\partial r}\right) \qquad (r_c < \mathrm{r} < \infty)\left(0 < |z| < \frac{h}{2}\right),$$

$$c\frac{\theta}{\partial t} = \frac{\partial}{\partial r}\left(\frac{\lambda}{r}\frac{\partial \theta}{\partial r}\right) \qquad (r_c < \mathrm{r} < \infty)\left(\infty < |z| < \frac{h}{2}\right).$$

here, T and θ are temperatures, respectively, in the reservoir and in the enclosing rocks; r, z, and t, respectively, are radial and vertical coordinates and time; h is the reservoir thickness; λ and λ_p are effective heat conductivities, respectively, of the reservoir and enclosing rocks; c and c_p are the specific heats, respectively, of the reservoir and enclosing rocks; and r_c is the well radius.

The edge conditions are as follows:

$$t = 0,\ T = \theta = T_0;$$

$$T = \infty, \quad \left(z = \frac{h}{2}\right);$$

$$t = \infty, \quad T = \theta = T_0;$$

$$\frac{\partial T}{\partial z} = 0, \quad (z = 0)$$

$$\theta = T_0 \quad (z = \infty);$$

$$\lambda_p \frac{\partial T}{\partial z} = \lambda \frac{\partial T}{\partial z} \quad \left(z = \frac{h}{2}\right),$$

where T_0 is the initial reservoir temperature.

Effective heat conductivity during vibration was determined as a function of its intensity:

$$\lambda = \lambda_0 + AJ_{ak}(r); \qquad J_{ak}(r) \geq 300 \text{ w/m}^2 \text{ for clay,}$$

$$J_{ak}(r) \geq 840 \text{ w/m}^2 \text{ for sand,}$$

$$\lambda = \lambda_0; \qquad 0 \leq J_{ak}(r) < 300 \text{ w/m}^2 \text{ for clay,}$$

$$0 \leq J_{ak}(r) < 840 \text{ w/m}^2 \text{ for sand}$$

Here, λ_0 is the effective heat conductivity of the reservoir without vibration. Vibration intensity is

$$J_{ak}(r) = J_c \frac{\exp[-\alpha(\mathrm{r} - r_{c)})]}{(r/r_c)^{0.5}}$$

where J_{ac} is the vibration intensity, J_c the vibration intensity at the well wall, and α is the vibration fading factor.

Effective heat conductivity of the media without vibration was determined by a method of the first kind regular regime. For the reservoir, the effective heat conductivity was equal to 1.28 and, for the enclosing rocks, 0.589 Bt/(m°C). Solutions of the above system of differential equations have been conducted by the method of finite differences using unidimensional pass. For increasing accuracy of the computation scheme, the step by radius was selected so that the nodes were getting into break boundaries of the heat conductivity factor. The approximation error was $\theta\,(\Delta\,\lambda + \Delta\,\lambda_{max}^2)$. Then, the calculated temperature field was compared with the experimental one and the constant A and media effective heat conductivity within the vibration field have been determined. With increasing heat intensity, the effective heat conductivity of the saturated sand was increasing by a factor of 5.5 and of the bentonite clay, by a factor of 3. Similar results have been obtained in field experiments.

The effect the properties of the saturating medium and a possibility of focused heating were studied using a reservoir model, which was a vertical cylinder, 420 mm in diameter and 730 mm high, filled with compacted sand. Along the entire cylinder length, an electric heater mounted within a quartz tube with a garland of electric vibration excitors has been positioned. The quartz tube has been tightly put on the heater. The model has been isolated from the surrounding medium by two layers of sound-absorbing rubber in order to eliminate a distorting effect of the reflected waves. The modeling medium was flushed-through sand with the grain size of 0.2 to 0.4 mm. Average porosity has been 34%. In order to understand the influence of a nature of the medium saturating the reservoir, the experiments have been conducted at 100% saturation with water, transformer oil, and air at various intensity of the elastic vibration field. In every experiment, a constant temperature was kept on the heater wall (130° to 160°C). The experiments lasted 1.5 to 2 hours until the onset of the quasi-stationary regime.

The experiments have been conducted in two stages:

- At the reservoir heating without vibration, the temperature has been measured at various points;
- Similarly, the temperature has been measured in time with the simultaneous heating and vibration.

Preliminarily, for each saturated medium, a possibility has been checked of increasing the temperature at the expense of vibration energy dissipation during a 3-hour-long vibration irradiation without heating. No temperature readings deflections have been observed within the measurement accuracy limits. Several series of experiments have been conducted for every type of the saturated medium. The largest temperature increase has been observed in the water-saturated sand due to a substantial elasticity of this medium. Somewhat smaller temperature increase has been observed in the oil-saturated sand. The explanation is that the oil has a much greater viscosity and vibration fading factor than water. The minimal temperature increase has been observed in the air-saturated sand.

The obtained results agreed with the results obtained from the above described physical experiments of direct observation of the effective heat conductivity. A nonlinear increase in the temperature growth has been observed at an increased intensity. A possibility of the focused heat action is illustrated by Figure 4.2. The position of vibration irradiator in the reservoir has been changed in these experiments in such a way that gradients of the temperature and vibratory fields were directed at different angles to each other. As may be seen, the creation of a focused heat field in the reservoir does not cause any particular difficulties.

The mechanism of vibration effect on the heat transfer in a saturated porous medium is poorly studied [17]. Most closely related with this issue are the works of acoustic drying-up of materials. However, usually indicated in those works is the prevalent role of gas vibration in the capillaries and a removal from the surface of moisture evaporated [19]. That is why for explaining the obtained results, we will review the heat transfer at the vibration action on the media of a very low permeability (like clays). An increase in the effective heat conductivity factor here occurs because the vibratory field activates the heat transfer through a saturated medium making the

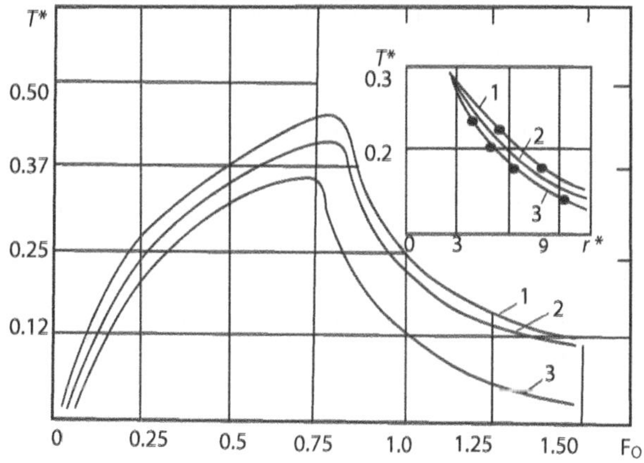

Figure 4.2 The temperature changes in the heat and vibration fields. 1. The heat and vibration sources are combined. 2. The heat and vibration sources are in different places. 3. There is no vibration source. $T^* = (T - T_0)/(T_c - T_0)$ is nondimensional temperature; $F_0 = \lambda_t/cr^2$ is nondimensional time (Fourier parameter); $r^* = r/r_c$ is nondimensional radial coordinate.

fluid within the closed pores vibrate. The duration of temperature leveling between phases in the reservoir is within 10^{-1} to 10^{-3} s. In the vibratory field, an elementary fluid volume within a closed pore may perform during this time from 3 to 2,000 vibrations. The effective heat conductivity in the elastic vibration field in the low-permeability media (like shale) is increasing apparently due to a declining duration of the temperature leveling between phases within closed pores. In the permeable media, an increase in heat conductivity factor in the vibration field may be caused by the fact that along with the vibratory fluid motions within the closed pores, a hydrodynamic flow is being formed by the saturation medium in the pore channels. Possibility of the formation of a hydrodynamic fluid flow under action of vibrations in the pore channels was indicated by Frenkel 1944 [35]. It has been noted that the pressure gradient generated by vibrations caused a directional movement of fluid (liquid and/or gas) particles, the so-called "sound wind". Thus, temperature increase at a fixed point is only possible at dynamic action of the heat and vibration fields and depends on the angle between the directions of heat and "sound wind" flows.

The formation of a hydrodynamic flow is affected substantially by the intensity, frequency, fading factor of the vibration field, permeability, porosity, matrix pore configuration, and properties of the saturating fluid. The hydrodynamic flow formation is visually demonstrated by experiments in temperature distribution at the heat carrier flow in the vibration field.

The study was conducted on a vertical vinyl plastic cylinder with the internal diameter of 143 mm, external diameter of 160 mm, and 300 mm high. Next to the cylinder's lower end, there was an organic glass chamber in which a constant temperature has been maintained. In the middle of the chamber, a heating coil has been placed through which the hot water from a pump has been supplied first in the second, smaller size chamber, and then through the filtration sieve in the reservoir. The filtration sieve served for creating uniform water flow over the cylinder area in the reservoir. A lid of organic glass with a tubelet for the fluid outflow has been placed at the exit from the reservoir.

A vibrato has been installed at a 10-mm distance from the entrance to the reservoir and a generator of sinusoidal vibrations. The reservoir model has been thermo-insulated by a cotton wool layer. The vibration intensity has been greater than 0.25 w/cm^2 next to the source. A quartz sand with medium-rounded grains, 0.2 to 0.3 mm in size, as well as a mix of sand and pebbles, 10 to 20 mm in size, have been used as a model medium. The experiments have been conducted in two stages for different flow rates. During the experiments:

- The temperature has been changed in the various reservoir points with a hot water flow through the reservoir during 3 hours;
- The temperature has been measured during simultaneous hot water flow and vibration during 3 hours. The hot water temperature at the input was maintained at 60°C. The total amount of water that has flowed through the reservoir was measured every hour for determination of average flow rate.

Figure 4.3 shows redistribution of the temperature profile in the reservoir under vibration. The temperature at the exit from the reservoir model increased by factor of 1.6 to 1.8 and its distribution nature approached rectangle. On increasing vibration field intensity, a relative temperature increment also increased. The results obtained have indicated a convective heating of the reservoir.

The issue of a joint heating and vibration on solid state hydrates is important for fighting hydrate formation [25].

Irradiating and receiving acoustic converters and an electric heater have been built-in in a crosspiece designed for the work pressure of 0.6 MPa. The crosspiece has been filled in with water (three quarters of its volume), vacuumed and in the remaining volume has been supplied a gaseous propane from a balloon at a pressure of 0.19 to 0.20 MPa. The crosspiece has

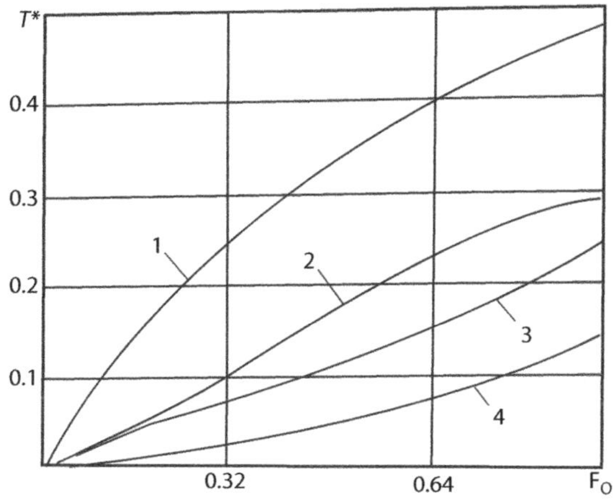

Figure 4.3 Injected hot water temperature vs. vibration power. 1, 2, and 3. Vibration fields of decreasing power. 4. No power.

been gradually cooled down to 0°C. When the ice crystals appeared on the crosspiece wall, the cooling has been stopped and the pressure has been raised to 0.49 MPa. The mixture has been continuously stirred by rocking the crosspiece in a horizontal plain using the vibrator with the 2 to 4 vibrations per second. As a result, a hydrate layer has emerged. Constant pressure has been supported in the crosspiece during the experiments. The stirring has been continued until a hydrate layer thickness reached 35 mm. The temperature has been raised to 2.5°C.

After the emergence of the hydrate layer, it has been subjected to one of the following:

- Vibration field;
- Heat with intensity of up to 84 w/cm^2;
- Combined vibration and heat fields with the same heat parameters.

The results are displayed in Figure 4.4. The figure shows that the vibration of the hydrate layer has caused its more intense decomposition, which has been increasing with increasing vibration. The maximum decomposition has been observed at joint heating and vibration. The hydrate gas destruction has accelerated twice or even faster. The hydrate decomposition process using a regular heating is least efficient.

Under a joint heat and vibration, the vibration power increase at a constant total power of both fields, has caused a more intense hydrate

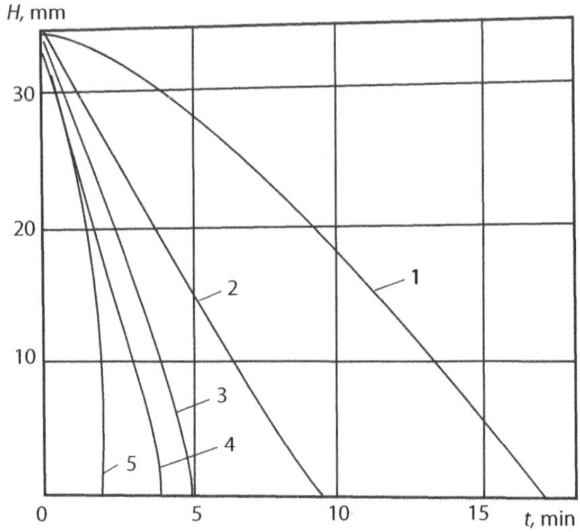

Figure 4.4 Gas hydrate decomposition in a vibration field. 1. Without vibration field. 2. Vibration field. 3. Heat field. 4 and 5. Joint fields and an increased power. H is the gas hydrate layer thickness.

decomposition. For instance, at the field, power increased by one order of magnitude (curves 4 and 5), and the time of hydrate layer destruction decreased from 3.5 to 2 min. Thus, the process was going on 8.5 times faster than at hydrate decomposition without the vibration.

4.3 Decolmatation of the Near-Bottomhole Zone in the Vibration Field

The vibration effect on decolmatation processes caused by contamination of the near-bottomhole zone by various materials has been studied on physical models. Decolmatation associated with paraffin or drilling mud entering the near-bottomhole zone was studied on an installation made of the reservoir model, diameter of 500 mm, with a well, 6 mm in diameter, in its center. A heater combined with vibration irradiator has been installed in the well. The reservoir thickness has been 200 mm. The reservoir has been modeled by washed up and graded sand, 0.2- to 0.4-mm fraction, with porosity of 37% and of bentonite clay layers, 130 mm thick with permeability 3 D, that have been used as top and bottom of the reservoir. Clay permeability has been close to zero with porosity of about 5%. A ring zone, 25 mm wide, completely saturated with paraffin has been made around the well.

The drilling mud has been prepared by a careful mixing of distilled water with 5% (by weight) bentonite clay powder. The rest of the reservoir has been totally saturated with the transformer oil. The zone's permeability (air) before the vibration has been equal to zero. Three series of experiments have been conducted:

- Reservoir was heated. The deteriorated zone has been heated above 60°C for 2 hours.
- Vibration for 2 hours.
- Joint heating and vibration for 2 hours.

Permeability of the damaged zone has been measured after each experiment. After each treatment, the reservoir has been cut, the deteriorated zone has been examined and its permeability studied.

The average permeability of earlier deteriorated zone (in Darcy) as a result of different treatment is shown. The numerator is the paraffinization zone, and the denominator is the clayed off zone.

Vibration	0/0
Heat action	0.6/0.45
Joint heat and vibration with intensity 0.25 w/cm^2	1.2/0.98
Joint heat and vibratory action with intensity of 1 w/cm^2	1.46/1.12

Visual examination of deteriorated permeability zones has shown, that after vibration only (w/o heating), no changes of earlier paraffined and clayed up zones have been observed.

After heating the paraffined zone, its upper part has been practically released of paraffin which has moved to the lower part of the reservoir under the action of gravity. Permeability of earlier deteriorated zone has been partially recovered (to 0.6 D). The changes after heating of the earlier clayed-up zone have been similar; however, permeability recovery has been lower. After the joint heating and vibration, a ring area 10- to 15-mm wide around the well has been observed, freed of paraffin or drilling mud. However, these substances have been moving deeper into the reservoir and have been positioned around the released zone as a 50-mm-wide ring with permeability respectively, 1.4–1.5 or 1.1–1.2 D.

Increased temperature resulted in melting of paraffin (and mud) in the reservoir pores. On application of heat and vibration, the melted medium under the vibration begun to move deeper into the reservoir [28].

These studies showed that the joined heat and vibration on a paraffin-soaked or clayed-up porous medium has resulted in the recovery of permeability on the average up to 40% to 50% of the initial one.

The described here studies are closely related to the studies on the effect of vibration on the movement of bound water. According to Frenkel (1944) [35], the bound water motion in a porous medium results in the emergence of eigenpolarization potential. For this reason, our studies have been based on measurement of eigenpolarization potential in the vibration field. The studies have been conducted on an installation in which a vibration source placed in a sand reservoir model has been a cylindrical ceramic irradiator screened by a copper foil and isolated from the medium by an epoxy layer. The vibration frequency and intensity in the rock have been registered by a quadrangular piezo-ceramic vibration probe. The reservoir model has been a parallelepiped, 60 × 40 × 25 cm in size, covered by a layer of sound-absorbing rubber and filled up with a compacted sand. A washed-through sand of medium-rounded grains (fraction 0.2 to 0.4 mm) and the initial porosity of 36%, served as a model medium. The sand has been saturated with a table salt solution with 0.005 n concentration. The eigenpolarization potential has been registered in *mv* by the contact method through platinum grid-electrode (45 × 45 mm in size) at distances of 150 and 200 mm from the vibration source (Figure 4.5). Zero of the device has been established before the beginning of an experiment by a short circuit of terminations, and then, the electrodes have been connected. The vibration source has been turned on under a quasi-stationary regime and simultaneous recording has been conducted of the eigenpolarization readings until reaching a new quasi-stationary regime. After that, the vibration source has been turned off and

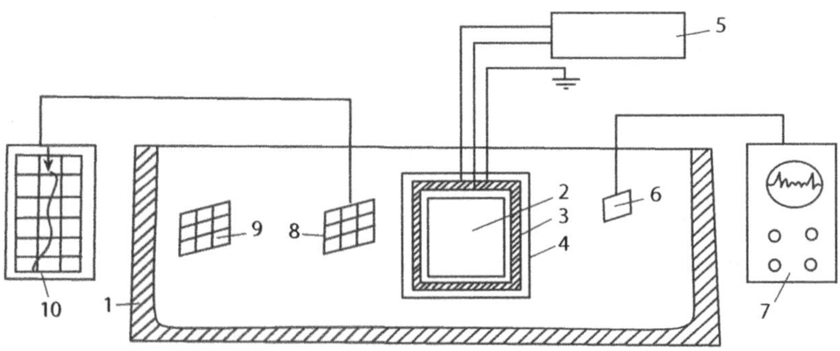

Figure 4.5 An installation scheme of measuring eigenpolarization potential of the reservoir in a vibration field. 1. Reservoir model. 2. Vibration source. 3. Foil. 4. Isolating layer. 5. Vibration generator. 6. Vibration probe. 7. Oscillograph. 8 and 9. Grid electrodes. 10. Potentiometer.

a decay of the eigenpolarization curve has been recorded. The potential has been registered at various intensity of the vibration field.

The eigenpolarization potential establishment time in the vibration field has been 10 to 15 min, which agrees with theoretical analysis of the duration of an acoustic flow establishment in liquids. The potential size has depended on the amount of bound water and intensity of the vibration field. Conducted studies have indicated a conceptual possibility of utilizing joint heat and vibration for removing bound water from the damaged zone, which may be among the reasons of declining reservoir permeability.

The intensification of the damaged zone cleanup from natural or artificially introduced colmatating deposits and improvement of its reservoir properties under the influence of vibration have received numerous field confirmations and have been implemented in various vibration techniques in production and injection wells [3, 5].

The greatest contributions in a theoretical study of the silting-up, colmatage, and efflux of small particles from a porous medium (suffosion) phenomenon has been done by Shekhtman (1961) [37]. He analyzed a one-dimensional problem of fluids with suspended solid particles flow through a porous medium without vibration. Called filtration deformations of the medium changes in the porosity, permeability, and volume weight occur in the process of flow. Part of the suspended solid is retained and deposited in the pore space and part of the deposited particles gets into the flow and is carried away.

As proposed by Shekhtman (1961) [37], in the case of mixed flow, one can denote as γ the volume fraction of a liquid matter of the pore space saturation with liquid/fluid, as β the pore space occupied by the precipitated solid particles, and

$$\beta = \zeta + \varepsilon \beta,$$

where ζ is the volume fraction of the solid matter proper, $\varepsilon\beta$ is the saturation with an immobile fluid bonded with the precipitated mass of solid particles, and ε is the porosity of the precipitated mass.

One can denote that, as $\delta = \alpha/(\alpha + \xi)$, the volume concentration of suspended solid matter in the moving mix, where α is the pore space occupied by suspended solid particles, and ξ is the pore space occupied by the free fluid. Thus,

$$\gamma + \alpha + \zeta = 1, \qquad \xi + \beta + \alpha = 1, \tag{4.1}$$

$$\gamma = \xi + \varepsilon \beta \tag{4.2}$$

Medium porosity in the filtration process is defined as $m = m_0\gamma$ where m_0 is the initial porosity. It is assumed that the size of suspended particles is small compared with the initial pore size and their motion velocity is proportional to the fluid velocity.

Let one assume that the porous medium with its suspension in the process of flow begins vibrating under the influence of the external vibration source. It is required to determine changes in the volume concentrations and saturations. As seen from Equations (4.1) and (4.2), it is sufficient to determine β and δ values as functions of the coördinates and time. Every particle in the process of flow may be in one of the two states: either in a state of quiescent or moving in a state of suspension along with the fluid. This allows introducing a status factor θ, which is describing the ratio of the amount of suspended solid matter to the entire colmatating matter in the pore space, which is composed of solid particles and bonded fluid:

$$\theta = \frac{\alpha}{\alpha+\beta} \tag{4.3}$$

The balance equation tying up the two sought-for functions in a differential format may be derived from the continuity equation for the liquid phase and the law of the matter conservation for the solid phase:

$$\frac{\partial\delta}{\partial x} = -B(t)\frac{\delta\beta}{\partial t} \tag{4.4}$$

where $B(t)$ at a constant pressure gradient is equal to:

$$B(t) = \frac{m_0(1-\varepsilon)}{Q(t)}, \tag{4.5}$$

where $Q(t)$ is the mixture throughput the process of flow. At a constant throughput ($Q(t) = Q_0 = \text{const}$)

$$B(t) = \frac{m_0(1-\varepsilon)}{Q_0} = \frac{1-\varepsilon}{u_0}, \tag{4.6}$$

where $u_0 = Q_0/m_0$ is the fluid motion velocity.

For solving the set problem, kinetic equation must be added to the balance equation, which would describe the processes of precipitation and tear-off of colmatating particles and would take into account the external vibration.

Disregarding the causes affecting the sticking of the suspended particles to the pore channel surface, it is possible to consider that the tear-off of a resting colmatant particle occurs when the flow velocity exceeds some critical velocity u_*. The critical velocity value depends in every specific case on the structure particulars of the pore channel and interaction nature between a solid surface and colmatating particles. Due to a nonuniformity and structural irregularity of natural porous media, in the process of flow, the field of the fluid flow true velocities emerges [38]. For a certain average value of flow velocity, in one or another number of cases of the flow velocity greater or smaller than the critical velocity. Beside the structural irregularity of the porous media, fluctuations of true fluid velocity in the pores are affected by the vibration and temperature. At a fixed value of the linear x coordinate, the status factor depends on its own initial value and on the solid particles transition from one status to another in a unit volume of the medium. For this reason, to account for changes in the status factor, methods of the probability theory should be used.

Let us review a general case when simultaneously tearing of and retention of particles may be occurring, which corresponds to a nonuniform deformation of the porous medium in the process of flowing.

Let us assume that in a unit volume of a porous medium per unit time there are θ_1 cases of exceeding critical velocity in pores and θ_2 cases when the fluid velocity is less than the critical one. Thus, θ_1 cases of maximum velocity amplitudes per unit time will cause $A\theta_1\beta$ cases of particles' transition from an immobile state to a suspended (per unit volume of the medium), and the status factor will change by

$$d\theta = A\theta_1 \frac{\beta}{\alpha+\beta} dt = A\theta_1(1-\theta)dt.$$

where A is the nondimensional parameter describing the properties of the medium and colmatant in their interaction and taking into account the physicochemical nature of the process.

During the same time and in the same volume, the θ_2 cases will cause $A\theta_2\alpha$ cases of sticking by solid suspended particles, and the status factor will respectively decline by

$$-A\alpha\theta_2 \frac{1}{\alpha+\beta} dt = A\theta_2\theta dt.$$

Thus, during elementary time interval the status factor will get an increment of

$$d\theta = [A\theta_1(1-\theta) - A\theta_2\theta]dt.$$

And from here comes a differential correlation between the status factor and cases of maximum velocity amplitudes which is

$$\frac{\partial\theta}{\partial t}+A(\theta_1+\theta_2)\theta=A\theta_1. \tag{4.7}$$

On assuming a linear correlation between the number of maximum velocity amplitudes and the critical flow velocity, one obtains

$$\theta_{1,2}=\frac{D}{2}\left(1\pm\frac{u}{u_*}\right), \tag{4.8}$$

where u_* is the critical velocity and D is the factor of maximum velocity cases for a given temperature and vibration field.

The physical meaning of the introduced D factor is the number of cases of maximum fluid flow velocity amplitudes in the medium's pores in unit volume per unit time at a given temperature and vibration parameters. A correlation between the critical velocity u_* and β according to experimental data may be written as follows:

$$u_*=u_0+\frac{(1-\beta)^2}{a_*}, \tag{4.9}$$

where u_0 is the initial velocity at which colmatant particle for the first time reaches the maximum equilibrium and begins moving in medium's pores and a_* is an experimental parameter on the order of 100 s/m.

Substituting Equations (4.8) and (4.9) in Equation (4.7), one obtains

$$\frac{\partial\theta}{\partial t}+AD\theta=A\frac{D}{2}\left(1+\frac{ua_*}{a_*u+(1-\beta)^2}\right) \tag{4.10}$$

Considering Equations (4.1), (4.2), and (4.3), one may write the status factor with β and δ as follows:

$$\theta=\frac{\delta(1-\beta)}{\beta+\delta(1-\beta)} \tag{4.11}$$

Therefore, Equation (4.10) together with the balance Equation (4.4) form a system of two equations with the two sought for functions $\beta(x,t)$ and $\delta(x,t)$.

In a description of the suffosion process, the porous medium permeability factor k and its porosity m may be written as follows:

$$k = k_1\beta + k_0(1 - \beta)^2, \tag{4.12}$$

$$m = m_1 + (m_0 - m_1), \tag{4.13}$$

where k_1 and m_1 are, respectively, permeability and porosity the beginning of the suffosion process ($\beta = 1$), and k_0 and m_0 are, respectively, permeability and porosity by the end of the suffosion process ($\beta = 0$).

Solution of the derived system enabled determination of changes in the volume concentration and saturation of the pore space with precipitated particles as well as the permeability and porosity of the porous medium at various time moments in the process of suffosion under vibration by the external vibration field parameters of which are assigned the factor D.

Of interest is a comparison of the solution of the derived equations with the experimental results for natural cores, 0.28 m in diameter and 0.1 m long with the initial porosity m = 25% and initial permeability k_0 = 0.25 μm^2. These cores have been initially colmatated by pumping through 150 pore-volumes of drilling mud until permeability has declined by two orders of magnitude and then subjected to a reverse flow at a constant velocity with a simultaneous vibration. In experiments, a relative permeability accrual in time has been studied. For studying the effect of different frequency vibration on the suffosion process has been the step-wise turning on different frequency vibration at the flow in the same core starting at high frequencies. For the avoidance of sharp irreversible changes in the pore system status, the flow velocity has been supported at two orders of magnitude below the initial u_0 = $0.1{\cdot}10^{-3}$ m/s and pressure vibrations have been set at near-threshold level.

Taking the experimental conditions into account, one can assume that the volume concentration δ of the matter suspended in the fluid has been constant in the process of the experiments. The particle tear-off occurs uniformly along the flow, does not depend on x coordinate, and changes only with time. As, during the entire experimentation, β values have not been much different from 1, we will disregard the value $(1 + \beta)^2$ in Equations (4.10). Then,

$$\frac{\partial \theta}{\partial t} + AD\theta = A\frac{D}{2}(1+u/u_0) \tag{4.14}$$

Integrating this equation at a constant flow velocity $u = V$ and initial conditions $\theta(0) = \theta_0$, we get:

$$\theta(t) = \theta_0 \exp(-ADt) + \tfrac{1}{2}\,(1 + V/u_0)(1 - \exp(-ADt)) \quad (4.15)$$

Now, the relative permeability changes, taking Equations (4.11) and (4.13) into account, will be

$$\frac{k}{k_0} = \frac{k_1}{k_0}\left(\frac{1-\theta(t)}{1-\theta(t)(1-\delta^{-1})}\right) + \left(1 - \frac{1-\theta(t)}{1-\theta(t)(1-\delta^{-1})}\right)^2. \quad (4.16)$$

In the calculation process based on the derived equations, the vibration nature (a step-wise switching different frequencies) has been considered by assigning appropriate values at certain time intervals in Equations (4.15) and (4.16). Diagrams of experimental functions k/k_0 at core cleanup of the colmatating material with the superposition of various frequency elastic vibration are presented in Figure 4.6. It also includes (dashed lines) k/k_0 values calculated using Equations (4.15) and (4.16). As shown, a quite satisfactory agreement has been achieved.

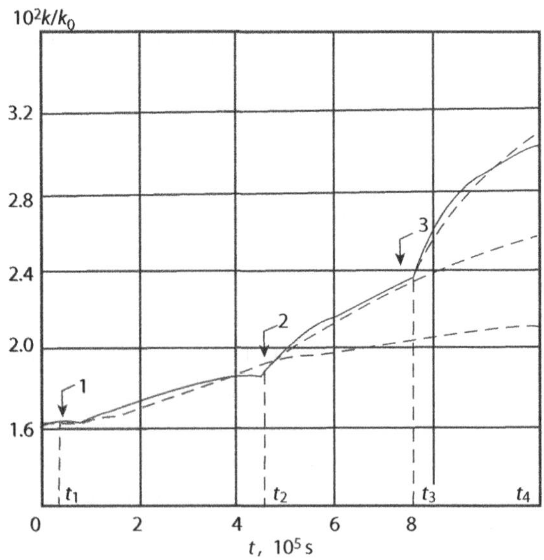

Figure 4.6 Permeability change in natural cores in the vibration field. 1. 11 kHz. 2. 6.8 kHz. 3. 1 kHz.

The model allows forecasting dynamics of the suffosion processes in consideration of real properties of the porous medium and of the colmatant as affected by vibration.

4.4 Cement Sheath Stability Around a Well in the Vibration Field

The petroleum production experience in general indicates the reliability of well cementing. However, in a case of a vibration, there are no substantiated quantitative criteria in the case of regulating vibration acceptable amplitudes and duration.

For this reason, of a great practical interest are studies of the conditions for a destruction caused by vibration and vibrocreep of the hardened cement in wells based on the available data of standard testing of concretes at periodical stressing.

Of interest is finding the causes of disturbance to casing isolation in the process of well production, which may occur due to the accumulation of irreversible deformations, vibrocreep in the hardened cement. For instance, if a well is under excessive internal pressure at simultaneous vibroaction, then the vibration load even at relatively low periodical stress notably accelerates the creep of the cement [12, 27]. There are numerous data about results of the standard testing for conditions of periodic stress (for the compression, extension, and bending) mostly for concretes of usual construction kinds. Oil well cements have not been studied in the existing range of a complexly stressed state of the cement sheath [8].

Hardened cement in the process of well production is in conditions of a nonuniform triaxial compression. Its strength is described by the tangential σ_τ and radial σ_r stresses in the cement lining and exceeds the standard parameter of uniaxial compression R. According to the experimental data, $\sigma_\tau - \sigma_r = \gamma R$ for the hardened cement of any age between 3 days and one year and more. γ is a factor assuming values of 1.2 to 1.45 correspondingly for "cold" and "hot" cements [26].

In the environment of periodic load, maximum stress of the cycle attributed to standard conditions of vibration load, $\sigma_* = \sigma_\tau - \sigma_r/\gamma$ must not exceed the limit of hardened cement vibro-endurance at uniaxial compression R^o, i.e.,

$$\sigma_* \leq R^o. \tag{4.17}$$

In this case, the material of a cement sheath is "adjusting" to the vibration load and no noticeable fatigue damage occurs within any arbitrary long time.

We will now identify major factors determining vibro-endurance of hardened cement in well conditions. Most substantial effect on the vibro-endurance breaking point has the characteristic of loading cycle $\rho = \sigma_0/\sigma^*$, where σ_0 is a minimum stress of the cycle. Depending on the ρ value, at equal average stress of the cycle $\sigma_m = (\sigma_0 + \sigma^*)/2$, the degree of fatigue damages is different. At small ρ, micro-fractures may emerge even at a stress less than R^0, whereas a periodic load $\rho \approx 0.9$ causes strengthening of the hardened cement [12]. As the correlation function of the vibro-endurance limit vs. ρ may the following expression proposed by Berg (1961) [8] be used

$$K_y = K_y^0 + (0.85 - K_y^0)\rho, \tag{4.18}$$

at $K_y^0 = 0.53$ it also describes experimental results of vibration loading of a fine-grained concrete, grade 500, with $R_p = 550$ MPa at loading frequencies of 10 to 100 Hz.

Here, $K_y = R^0/R_p$ is the relative limit of vibro-endurance, R_p is the threshold strength, and K_y^0 is the relative threshold of vibro-endurance at zero characteristic of the cycle amplitude.

Studies of the concrete strength and age and of some other factors give contradictory results. For instance, according to Berg [8], with increasing strength the relative threshold of vibro-endurance increases, because the threshold of fracturing increases. However, Gvosdev *et al.* (1972) [12] obtained opposite results. The elevated-plasticity concretes somewhat better resist the multiple repeated loading; however, there is no direct connection between concrete creep and its strength at vibro-loading. As for humidity, it is known that water may lower the vibro-endurance of the hardened cement if its humidity is high, because in this case water's wedging effect shows up. However, if the hardened cement has sufficiently high water saturation, then water creates a certain "healing" action. An increase in the water/cement ratio results in a substantial deterioration of the hardened cement's cyclical strength.

As for the effect of loading frequency, it is most dangerous in cases of vibro-fatigue as well as vibro-creep have low frequencies. The reason is that under effect of the deformation speed, if temperature conditions and the number of loading cycles coincide, then the high-frequency endurance threshold is always higher than its low-frequency values. The value of high-frequency endurance may be equal or even lower than low-frequency only in a case of an intensive heating of the sample at the ultrasound deformation frequency associated with the heat exchange breakdown leading to the lowering fatigue strength. It was found that at low frequencies, with the constant load amplitude, the creep speed linearly increases with the increase in frequency. However, as at fixed source power the vibration amplitude sharply decreases with increasing frequency, the amplitudes sufficient for creep appearance emerge only at the resonance regime of exciting vibrations.

The authors model a well by a system of round, equal wall thickness tube with the internal radius *a* and an external radius *b* and a shell of unlimitedly large loading radius *d*, which includes the cement sheath and the enclosing rocks. Such, a merger is justified by closeness of elastic properties for the hardened cement and the enclosing rocks. A cylindrical coordinate system where *z* axis is coincident with the well axis is associated with this model. Pressure is applied from the side of the cement sheath to the surface of contact (*b*). This pressure is external to the pipe and internal to the sheath. The hardened cement is under a triaxial uniform compression in a case when the casing is loaded by the internal pressure P^0 coinciding with the static bottomhole pressure P_s, which is uniformly distributed in the hardened cement after the cement setting. In this case, $\sigma_\tau = \sigma_r$. At a change of the internal pressure by $\pm\,\Delta P$, the state of stresses becomes complex and the change of tangential and radial stress in the hardened cement reach maximum value on the internal surface of the cement sheath and are equal:

$$\Delta\sigma_\tau = \pm K\Delta P, \quad \Delta\sigma_r = \mp K\Delta P.$$

Thus, the difference of total main stresses, which is describing the strength of the cement sheath, will be equal to the difference of these additional values. The *K* factor is determined on the assumption of a plane deformation when the displacement $u_z = 0$ on condition of equal radial displacements and also tangential and radial stresses of the casing and cement sheath at the *b* boundary and is equal to

$$K = \frac{2(1-\mu_1^2)a^2}{[(1+\mu_1)-\varepsilon(1-\mu_2)]a^2 + ([1-\mu_1-2\mu_1^2)+\varepsilon(1+\mu_2)]b^2}.$$

where $\varepsilon = E_1/E_2$, and μ_1 and μ_2 and E_1 and E_2 are Poisson coefficients and elastic moduli, respectively, for the materials of the casing and the cement sheath.

Periodic loading with the internal pressure at the bottomhole may be subdivided in two cases that differ from one another by the values of maximum reduced stress of the cycle and cycle amplitude characteristic.

1. A pulsating pressure of amplitude A *is created* in a well and the static bottomhole pressure is greater than P_s, i.e., $\Delta P = P^0 - P_s$. In this case there is a maximum change of stresses:

$$\Delta\sigma_\tau = K(\Delta P + A), \quad \Delta\sigma_r = -K(\Delta P + A),$$

and therefore, the maximum reduced stress of the cycle is equal to

$$\sigma_* = 2K(\Delta P + A) \tag{4.19}$$

and the cycle amplitude characteristic is determined as

$$\rho = \Delta P/(\Delta P + A) \tag{4.20}$$

Combining Equations (4.17) and (4.18) and introducing safety factor n, the condition of long-time vibration strength may be expressed as follows:

$$\sigma_* \leq [\gamma(0.53 + 0.22\rho)R_p]/\text{n}.$$

Considering Equations (4.19) and (4.20), the final expression of acceptable amplitude of pulsating pressure in well is

$$A_* \leq \frac{\gamma R_p}{2nK}\left[0.53 + 0.22\left(\frac{\Delta P}{\Delta P + A_*}\right)\right] - \Delta P \tag{4.21}$$

2. Pulsating pressure has an amplitude A and the bottomhole pressure is lower than P_s. In this case, similar calculations give:

$$A_* = \frac{\gamma R_p}{2nK}\left[0.53 - 0.22\left(\frac{\Delta P}{A_* - \Delta P}\right)\right] + \Delta P \tag{4.22}$$

Figure 4.7 is a nomogram for determination of the amplitude A_* of the acceptable pulsating pressure based on a given excessive pressure ΔP in the well obtained by the solution of Equations (4.21) and (4.22). It is estimated for various values of a/b parameter, the values of used in the oilfield practice for choosing the casing thicknesses. For a casing material with $\mu_1 = 0.2$ and $E_1 = 2.1{*}10^{-5}$ MPa, as hardened cement constants data for the grade 500 standard oilwell cement. For the hardened "cold" cement, $R_p = 50.8$ MPa, $E_2 = 16.8{*}10^3$ MPa, and $\mu_2 = 0.4$; for the hardened "hot" cement, $R_p = 20.3$ MPa, $E_2 = 10.9{*}10^3$ MPa, and $\mu_2 = 0.3$. The safety factor n for technical constructions subject to vibration is selected as 1.4 to 1.5.

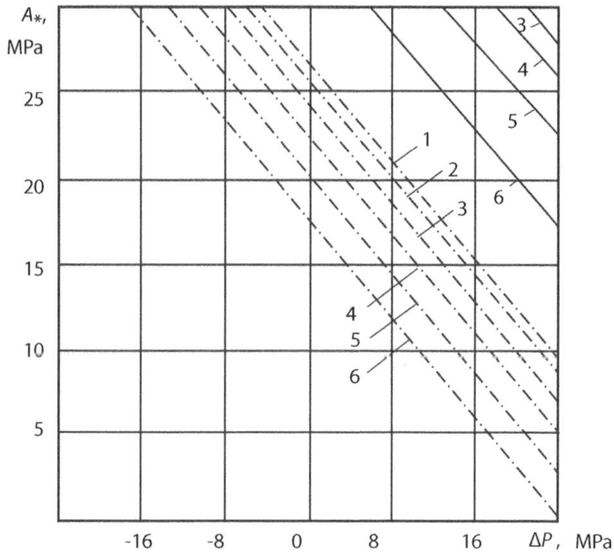

Figure 4.7 A nomogram for evaluation acceptable vibration pressure. a/b =: 1. −1.167. 2. −1.16. 3. −1.15. 4. −1.14; 5. −1.125. 6. −1.105. Dash-dot lines are for hot cement, and solid lines are for cold cement.

As the geostatic or formation pressure P_{form} acting on the external surface of the cement sheath is practically completely passed on the casing and, in most cases $P_s \leq P_{form}$, it is possible to assume $P_s \sim P_{form}$. The obtained result allows to evaluate acceptable amplitudes of pressure variations at the well bottomhole A_* under various conditions of fluid injection in the reservoir ($\Delta P > 0$) or its production from the well ($\Delta P < 0$). Figure 4.8 shows the correlations computed using the constants R_p, E_2, and μ_2 for hardened cement of various age (3 days and longer). As the derived correlations indicate, the danger of a fatigue destruction for a "cold" cement is the greatest during the first 3 to 30 days, whereas the "hot" cements, on the contrary, at this age have maximum vibration endurance.

Another cause of a possible hardened cement destruction is the formation of a ring channel. This may happen in a case of loading the casing with an overpressure ($\Delta P > 0$) and with pulsating pressure of amplitude A. The hardened cement creep strengthens if in the process of vibration pressure bleedoff occurs in a well ($\Delta P < 0$).

The conducted computations of a viscoelastic behavior of the cement sheath indicate that the size of a potentially "dangerous" gap between the casing and the cement sheath is (80–100)*10^{-6} m [7]. Knowing this value, it is possible to estimate the amplitude variations for different conditions.

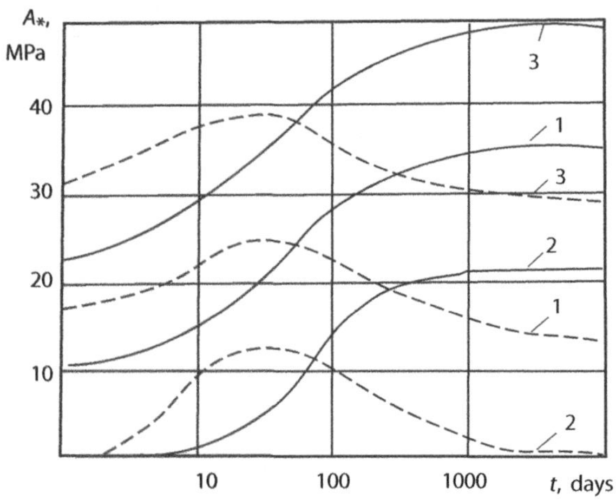

Figure 4.8 Amplitudes of maximum acceptable vibration pressures vs. the hardened cement age at a/b = 1.105 and various static pressures. Static overpressure: 1. 1 MPa. 2. 20 MPa. 3. 20 MPa. Dashed curves: hot cement; solid curves: cold cement.

Correlation of vibration times vs. the pulsating pressure variations amplitude in a well computed for various values of the static pressure P^0 in a well (for various values of parameter $D = P^0/P_{form}$) showed that with the static pressure increase, the cement sheath vibrocreep increases to a certain limit. The cement sheath's vibrocreep is at a maximum when the formation pressure value is close to the static pressure in a well. The D parameter is close to one. However, pressure gradients $P^0 - P_{form} \approx 5$ to 10 MPa and D parameter close to 0.4–0.5 are better appropriate with the real vibratory conditions. For instance, when the duration of vibration does not exceed 24 hours, a noticeable manifestation of vibrocreep is observed at pulsating pressure amplitudes greater than 8 to 10 MPa. The 3- to 5-MPa amplitudes developed by the currently existing downhole vibrators at action on the near-bottomhole zones practically cause no manifestations of vibrocreep in the cement sheath and, therefore, do not cause disruptions in the well isolation.

References and Bibliography

1. Abdullin, F.S., *Improvement of well performance*, p. 262, Nedra, Moscow, 1973.
2. Allakulov, P.E. and Belonchenko, V.N., *Effect of acoustic vibration on the low in cxapillaries and porous media*, vol. 1, pp. 44–46, DAN Uz, SSR, 1991.

3. Allahverdiyev, R.A., The flow rate intensification by the cyclic impulse action technique on the reservoir near-bottomhole zone. *Oil production transport*, 3, 10–12, 1985.
4. Amiyan, V.A., A possibility of forming emulsions on the near-bottomhole zone. *Oil production*, 11, 18–21, 1959.
5. Akhmetshin, E.A., Nurgaleyev, R.M., Mavlyutov, R.M., Applicaion of vibratory action on well near-bottomhole zone. *Oil production*, 8, 23–24, 1970.
6. Babalyan, G.N., *Physicochemical processes in oil production*, p. 200, Moscow, Nedra, 1974.
7. Babeshko, V.A., Glushkov, E.B., Glushkova, N.V., *On calculating a seismic source with a preset direction of irradiation*, vol. 262, pp. 831–834, DAN USSR, 1982.
8. Berg, O.Ya., *Physical fundamentals of concrete and reinforced concrete strength*, p. 281, Leningrad, Gosstroyizdat, 1961.
9. Briedis, I.P., High-frequency periodic deformation of viscoflowing polymers. *Polym. Mech.*, 4, 13–16, 1978.
10. Vakhtiyev, G.G., Kuznetsov, O.L., Simkin, E.M., *Thermodynamics of oil reservoir near-bottomhole zone*, p. 215, Nedra, Moscow, 1978.
11. Vakhtiyev, G.G. and Simkin, E.M., *Using geophysical field for oil recovery from reservoirs*, p. 231, Nedra, Moscoiw, 1985.
12. Action of static, dynamic and numerously repeating loads on the concrete and on elements of reinforced concrete constructions, p. 235, Stroyizdat, Moscow, 1972.
13. Ganiyev, R.F., Ukrainsky, A.E., Frolov, I.V., *Wave mechanism of accelerating liquids motion in capillaries and porous media*, vol. 306, pp. 803–806, DAN USSR, 1989.
14. Danilov, I.Z., Prokhorenko, P.P., Zhuravleva, V.P., Ultrasound effect on the mass transfer process in cement grout and in concrete. *Concrete*, 29, 4, 638–645, 1976.
15. Egorov, A.G., Kosterin, A.V., Skvortsov, E.V., *Consolidation and acoustic waves in saturated porous media*, p. 102, Kazan University, Kazan, 1990.
16. Efimova, S.A. and Shubin, A.V., Effect of the acoustic field on rocks' phase permeability in the reservoir near-bottomhole zone. Nuclear-geophysical and geoacoustic techniques in the determination of the rock permeability-porosity properties in oil and gas accumulations. *Proc. VNIIYaGG*, pp. 104–106, 1989.
17. Kalashnikov, N.V. and Chernikin, V.I., *Vibroheating of viscous oil products*, p. 120, Hedra, Moscow, 1961.
18. Karus, E.V., Efimova, S.A., Kuznetsov, O.L., Regulating effective heat-conductivity of reservoirs in acoustic field. *Oil husbandry*, 4, 52–55, 1975.
19. Konovalov, E.G. and Germanovich, I.N., Temperature effect on liquid rise height in capillaries under action of a high-frequency vibration. *IFZh*, 6, 3, 54–56, 1963.

20. Kuchumov, R.Ya., Application of vibroaction technique in oil production, p. 111, Ufa, Bashknigizdat, 1988.
21. Lyushin, S.F. and Khabibullin, R.M., On a possibility of inorganic salts deposition in reservoir conditions. *Proc. BashNIPIneft*, vol. 145, Ufa, pp. 120–122, 1975.
22. Mirzadzhanzade, A.Kh., Devlikamov, V.V., Khabibullin, Z.A., Study of the temperature of acceptable oil cooling. *Oil Husbandry*, 7, 57–59, 1982.
23. Mikhaylov, V.M. and Neretin, V.D., Kuznetsov O.L. A study of ultrasound action on the flow processes in porous media. *Proc. VNIIYaGG*, vol. 24, pp. 78–87, 1975.
24. Popov, A.A., *Strike action on the well near-bottomhole zone*, p. 157, Nedra, Moscow, 1990.
25. Kuznetsov, O.L. *et al.*, Decomposition of gas-hydrates at joint heat and acoustic action. *Dev. exploitation gas-condensate fields*, 8, 17–21, 1974.
26. Sarkisov, G.M., *Calculations of drill pipe and casing*, p. 217, Nauka, Moscow, 1971.
27. Sarkisov, G.M., Saroyan, A.E., Burmistrov, L.G., *Cementing strength of oil wells*, p. 256, Nedra, Moscow, 1977.
28. Simkin, E.M., Kuznetsov, O.L., Efimova, S.A., On a possibility of permeability recovering in paraffined and clayed-out reservoir zones by heat and acoustic action. *Oil Husbandry*, 10, 80–82, 1975.
29. Simkin, E.V. and Bernstein, M.A., Dynamics of reservoir paraffining in the oil flow process. *Oil Husbandry*, 2, 44–46, 1975.
30. Simkin, E.M., Pogosyan, A.V., Stremovsky, E.V., Forms of residual oil occurrence in water-flooded reservoirs and a possibility of its mobilization using the vibroaction. *Current methods of increasing reservoir oil yield. Proc. VNIINeft*, pp. 25–26, 1998.
31. Sokolov, A.V. and Simkin, E.M., Study of acoustic action effect on rheologic properties of some oils. *Proc. VNIIYaGG*, pp. 104–106, 1981.
32. Sultanov, S.A. and Svinushin, N.M., *Using methods of oilfield geophysics for improving reservoir oil yield*, p. 200, Nedra, Moscow, 1967.
33. Surguchev, M.L., Kuznetsov, O.L., Simkin, E.M., *Hydrodynamic, acoustic, heat cyclical action on oil reservoirs*, p. 184, Nedra, Moscow, 1975.
34. Surguchev, M.L., Yu.V., Zheltov, Simkin, E.M., *Physicochemical microprocesses in oil and gas reservoirs*, p. 215, Nedra, Moscow, 1984.
35. Frenkel, L.I., On the theory of seismo-electric phenomena in moist soil. *Proc. AN USSR, Geophysical series*, vol. 4, pp. 134–136, 1944.
36. Chersky, N.V., Tsarev, V.P., Konovalov, V.M., Ultrasound field effect on rock permeability at waterflow. *DAN USSR*, 232, 1, 201–205, 1977.
37. Shekhtnman, Yu.M., *The flow of low-concentration suspensions*, p. 208, AN USSR Publ., Moscow, 1961.
38. Shwidler, M.I., *Statistical hydrodynamics of porous media*, p. 287, Nedra, Moscow, 1982.

39. Fairbanks, H.I. and Chen, N.Y., Ultrasonic acceleration of liquid flow through porous media. *Chem. Prog. Syst.*, 61, 109, 108–117, 1971.
40. Larson, R.R., Scriven, L.E., Davis, H.T., Percolation theory of two-phase flow in porous media. *Chem. Eng. Sci.*, 36, 1, 57–64, 1981.
41. Simkin, E.M. and Surguchev, M.L., Advanced vibroseismic technique for water flooded reservoir stimulation, mechanism and field test results. *Proc. 6th European Symp. On Improved Oil Recovery*, vol. 1, Stavanger, Norway, pp. 213–241, 1991.
42. Chilingarian, G.V. and Vorabutr, P., Drilling and Drilling Fluids, p. 767, Elsevier, 1981.
43. San Joaquin Valley Section of Society of Petroleum Engineers. Society of Petroleum Engineers Proceeding, Eighth SPE Sym. On Formating Damage Control, Bakersfield, CA, San Joaquin, pp. Feb 8–9, 1988.

5

Effect of Vibration on Improving Oil Yield and Various Tertiary Recovery Technologies

5.1 Major Causes of Incomplete Oil Recovery From the Subsurface

When reviewing the reasons obstructing complete oil recovery from the reservoirs, the first reason to be noted is the incomplete oil displacement by water (poor sweep efficiency) at using the waterflooding processes. Whenever a uniform nature of a porous medium is combined with capillary effects, it is impossible to displace oil by water from some part of the porous space. It is believed that in the hydrophilic reservoirs this residual oil is concentrated mostly in the largest pores. The completeness of oil displacement from a porous medium is usually determined by the displacement factor. This factor is a ratio of the difference between an initial and residual oil saturation. The oil by water displacement factor varies under different conditions between 0.4 and 0.75. Thus, residual oil is taking 15% to 40% of the total volume in a waterflooded space.

The factor defining the reservoir oil yield at waterflooding is an incomplete reservoir sweep by the displacement process. Progressively, as the displacement front advances in the injection from injection to production wells, the waterflooding involves not the entire productive reservoir area included in the development. This waterflooding parameter is usually described by the waterflooding sweep factor of the reservoir area. It represents the ratio of the water-flood-covered reservoir area to the entire reservoir area included in the development. The sweep efficiency increases with the increasing water pumped in to the reservoir. To a significant extent, the area sweep factor depends on the interrelation between mobilities M of water injected into the reservoir and the displaced oil. The mobility is the ratio of the relative permeability of a respective fluid to viscosity:

George V. Chilingar, Kazem Majid Sadeghi, and Oleg Leonidovich Kuznetsov. Acoustic and Vibrational Enhanced Oil Recovery, (123–180) © 2022 Scrivener Publishing LLC

$$M = \frac{k_w}{k_o}\frac{\mu_o}{\mu_w}$$

where k_w is the relative permeability to water ahead of the displacement front (at residual oil saturation); k_o is the relative permeability to oil ahead of the displacement front (at the initial water saturation); and μ_w and μ_o are, respectively, water and oil viscosities at reservoir conditions.

Numerous studies found a strong correlation between the areal waterflooding sweep factor and ratio of mobilities. Based on these patterns, the ratio of mobilities close to one and lower number is considered favorable. At these M values, a significant waterflooding areal sweep may be reached with relatively small water amount injected into a reservoir (relative to the produced oil amount). The waterflooding areal sweep substantially declines at $M > 10$. Thus, at $M = 1$ and maximum water content of the production on the order of 95%, the areal waterflooding sweep of a uniform reservoir will be close to 100% both at areal and linear action systems. At the same time, at $M = 10$ and the same water-content in the production, the areal sweep will decline to 85% at the areal system of action and to 90% at linear. It is also important to emphasize that, under favorable interrelation of mobilities ($M = 1$), the indicated high percentage of the areal sweep is reached with pumping through the reservoir the water volume double the recovered oil volume. Under unfavorable interrelation of the mobilities (for instance, $M = 10$) for achieving even a smaller areal sweep it will be necessary to inject a significantly greater volume of water: triple compared to the recovered oil at the areal system and 2.5 times, at the linear one.

Analysis of the quoted correlation indicates that the waterflooding areal sweep to a substantial measure is defined by oil and water viscosity ratio. That is why it is believed that it is inefficient to develop high-viscosity oilfields using waterflooding. Indeed, with mobility ratio of about 100 and higher (which corresponds to the oil viscosity over 100 cP), it is a major issue to reach waterflooding areal sweep of 50% even at very high water content of the produced oil. What it means is that waterflooding even of reasonably uniform reservoirs containing high-viscosity oil will result in a low oil yield, rarely above 15% or 20%.

Another important factor defining the sweep by the oil displacement process at waterflooding is geological nonuniformity of the productive reservoirs (the presence of layers with different permeability, fractures, caverns, lenses, semi-lenses, lithological variability, etc.) (Simkin, *et al.*, 1981 [1]; Simkin and Surguchev, 1981 [9]). The geological nonuniformity usually results in that not in the entire reservoir volume, within the area involved in the development, an active oil with water displacement process occurs [8]. In every cross-section of a waterflooded reservoir segments may be present, to which the displacement front either have not reached or they were bypassed by displacing water. The effect of geological nonuniformity on the productive reservoir sweep by the displacement process is usually described by the so-called sweep factor by waterflooding of the reservoir thickness. This parameter is defined by the ratio of the reservoir volume where the flow of injected water and displaced oil occurs to the total reservoir volume within the waterflooded area. Obviously, in uniform reservoirs, the waterflooding sweep factor of the thickness may reach 100%. At waterflooding of a laminated-nonuniform reservoir, high waterflooding sweep factor of the thickness may be achieved by way of pumping through the reservoir substantial amounts of water, which may cause high water content of the produced oil. Along with this, it has been established in studies that a high waterflooding sweep factor of the thickness increases with a declining mobility ratio of the injected and displaced fluids. This means that, at waterflooding of the high viscosity oilfields (i.e., with unfavorable mobility ratio), geological nonuniformity causes a larger decline in the oilfield than at waterflooding of low-viscosity oil accumulations.

The oil yield at waterflooding is determined by the product of aforementioned displacement and sweep factors. From the viewpoint of applying one or another oil yield improvement methods, it is important to know where the oil remaining after waterflooding is concentrated. It is believed that, due to an incomplete oil displacement by water, 60% to 90% of all oil is not recovered with waterflooding. The remaining 10% to 40% of the oil arc not recovered because of low reservoir sweep efficiency. Thus, at an unfavorable rate of the mobility ratio as well as due to an extensive reservoir nonuniformity, the fraction of unrecovered oil within the reservoir swept by the waterflooding increases.

Thus, following are the major causes of low oil yield at waterflooding:

- Capillary forces obstructing oil displacement from some pores of poorly sorted rocks;
- Unfavorable mobility ratio of the displacing and displaced fluids;
- Heterogeneity of reservoir rocks.

In specific field geology conditions, one of the aforementioned factors turns out dominating. For example, at the development of low-viscosity oilfields, the major cause of an incomplete oil recovery is the manifestation of capillary forces. Obviously, methods of the oil yield improvement from such fields must be the exclusion (partial or total) of the capillary forces in the process of oil displacement.

If the field under development is characterized by a strongly developed geological heterogeneity of the productive reservoirs, then the ways of improving their oil yield should include in the first place the increased sweep efficiency by the action process. At developing fields of high-viscosity oil, a substantial fraction of unrecovered oil is due to unfavorable mobility ratio. For this reason, for improving oil yield from such fields the attempt should be made in the first place to increase viscosity of the displacing fluid and/or to lower viscosity of the displaced oil.

Scientific research in recent decades resulted in the creation of a whole series of methods conceptually allowing to reach a more complete oil yield than at a regular waterflooding.

Most of the known methods of improving the oil yield are envisioning the application of means completely or in part eliminating the output causing a decline in efficiency of the oil displacement from the reservoir. Such methods are directed to an increase in the sweep efficiency during flooding. Three groups of oil yield improvement methods differing in their way of achieving the desirable result are as follows:

- Methods of increasing the reservoir sweep by the oil displacement processes;
- Methods of increasing the oil displacement extent;
- Methods of increasing the reservoir sweep and increasing the oil displacement extent.

Also common is the following classification of the oil yield improvement methods related to the kinds of means of improving the oil recovery process:

- Physico-hydrodynamic methods (changing the direction of the fluid flow in the reservoir, cyclical alternation of fluids injection and withdrawal, utilizing high-pressure injection, increasing pressure gradients, etc.);
- Physicochemical methods (waterflooding with the application of surface-active substances, viscosity increasing agents, sulfuric acid, etc.);
- Methods of the oil displacement by agents miscible with the oil (high pressure hydrocarbon gas, liquid solvents, carbon dioxide, etc.);

- Heat methods (oil displacement with heat carrying agents or using in-reservoir oxidation reactions);
- Vibration and vibroseismic methods.

Physico-hydrodynamic methods of improving the oil yield emerged as a result of long-lasting job of perfecting the waterflooding. They all are directed to increasing the sweep efficiency.

Currently, numerous new methods of improving oil yield are being tested and applied. Due to these methods, dozens of thousands of tons of oil are already being produced. Each new method is not universal and has its own range of use. It is defined to a substantial extent by geologic-physical conditions of the accumulation.

Requirements to a geological study of the object for a purpose of applying a new development method are basically the same as at the application of traditional methods. At the same time, specific features of the new methods of improving reservoir oil yield should be taken into consideration at geological studies and associated with them need in obtaining additional information about the accumulation. A result of the geologic-physical study should be as complete as possible idea of the accumulation features, its structure, physical properties of rocks and fluids, temperature field within the field limits, and its operation regime. Preliminary recommendations about the exploitation targets should be issued.

A general geological description of a field includes major features of the regional orohydrography and geomorphology, specifics of the field's geological structure (tectonics, stratigraphy, lithology, and oil saturation), and geological oil and gas reserves (initial and current in case a development is conducted).

The accumulation's structure means the following:

- The identification of isolation extent of production targets, the presence of faults;
- The identification of structural features of production targets by way of delineating their top and base, preparing structural maps;
- Preparing isopach maps (general, effective, and oil-saturated); the determination of average thickness of the production targets and their areal changes;
- The establishment of internal and external gas water and oil water contacts.

The following tasks are identified for the study of physical properties of productive reservoirs and the contained fluids:

- The porosity determination and its variation in the section and over the area;
- Permeability determination and its variation in the section and over the area; preparing maps of permeability and hydraulic conductivity of the area;
- Identification of productive reservoir's nonuniformity in the section and over the area;
- Identification of thermophysical parameters of the reservoirs and of the reservoir-enclosing rocks;
- Determination of lithological composition of the oil reservoir rocks;
- Determination of the oil by gas saturation pressure, and the reservoir oil formation volume factor;
- Determination of oil density and viscosity, their variation in the section and over the area and also at the temperature change;
- Determination of the formation water properties and its salt composition;
- Determination of porous medium fluid saturation, its behavior in the section and over the area.

In the process of a geologic-physical study of the accumulation, the initial temperature of the field and the appropriate maps prepared (initial isotherms) should and must be established.

Most new methods of improving the oil yield require large capital investments in the hope for additional oil production. The authors will review below some efficient methods of improving the oil yield whose application results in substantial increase of recoverable oil reserves.

5.1.1 Oil Displacement by Miscible Hydrocarbons

Achieving conditions of the total oil displacement is possible if the displacing agent is miscible with the displaced oil forming a uniform phase with it. As the oil is a multicomponent system, its phase status is defined by its pressure, temperature, and composition. Usually, this circumstance is demonstrated by a triple diagram of the oil status where three components are tentatively identified: methane (C_1), ethane-hexane (C_2-C_6) (ethane-pentane), and octane and the rest of heavier components (C_7).

Within a substantial range of formation pressure and temperature, methane may be only in a gaseous state and the C_7 fractions, only in a liquid state. As for the intermediate C_2-C_6 fractions, their phase state is defined by specific values of the pressure and temperature. On a triple diagram of state of these components' mix, five characteristic areas are identified, every one of them reflecting a certain phase state and component composition of the hydrocarbon mix. Diagrams of state also allow to come up with a judgement regarding mutual miscibility of hydrocarbons belonging with different areas. A condition of hydrocarbons' miscibility are basic theoretical preconditions for the application of a number of varieties of the oil displacement processes by hydrocarbons miscible with it.

5.1.2 Oil Displacement by a High-Pressure Dry Gas

This method is based on miscibility of the gas injected into a reservoir containing the oil. In the process of oil displacement by a dry gas, the intermediate C_2-C_6 fractions in the dual phase flow area pass from oil to gas. Benefication of the injected gas, which usually is mostly methane, occurs according to the saturating vapor curve until reaching critical composition. In those parts of the two-phase flow area where the gas is beneficiated to the critical state, the separation boundary between phases vanishes and oil and gas become totally mutually miscible. The mix zone gradually grows along with the increasing volume of a beneficiated gas. The gas beneficiation occurs at the expense of logs of intermediate fractions by the oil. As a result, the oil becomes heavier and more viscous and its displacement by the gas becomes less efficient. In the process of subsequent contact of this oil with the injected gas, a substantial part of the oil remaining behind the displacement front is evaporated. Eventually, in the zone, through which the front of mixing displacement occurred about 5% of the oil remained, mostly represented by asphaltenes and other heavy components.

This method is usually applied to the development of light oilfields with a substantial concentration of intermediate components C_2-C_6. The oil displacement by a high-pressure gas is recommended for application at initial field development stages. An important parameter of the process implementation efficiency in this case is the formation pressure level. With the pressure increase, the size of the two-dimensional area decreases; therefore, the miscibility conditions of the injected gas with the formation oil improve. For this reason, the formation pressure is among the major criteria for the application of oil displacement with dry gas. It is believed that the miscibility of the injected gas with the oil the formation pressures above 20 MPa is required.

Injection of a high-pressure gas into the reservoir results in high oil displacement factors. However, due to an unfavorable viscosity interrelation of the injected gas and the displaced oil, the displacement process often covers a smaller area than waterflooding. This circumstance substantially lowers the application efficiency of this method. For increasing sweep efficiency, the produced gas may be reinjected into the reservoir. Such gas recirculation may increase the sweep factor at the expense of increased volume of gas pumped into the reservoir. What should also be kept in mind is that, with the pressure increase, gas viscosity grows and the mobility ratio improves. It also indicates advantages of the method application at high formation pressure levels.

Upon the method's implementation, the gas and liquid are simultaneously flowing in the reservoir, which may cause a gravity segregation. The sweep deterioration of the miscible displacement process due to the segregation phenomena is observed in reasonably uniform high-permeability reservoirs of a large thickness. For this reason, the application of the oil displacement method by high-pressure gas is usually not recommended in such reservoirs. For decreasing the oil yield loss due to unfavorable viscosity interrelations between the injected gas and the displaced oil, water is injected together with the gas. Usefulness of this measure is obvious as in conditions of insubstantial segregation phenomena effect, the sweep by the oil-by-gas displacement process may be on the same order of magnitude the waterflooding. Volume interrelations of the injected gas and water should be selected so that the water-gas mixture displacement front lags behind the zone of mixing hydrocarbons.

5.1.3 Oil Displacement by an Enriched Gas

The application of this method is based on a fact that the intermediate fractions pass from the injected gas into the oil, gradually enriching it. The oil beneficiation process occurs according to the liquid saturation curve. The so beneficiated by intermediate fractions, oil is mixing both with the formation oil and with the injected gas. The emerging mixture zone assures a total oil displacement. With an increase in volume of the beneficiated gas injected into the reservoir, the mixture zone grows. Usually, for saving the expensive intermediate fractions, the beneficiated gas is injected in the amount of 10% to 20% the pore volume. After that, the created fringe of the beneficiated gas and the mixture zone is pushed through the reservoir by a cheaper agents (dry gas or a dry gas with water). Gas beneficiation may be implemented by way of adding ethane, propane and butane. In some cases, miscible displacement is conducted by way of injecting into the reservoir of the associated gas after the separation. Propane is most often used as a beneficiated agent. The beneficiating components' concentration in the injected gas for assuring the miscibility conditions depends on the oil composition, formation pressure and temperature. If propane used as the beneficiating component, in many cases in field conditions to assure the miscibility with the formation oil, then the propane concentration in the injected gas must be around 20% to 50%, at the formation pressure in the range of 10 to 20 MPa. The beneficiation component concentration needed for providing for miscibility of the injected gas with the formation oil is decreasing with the increase of the formation temperature and pressure. It is preferable to use this method in the fields of low viscosity (up to 5 to 10 Cp) and low density (up to $0.88/cm^3$) oil. On decreasing the intermediate component contents in the oil, it becomes necessary to increase the use of expensive enriching components.

The use of the method of injecting a reservoir with the beneficiated gas as well as high-pressure gas results in achieving a high oil displacement factor. However, an unfavorable mobility relationship demands the measures of increasing the sweep of the displacement process. It may include water injection along with gas. For preventing the oil yield loss due to gravity segregation, the method of oil displacement with a beneficiated gas is recommended for low-thickness reservoirs of low permeability. Wherever the geological structure allows, it is desirable to inject gas in the upper part of the structure and displace the oil in its lower portions.

5.1.4 Oils Displacement by Liquefied Petroleum Gas

The reason of this method application is that, with an increase of intermediate components in the injected gas, its miscibility with the formation oil improves. If the injected agent would be totally composed of intermediate components, then it would be mixing with the formation oil without its enrichment by intermediate components. This circumstance is the basic theoretical precondition for the application of the method of liquid components' injection in the reservoir. As solvents, liquefied hydrocarbon gases (ethane, propane, and butane), unstable gasoline, wide fractions of light hydrocarbons obtained at oil stabilization and processing, kerosene, gas condensate, organic alcohols, etc., may be used. Most commonly, propane is used as the solvent. Usually, for a purpose of economizing, a solvent fringe is created in the reservoir, which is then pushed by gas or by gas with water. In the case that the solvent fringe is being displaced by the gas and water, the ratios of the injected volumes should be such that water lags behind the displacement front of the solvent by gas. The solvent fringe usually constants 2% to 10% of the reservoir pore volume. With increasing reservoir nonuniformity, the needed volume of the solvent fringe increases. The fringe moving in the reservoir displaces the oil in front of it. Between the fringe and a displaced oil a transition zone emerges of mixed oil and solvent. If the solvent is injected in already waterflooded reservoir, then the solvent fringe displaces both the water and oil. In as much as the water mobility is higher than that of the oil, water moves faster to the production wells. As a result, an oil wave is formed. If the method is applied from the very beginning of field development, then the oil wave will be absent and the flow will occur due to water injection.

The oil displacement by liquefied petroleum gas method is recommended to apply if the formation pressure ranges between 8 and 14 MPa. At such a pressure, the solvent (propane and butane) is maintained in a liquid state as well as its miscibility with the pushing gas (if the natural gas is used as such). The method may not be applied in the fields with the temperatures over 96°C to 97°C because at such a temperature, propane will be in a gaseous state regardless of the formation pressure.

An important advantage of the oil displacement by liquefied petroleum gas compared to the oil displacement processes by dry and an enriched gas is that it may be conducted at lower formation pressure, i.e., in shallower reservoirs. The injected solvent is displacing practically the entire oil from the reservoir portions involved in the displacement. However, oil displacement by a propane fringe causes problems with increasing sweep by the displacement process. For this reason, the method is recommended for fields with low viscosity of oil (no greater than 5 to 10 cP), with relatively thin reservoirs (up to 8 to 10 m) and relatively low permeability (less than 50 mD).

5.1.5 Oil Displacement With Carbon Dioxide

This method is based on a capability of carbon dioxide at positive temperature to be in a liquid state. As the temperature rises above the critical, the carbon dioxide may only be in gaseous form. Most commonly, the field conditions are such that the carbon dioxide within the formation is gas. At the same time, the cases are known when the carbon dioxide within the formation is liquid.

The oil yield improvement mechanism to a substantial measure is defined by the state of the carbon dioxide within the formation. The oil displacement process by a liquid carbon dioxide depends on a high degree of their mutual solubility. At the carbon dioxide dissolving in the oil, the oil volume substantially increases, its viscosity declines, with capillary forces manifestation. All these result in improved oil displacement. In the same direction, the formation near the oil displacement front of a wave of light hydrocarbon fractions extracted from the oil is working. Usually, it is recommended to apply a technology of the oil displacement by the fringe of a liquid carbon dioxide which is pushed within the reservoir by water. As the carbon dioxide is well dissolvable in water, a zone of carbonized water, preventing destruction of the carbon dioxide fringe, forms in the rear part of the fringe.

Usually, the injected carbon dioxide is within the formation in a gaseous state. In this case, the oil displacement process looks like a described above process of the oil displacement by a high-pressure gas. At injection, a mixture zone is formed due to a transition of the oil's hydrocarbon components in the carbon dioxide. As a result, it is capable of mixing with the oils that have an insignificant content of intermediate components. Also important is the fact that the pressure required for the miscibility of the carbon dioxide with the oil is much lower (by a factor 1.5 and greater) than at the dry gas application. That is why the method of the oil displacement by the carbon dioxide may be used in oilfields with the oil of medium and low viscosity, with the density lower than 0.9 g/cm^3 and pressure 10.5 MPa.

The carbon dioxide viscosity is higher than that of a dry gas by the factor of 2 to 4. This gives certain advantages to this oil displacement method. At the same time, the mobilities relationship here still remain unfavorable. That is why it is recommended to alternate the injection of gas with water. The most efficient variety of this method envisions the creation of a fringe on the order of 5% to 6% of the pore volume. The fringe is pushed into the reservoir by way of alternating injections of water and gas until the accumulated volume of the pumped-in gas reaches 15% to 20% of the pore volume. After this, a continuous water injection is conducted.

Carbonate reservoirs with a low-viscosity oil are important perspectives of an improved oil yield. However, when applying this method, it is necessary to protect the well equipment from corrosion by the carbon dioxide.

5.1.6 Oil Displacement by Polymer Solutions

It has been already noted above that poor reservoirs are to substantial measure determined by the mobility ratios of the injected agent and of the displaced oil. The studies show that, at lowering the mobility ratio from 10 to 1, the areal sweep by the displacement process (at five-point well system) doubles up to the breakthrough of the displacing agent. On lowering M to 0.25, the sweep efficiency increases more than 2.5 times. Lowering of the mobility ratio results in an improved areal sweep. Thus, at the M change from 10 to 1, the areal sweep efficiency displacement process (by the time a 95% water content is reached) increases by 18% to 20%. In laminary-nonuniform reservoirs, lowering of the mobility ratio facilitates a more uniform advance of the oil displacement front due to the water redistribution into the lower-permeability interbeds. The mobility ratio is among the most important parameters determining the oil yield.

The oil displacement by polymer solutions method involves for lowering mobility of water injected into the reservoir (water-soluble polymers are added to water). The presence of even very low polymer concentration in water (a fraction of a percent) substantially increases its viscosity. Most commonly used among such substances is polyacrylamide. This is a polymerization product of amid monomers composed of carbon, hydrogen, oxygen, and nitrogen. This polymer is water-soluble and has a large molecular weight, on the order of a few millions. A high viscosity is reached by polyacrylamide solution upon hydrolysis. As a result, amid groups convert to hydroxyl ones. The polyacrylamide solution viscosity to a substantial is determined by the extent of hydrolysis (by the number of amide groups converted to hydroxyl ones). Polymers used for the injection (in particular, polyacrylamides) are non-Newtonian liquids characterized by a nonlinear relation between the flow velocity and pressure gradient. It was found that a flow of the polyacrylamide solution within a porous media may be:

- Newtonian, when the medium resistance does not depend on the flow velocity;
- Plastic, when the medium resistance decreases with the increasing flow velocity;
- Dilatant, when the medium resistance increases with increasing flow velocity.

Thus, at the flow of polymer solutions, the pore medium resistance may undergo substantial changes. In the case of dilatant flow, the factors show up favoring the process of oil displacement because in decreasing resistance against the injected water as increasing permeability of the medium's facilitates a more uniform performance of reservoirs. An important feature in the flow of polymer solutions is adsorption and mechanical increase of the porous medium's dissolved amount. These sorption phenomena increase resistance against water injected into the reservoirs which also improves the oil displacement process. Therefore, the reservoir sweep by the oil displacement by polymer solutions process compared with the waterflooding is due mostly by increasing viscosity of the injected water and decline of permeability for this water due to sorption phenomena as well as by possible non-Newtonian flow nature of polymer solutions.

For injection into oil reservoirs, the applied solutions usually have polymer concentration of about 0.025% to 0.15%. Usually, the reservoir created a polymer solution fringe (20% to 30% of the total pore volume), which then is pushed within the reservoir by a regular water.

Due to sorption phenomena, an intensive decrease of polymer content in water occurs in the frontal part of the fringe. As a result, in the head-portion of the fringe the mobility ratio of the displacing water and displaced oil deteriorates. As a result, the idea of a so-called programmed injection of a polymer solution became common. Under this approach, a solution with an elevated polymer concentration is injected in the head portion of the fringe and in the rear portion, with a lower concentration.

Polymer solutions are most commonly recommended for developing elevated oil viscosity fields with a moderate nonuniformity of productive reservoirs. The oil-by-polymer solutions displacement method is usually applied to high permeability reservoirs (over 100 mD). The reason is a need to compensate for possible loss of the development tempo due to the injected fluid's viscosity increase and decrease of the injection wells' intake rate. A danger in the application of this method is that in reservoir with high clay content (over 10%) of aggregates blocking the porous medium may emerge, besides, the polymer adsorption increases. Similar problems emerge also in the case of significant contents of calcium, magnesium, and other cations in the formation water.

5.1.7 Oil Displacement by Micellar Solutions

This method is based on a high displacing capability of micellar solutions. Micellar solutions used for oil displacement may be described as micro-emulsions composed of aggregates (micelles) of water molecules and hydrocarbons solubilization (mutual solution) of which is achieved by the action of emulgators—surface active agents. Depending on which of the two phases (water or a hydrocarbon fluid) is a dispersion medium, there are direct micellar solutions (with the external water phase) and inverse micellar solutions (with the external hydrocarbon phase). In micellar solutions with the external water phase, droplets of a hydrocarbon fluid are dispersed in a solution of a surface-active agent in water. In micellar solutions with the external hydrocarbon phase, which is sometimes called "soluble oil" in publications, water droplets are dispersed in a solution of a surface-active agent in a hydrocarbon fluid. Currently, both types of the micellar solutions are being tested in field conditions. For making micellar solutions more aggregate-stable and for controlling their properties (in particular, viscosity) in the reservoir conditions the so-called stabilizers (inorganic electrolytes and organic surface-active substances) are introduced in the dispersion medium. The addition of stabilizers decreases adsorption of surface-active agents in the porous medium.

The micellar solutions tested in the oil displacement process in the field and in the lab have a broad range of change in the composition of individual components. According to published data, the micellar solutions are usually composed of surface-active substances (4% to 15%), hydrocarbons (4% to 80%), water (10% to 95%), organic stabilizers (up to 4%), and electrolytes (up to 1%).

Usually used as surface-active substances are oil sulfonates (produced from various hydrocarbons). As the hydrocarbon phase, it may be used light oil fractions and, sometimes, just oil. For making a micellar solution, fresh or low-salinity water is used. As organic stabilizers are used various alcohols (isopropyl, butyl, amyl, oxyethylated, etc.). Electrolytes added to a micellar solution may be ammonium sulfate, sodium chloride, sodium hydroxide, sodium sulfate, etc. The optimum micellar solution's composition recommended is selected experimentally in a lab environment for every productive interval at maximum closeness to the formation conditions (reservoir type, formation fluids' properties, temperature, pressure, etc.) For the achievement of efficient technological results, a micellar solution is injected as a fringe 5% to 10% by volume of the pore space. The micellar solution fringe is pushed within a reservoir by a buffer fluid—a polymer water solution, which, in turn, is pushed by a regular water. Usually, the buffer fluid fringe size reaches 50% and more of the reservoir pore volume. The mechanism of this process is not yet totally clear. It is suggested that its efficiency is determined by a mixing displacement of the reservoir oil with the micellar solution.

A high displacement extent of the formation fluids (oil and water) is due mostly to dropping practically to zero (less than 0.01 dye/cm) of interfacial tension at the oil/water boundary. The same factor assures a high degree of the displacement by water solution of the micellar solution and the preservation of a relatively small size micellar solution fringe in its advance within the reservoir.

Regardless of the interpretation of the oil displacement by a micellar solution, a high efficiency of this process is proved by the experimental studies and field data. A consequence of the plunger-like displacement of the reservoir-saturating fluids by a micellar solution is the formation of in front of the micellar solution fringe, the so-called water-oil wave. It needs to be emphasized that a high displacement by a micellar solution of oil and water does not depend on the initial reservoir saturation values by these fluids. The flow area in front of the water-oil wave is defined by the reservoir fluid-saturation conditions prior to injection of the micellar solutions. That is why the oil displacement by *s* micellar solution is accompanied by a high water-content of production until such time when the water-oil wave approaches the production wells.

An important feature of the oil displacement by micellar solution method is a capability to control its viscosity. This allows to conduct the displacement process at favorable mobility ratios between, on the one hand, the micellar solution and the displaced fluid and, on the other hand, between the micellar solution and the buffer fluid. A result is a wider reservoir sweep by the oil displacement process by a micellar solution compared with waterflooding.

Therefore, a high degree of oil displacement together with favorable conditions for reaching a significant reservoir sweep by the displacement process advances the process of micellar-polymer flooding among most promising methods of improving the reservoir oil yield. At the same time, it should be kept in mind that the micellar solution contact with a mineralized formation water containing plenty of such bivalent cations as calcium, magnesium, and iron most commonly results in a deterioration of solution's displacing properties. The mineralized water may negatively affect the nature of a micellar solution by a buffer fluid (by a polymer solution) due to the viscosity lowering of the latter and by the increase in interfacial tension between the polymer and micellar solutions. An obvious consequence of lowering displacement effectiveness of the micellar solution by the polymer one may become the destruction of a relatively small fringe of the micellar solution and a significant deterioration of the entire oil recovery process. That is why prior to creating a micellar solution fringe in the reservoir, it is recommended preliminarily to pump into the reservoir a fresh or slightly mineralized water. The amount of such preliminarily injected water should be no smaller than the micellar fringe volume. However, a danger of the contact between a strongly mineralized water and a micellar solution and a buffer fluid with relatively insignificant amount of a preliminary injection of small amounts of a regular water still remains. As a result, major potential for the micellar solution application is intended in the reservoirs after the waterflooding.

Application of the method of oil displacement by a micellar solution is mostly defined by the areas of a successful waterflooding application, i.e., is associated with fields with moderately nonuniform clastic reservoirs containing a low-viscosity oil (10 to 15 cP). Its application in the fields with elevated viscosity oil is associated with substantial cost increase for increasing viscosity of the micellar solution and buffer fluid. The application domain of the micellar-polymer flooding method is limited to reservoirs with temperature of up to 90°C, because at higher temperatures, the polymer solution may be destroyed with all ensuing consequences.

5.1.8 Thermal Methods

The methods of heat action on oil reservoirs are based on a favorable effect of the temperature on the mobility ratio of the displacing agents on the oil displacement extent. Two major classes of these methods are known. The first one encompasses methods by which heat is introduced in the reservoir from the surface (for instance, the injection of hot gases, water, and steam). In the methods belonging to the second class, the heat is formed directly in the reservoir due to inter-reservoir exothermal reactions of hydrocarbon oxidation. For the implementation of the first-class methods, water and water vapor are used as heat carrying agents. They have a maximum among the known work agents heat content and, therefore, allow to assure a better efficiency of thermal action on the reservoir. At the development of fields with relatively shallow oil horizons, the steam is preferred as its heat content (enthalpy) is much higher than water's.

With increasing pressure, the difference in the heat content of water and steam declines. This means that, with the formation pressure increase (i.e., the depth of oil reservoirs), the advantage of steam over water as a heat carrier declines. At the heat carrier motion to the bottomhole and then within the reservoir, heat loss is unavoidable. If the water is used as a heat carrier, with its motion in pipelines, then the heat carrier temperature declines. If the steam is used as a heat carrier, then due to the internal latent heat (additional amount of heat expended for transferring boiling water into the steam), the heat loss, until it is small, do not result in heat carrier temperature decline. That means that, using steam as a heat carrier, it is possible to assure the effect on the reservoir of a higher temperature than in a case of injection of a hot water in the same conditions.

Among the major problems in the implementation of oil production method by way of injection, heat carrier into the reservoir is bringing heat to the reservoir with a minimum loss and then the use of this heat in the productive reservoir with minimum possible loss into the enclosing rocks. The heat loss occurs in the generating equipment, in the surface lines, in the well stem, and, at last, at energy loss for heating the enclosing rocks. The heat loss in the generating equipment is defined by its efficiency which is on the average 80%. The daily heat loss on the surface varies between 0.08 and 0.8 million kcal/day per 100 m of the heat carrier length. As an approximate estimate, the heat loss in the well stem is 0.4 million kcal/day per 100 m of the well stem. Special measures of lowering heat loss in wells allow to bring its value per 100 m of the well stem to just 2% to 5% from the total heat amount introduced into a well. Heat loss in the well stem is defined as one of important criteria in the application of oilfield development methods with the help of injecting a heat carrier into well which is expressed in a limitation on the oil reservoir depth.

At heat carrier motion within the oil reservoir, the heat is transferred to the productive reservoir rock and fluids it contains as well as to the rocks enclosing productive reservoirs. Heat transfer in the reservoir is implemented by convection (the flow of fluids in the reservoir) and diffusion (heat conductivity). Heating of the rocks enclosing the reservoir occurs due to their heat conductivity. The heat efficiency of the process of a heat carrier injection in the reservoir is usually estimated by the ratio of amount of heat accumulated in the reservoir and the amount of heat introduced in it (heat utilization factor) or by the ratio of integral heat loss to the reservoir enclosing rocks and the amount of heat introduced in the reservoir (heat loss factor). Heat loss is defined to a substantial degree by the process implementation time. Therefore, a quite important requirement of a high tempo in the field development by the method of injecting heat carriers. Heat loss to the enclosing rocks also substantially depends on the reservoir thickness and decreases in proportion with squared reservoir thickness. This is a reason of selecting reservoirs thicker than 10 to 15 m for limiting heat loss.

Due to the heat loss to the enclosing rocks and the need to heat large masses of a porous medium matrix, the reservoir heating front lags behind the oil displacement front. When hot water is used as a heat carrier, the reservoir heating front advances slower than the oil heating front by the factor of 4 to 6. The result is that for a substantial reservoir heating and increasing its oil yield it is required to inject reservoir with substantial hot water amounts exceeding the reservoir pore volume by three to four times. When the water steam is used as a heat carrier, the heating front is lagging behind the displacement front two to three times greater than in a case of hot water. The explanation is, as have already been mentioned, a much greater steam's heat content and the fact that decline in steam heat content when advancing in the reservoir up to its total condensation occurs without a temperature decline. Due to an important advantage of steam as a heat carrier, for heating the reservoir and increasing its oil yield it is required to inject much less steam than the hot water. Of course, with increasing pressure (i.e., increasing productive reservoir depth) the indicated advantages of steam before hot water decline because the heat content difference between the both agents decreases. However, at heat treatment of shallow reservoirs (down to 500 to 700 m), steam has a clear advantage as a heat carrier. Exactly this is the reason that, currently, the saturated steam is most commonly used as a heat carrier.

As a result of different speed of heat and mass transfer in the reservoir and heat loss to the enclosing rocks, a number of characteristic zones form at the saturated steam injection in the reservoir. This is a steam zone of approximately equal temperature determined by the pressure level; a zone of condensed steam (hot water) where the temperature decline is observed from the saturated steam temperature to the initial formation temperature; and a zone not involved in the heat action. Temperature environment in the reservoir has a decisive influence on the oil displacement mechanism. Thus, in a zone not involved in the heat action, a mechanism is implemented of the oil by water displacement under isothermal conditions and at formation temperature. The oil displacement in the transitional temperature zone (condensed steam zone) is performed by water under non-isothermal conditions at the temperature change from the saturated steam temperature to the initial formation temperature. At last, in the steam zone, the oil is displaced by steam. Thus, the processes within each zone are not isolated but affect each other.

The oil by hot water displacement mechanism plays an important role both at the injection of steam and of the hot water. Additional oil production occurs due to the action of the following three factors:

- Decline of oil viscosity which improves the mobility ratio of the displaced oil and displacing water;
- Strengthening of the molecular surface strength in the reservoir resulting in a change of relative permeabilities to water and to oil;
- Heat expansion of the porous medium matrix and its contained fluids (especially oil) resulting in a decline in the residual oil saturation.

The role of each indicated factor substantially depends on oil thermal parameters. If the oil viscosity sharply declines with the temperature growth, which is typical of heavy high-viscosity oils, then the major contribution in the increase of oil yield improvement comes from improvement of the mobility ratio of the displaced oil and displacing water. However, if the oil viscosity moderately or weakly changes with the temperature change, then the oil displacement by hot water is caused mostly by changes in molecular surface forces and oil heat expansion. As for the reservoir nonuniformity, its negative effect on the oil yield at heat action is smoothed out to some extent due to a possible heating through heat conductivity of less permeable reservoir zones from the heated high-permeability zones and the aforementioned associated factors improving the oil yield. Thermal interaction between zones of different permeability, especially in the reservoir thickness, results in improved reservoir sweep efficiency by the heat action.

The indicated major factors of increasing oil production are implemented at hot water use. The same factors operate at steam injection in the reservoir. However, at steam injection, the boundary of a zone where the mechanism of oil displacement by hot water is efficient is to a smaller degree lagging behind the water/oil contact than at the hot water injection in the reservoir. As a result, the oil yield improvement at steam injection (and comparable volumes of heat carrier injection) is greater than at hot water injection. At the same time, the oil recovery at steam injection into the reservoir is affected by processes occurring in the steam zone. There, the major additional factor is the oil by steam distillation, the distillation of relatively light components of the oil remaining in this zone (after the oil displacement by cold and hot water). The presence of a gas phase (saturated steam) and a high temperature result in the evaporation of light fractions. They are carried away from the steam zone and condense within the cooler parts of the reservoir. An increase in oil displacement extent at the expense of its distillation by steam is determined by the oil composition. The lighter the oil, the more of the light components may be extracted from it. It is believed that this increase may reach 20% of oil reserves. As for absolute values of the residual oil saturation within the steam zone, they change in a wide range of 3% to 20%. As a result of displacing light hydrocarbons in the flow direction and their condensation in the "cold" zones of the reservoir, in front of the steam zone conditions are formed for the emergence of a fringe of lighter hydrocarbons (possibly, along with the formation oil). This may result in a mixed displacement (partial or total) of the oil originally contained in the reservoir and in an improvement of the oil yield. Currently, the steam injection method is used mostly in the development of heavy high-viscosity oilfields where the miscible displacement mechanism is of a subordinated value in the reservoir oil yield.

Under the influence of the heat action, oil waves formed in a heated reservoir part rush into the unheated reservoir portions. They represent a clear example of mutual influence between the processes occurring in various temperature zones. At continuous heat carrier injection, in particular steam, substantial volume of reservoir may be heated. However, the efficiency of such heating even after a relatively long heat injection those rapidly declines. Thus, there is practical and economic inexpediency of a continuous heat carrier injection. At a certain stage of a continuous injection, a substantial decrease in the oil production accrual occurs and, therefore, an increase in a heat carrier use for production. At this stage, the heat carrier is used mostly as a pressure support in the reservoir. This indicates the possibility of passing at a certain stage of the injection to injection of a less costly working agent, i.e., water.

The studies established that, due to a heat interference between various permeability reservoirs (areas), the negative effect of nonuniformity over the reservoir sweep by heat action and oil displacement processes is smoothed out to some extent. In the reservoirs with the most strongly developed heterogeneity (for instance, fractured-porous and laminated-nonuniform sections with typical size of blocks and layers on the order of 3 to 5 m), their sweep area by heat action is practically independent of the penetration ways by the heat carrier in the reservoir. In such a case, the reservoir may be considered thermally uniform. This is an important characteristic feature of applying a heat fringe. It allows first, to make the fringe creation easier and second, to decrease its "blurring" at advancing in the reservoir due to its nonuniformity. The size of initial reservoir heating zone is usually selected so that on approach to the fluid withdrawal zone certain temperature is preserved (70°C to 100°C). That assures the needed displacement efficiency to the extent of the heat fringe advancing in the reservoir. With increased distance between the wells, the initial size of steam heat fringe must be increased. For this reason, advantages of oilfield development technology with the application of moving heat fringe show up mostly at dense well grid. At heat injection in the reservoir, currently applied well density usually does not exceed 3.5 to 4.0 hectares per well. Under these conditions, the initial volume of a heat fringe is 0.15–0.2 to 0.6–0.7 relative to the total volume of the "heat-treated" reservoir. Obviously, with increasing distance between the wells, the size of the fringe should increase.

The method of oil displacement by steam is substantially tested in field conditions and is at a stage of industrial implementation in the high-viscosity oilfields (over 50 to 100 cP) with shallow productive reservoirs (up to 500 to 700 m) of large thickness (10 to 15 m). The method assures a high oil recovery (40% to 60%, sometimes higher) and may compete with conventional methods. The disadvantages of the method of oil displacement by heat carrier are as follows:

- Heat loss at the heat carrier delivery to the reservoir in steam pipelines on the surface and in the well stem. As a result, the application is restricted to relatively shallow reservoirs;

- Decline in the heat action efficiency in oil reservoirs with declining reservoir thickness and heat carrier injection in the reservoir due to the increase in the heat loss for heating the rocks enclosing the reservoir. As a result, the application is restricted to reservoirs thicker that 10 to 15 m with relatively high permeability (over 300 to 500 mD);
- More complex well design and well completion in consideration of the work under high temperature;
- Need of taking measures against sloughing of sand;
- A possible formation of stable water-oil emulsion;
- A possible manifestation of gravity effects and decline because of that the reservoir heat action sweep.

The oil displacement technology with the application of *In Situ* Combustion (ISC) has been for the first time theoretically substantiated in the early 1930s by the Russian scientist A. B. Sheinman. He conducted the first successful testing of this technology. It is based on capability of hydrocarbons (oil) to enter reaction with oxygen with the release of large amounts of heat. The heat generation directly in the reservoir is a major distinctive feature of yield improvement with ISC application. The heat generation directly in the reservoir allows to eliminate technical complications and heat loss at its generation on the surface and delivery in the reservoir by way of injecting heat carriers. The combustion is initiated in the firing well. This important operation boils down to the injection of an oxidizer (usually air) in the well at simultaneous heating of the reservoir near-bottomhole zone using a downhole heater, gas burner, firing chemical mixes, etc. A result is an acceleration of isothermal reactions of air with oil mix oxidation eventually resulting in the emergence of the process of burning in the reservoir near-bottomhole zone. In some cases, spontaneous oil inflaming occurs without supplying an additional fuel. After the burning initiation, a continuous air injection provides for a support of the ISC and advancing of the burning zone in a reservoir. The burning zone size is small compared with the distance between the wells so it is often called the burning front. When the air for burning support gets in the firing well, the burning front is advancing from this well to a production well in the movement direction of the injected air. This burning process is called a direct flow as opposed to a counter flow when the burning front is advancing from the production firing well toward an injection well, i.e., against the flow of the injected air. So far, a counter flow burning did not get any noticeable application, so the following description relates only to a direct flow burning process.

At advancing of the burning front, as a fuel is expended, some oil remaining in the reservoir after its displacement by the burning gases, water steam, and water evaporated oil fractions ahead of the burning front subjected to alteration due to the distillation, cracking, and other complex physicochemical processes. As a result, burn most heavy oil fractions, the so-called coke. In various geological field conditions, the coke concentration may be 10 to 14 kg/m^3 of the reservoir. This important parameter of the burning process is usually recommended to be experimentally determined in a lab. It is found that the coke concentration increases with the increasing oil density and viscosity and decreases with increasing permeability. It is believed that at coke burning is releasing heat on the order of 7,000 to 10,000 kcal/kg. Whereas it is *in situ* burning is supported by injecting in the reservoir of only gaseous oxidizer (air), the heat removal from the burning-heated rock by the air flow, due to its low heat-absorbing capacity, is occurring much slower than the rock heating by the advancing burning front. This circumstance results in that, at a dry *in situ* burning (this process is called so when is used for burning support, only air), most of the heat generated in the reservoir (80% and more) remains behind the burning front and gradually dissipates in the reservoir enclosing rocks. Conceptually, this heat positively affects the oil displacement process from the burning reservoir portions. However, that passing of the heat in the area ahead of the burning front will result in approaching of the heat generated in the reservoir to zones of the oil displacement and, therefore, in a substantial increase of the oil production process efficiency. Passing of the heat in the area ahead of the burning front is caused by an acceleration of the heat transfer in the reservoir by adding water to the injected air. In this connection, intense studies are conducted for creating methods combining the *in situ* burning and waterflooding. The major distinction of such methods is acceleration of heat transfer and passing heat in the area ahead of the burning front. In particular, ever greater recognition in the world practice acquires the moist burning method.

The substance of the moist burning is in that water injected in certain amounts together with the air evaporates near the burning front and passes the generated heat in an area ahead of it. As a result, expansive heated zones expressed mostly in the saturated steam and condensated hot water develop in this area. The *in situ* steam generation process is among the most important distinctions of the moist burning. To a large extent, it defines the oil displacement from the reservoir mechanism. The ratio of the water to air injected into the reservoir is about 1 to 5 m^3 per 1,000 nm^3 of the air, i.e., water/air ratio on the order of $(1–5)*10^{-3}$ m^3/nm^3. The specific values of the water/air ratio are defined by geologic-field conditions.

With increasing oil density and viscosity (or rather with increasing coke concentration), the value of needed water/air ratio declines. If this value is less than the indicated limits, then passing of the heat ahead of the burning front declines. If, however, larger water amounts are injected, then the moist burning method converts into other modifications of joint action on the reservoir by burning and waterflooding. It is important that an increase in the value of water/air ratio does not result in discontinuation of oxidizing exothermal processes in the reservoir even if there is no longer a high-temperature burning zone. Whereas a decline in the value of water/air ratio results in an efficiency decline of the heat action on the reservoir and of the oil recovery. That is the reason why the moist burning process is conducted with a maximum possible value of water/air ratio.

The highest temperature is associated with the burning front where it reaches 370°C and higher. As the burning front advances, several clearly identified temperature zones form in the reservoir. Two temperature zones emerge in the burned-out zone behind the burning front. The temperature in the transition zone varies from the work agents' (water and air) temperature to the evaporation temperature of injected water. Directly adjacent to the burning front is the overheated steam zone. This steam forms at evaporation of water injected together with air in a rock heated to a high temperature. This zone is positioned ahead of the burning front.

The heat transfer in the area ahead of the burning front is implemented at the moist burning mostly by way of convective transfer with the flows of evaporated injected water and burning products as well as through the heat conductivity. As a result, several temperature zones develop ahead of the burning front. First is a zone of the overheated steam within which the temperature declines from the frontal temperature to the steam condensation (evaporation) temperature. The size of this zone is relatively small as the heat loss to the reservoir-enclosing rocks results in a rapid cooling down of cooling there gaseous steam of water and low heat absorbing capacity burning products. The major fraction of heat passed in the area ahead of the burning front is concentrated in the saturated steam zone. This is a zone of the steam "plateau" where the heat loss to the enclosing rocks is accompanied by the steam condensation and not by a decline of its temperature. The major fraction of heat is also passed in the transition zone—a hot water zone formed as a result of a complete condensation of the saturated steam. Temperature in the saturated steam zone is mostly defined by the formation pressure level in consideration of steam fraction in the gas flow. Usually, it is changing insignificantly there and is equal approximately to 80% to 90% of the saturated steam temperature. The transition zone temperature varies between the steam condensation temperature and the initial formation temperature. At last, ahead of the transition zone is positioned a zone not involved in any heat action, i.e., a zone with initial temperature. The size of the reservoir heating area ahead of the burning front is to a great extent defined by the heat generation rate at the burning front (therefore, also by the rate of the air injection). It is also defined by the water/air ratio. With its increase, the size of the reservoir heating area increases. If the moist burning process is implemented at maximum possible (or close to it) water/air ratio, then, practically, the entire accumulated heat will be concentrated in the area ahead of the burning front and the size of this area will be maximal.

The temperature field at moist burning is mostly defined by the steam generation at the burning front and by heating by this steam of the reservoir area ahead of the burning front. That is why it is not accidental that at moist burning the temperature environment ahead of the burning front is much similar to the temperature environment at steam injection in the reservoir. At moist burning, the same oil displacement mechanisms will be implemented as at the steam injection in the reservoir. These mechanisms are the oil displacement by the steam and hot water and the miscible displacement by light oil fractions evaporated in the steam zone. As for the implementation of the ISC, air and water are injected into the reservoir, and the mechanism of oil displacement by water/gas mixtures also shows up. Besides, the oil recovery may be substantially affected by the products of oil burning and oxidation in the porous medium as well as by physicochemical transformations of the reservoir rock itself. In particular, a substantial carbon dioxide gas amounts emerge at burning. This creates conditions for the oil displacement by carbon dioxide. This mechanism may be substantially reinforced at ISC in carbonate reservoirs. There, additional amounts of carbon dioxide form due to the thermal and chemical decomposition of carbonates. Carbon dioxide together with the oil and water may form suds which also positively affects the oil displacement. It also forms at burning surface active substances, aldehydes, ketones, and alcohols which facilitate emulsion formation. Same as in a case of the oil by steam displacement, the formation of the oil wave is possible at moist burning ahead of the heated reservoir portion.

The size of the reservoir heating zone ahead of the burning front at the implementation of moist burning is about the same as the burnt zone and in most practically important cases may reach 100 to 150 m and greater. This indicates a possibility to apply moist burning method at loose well grids (like 16 to 20 hectares per well and greater). On the other hand, there is no need to bring the burning front to the production wells, which decrease the air expense for producing the oil. Only on the account of a reservoir heating zone ahead of the burning front the air usage decreases on average 1.5 to 2.0 times. Additional saving of the air for the oil production may be achieved by moving in the reservoir of the heat fringe created by the moist burning. This is implemented by way of injecting cold water. Overall, in this case (at moist burning), the air usage for oil production is lower by the factor 2.5 to 3.0 and greater than at dry burning. As for the air use at dry burning, it varies (according to the field data) between 1,000 and 3,000 nm^3/m^3.

A substantial decrease in the air use at moist burning is a precondition for broadening the heat action domain application to deeper reservoirs than currently accepted. Lab studies and field tests indicate that the moist burning method is implementable to the targets with substantial variation range of geologic-physical conditions. In particular, a possibility appears of developing using this method the fields with medium and low viscosity oils, including after the flooding. At elevated water/air ratio values emerge the other technological varieties based on a combination of waterflooding with *in situ* oxidizing reactions. In this case, the burning front (and its adjacent zones) disappears and oxygen from the injected air enters the saturated steam zone where it enters exothermal reactions with the oil. It should be noted that the rate of oxidizing reactions is quite high even at temperatures of the steam zone (200°C and higher). Such process is called a super-moist burning. At super-moist burning a "cold" water invades the burning zone even prior of the moment as the entire remaining oil burns down. The specifics of this burning is in that the heating and evaporation of water, heat regeneration, and its emergence as a result of oxidizing reactions are concentrated in one zone advance rate of which is defined mostly by the water injection tempo. It is substantially higher than the burning front advance rate at dry and moist burning. Thus, as the super-moist burning is reached, there is a substantial intensification of the heat action on the reservoir as well as a significant decrease of the air usage for the oil production. Small fuel usage is required for supporting the super-moist burning (on the order of a few kg per 1 m^3 of the reservoir). That is why the super-moist burning method promises great increase in the case of low-viscosity oilfields, especially after the waterflooding.

The major drawbacks of the oil displacement methods with the ISC application are as follows:

- The need in the environmental protection measures and utilization of the products of burning;
- The need in taking measures of preventing equipment corrosion;

- Possibility of gravity effects manifestation and as a result a decline in the reservoir sweep by a heat action

5.1.9 The Vibroseismic Method

An idea of using vibration for improving the oil yield was first proposed by Kuznetsov, Simkin, and Surguchev in early 1970s [7] after correlation was identified between an earthquake and the oil production level from nearby fields. For instance, in the Starogroznenskoye oilfield, as affected by the earthquake of January 7, 1938 (the epicentral distance about 30 km, the magnitude 4.8, seismic intensity 6 balls), the daily oil production increased by 45% and then normalized. The cases are known when the production grew each time after near a field (in the Ukraine) were running railway trains and the regime of changing ground water well production flow rate depended on the level of technogenic seismic noise. It was also noted that 1.5 to 2 years prior to the earthquake, the formation pressure in the Kolodeznoye oilfield began rising and by the moment of a jolt increased by 3 MPa.

The cases are also known of an inverse connection between the water injection in a reservoir and earthquakes. For instance, a Dagestan earthquake of May 14, 1970 (magnitude 6.7) has been provoked by an oilfield development and has been accompanied by a great number of jolts with the epicenter near the injection well. Because of the identified tie between the earthquakes and water injection in the well, the injection was discontinued.

In the case of strong earthquake in Denver, Colorado (USA), when the water injection has been stopped, the frequency of earthquake occurrence decreased. However, more than a year after stopping the injection, the seismic activity again increased. Three earthquakes have occurred with the magnitude 5 to 5.2 that caused some destruction. Quite strong jolts continued even 3 years after the injection has stopped (two jolts with the magnitude of 2.5 have occurred). Using seismometers positioned at a distance of 0.5 km from one another, it was found that epicenters lay within a 2-km-wide band running through the wells.

In the Raingly Field (USA), the oil was being produced from the depth of 1,830 m from the Late Paleozoic quartz sandstones. The field was a closed anticlinal fold. After 12 years of production, the formation pressure strongly declined. For its recovery, peripheral waterflooding has been conducted. Five years after the injection has begun, 65 km northeast of Raingly a seismic station has been constructed, which immediately began recording numerous weak earthquakes in its vicinity. By the end of 8-year long period of its existence, the station has recorded 976 jolts out of which 320 had a magnitude over 1. The annual number of earthquakes correlated with the volume of an annual water injection into the reservoir. Based on a network of mobile seismic stations, earthquakes have been concentrated in two areas: in the northwestern area and in the south of the field's central part. It has been noticed in both cases that, due to injection, pore pressure exceeded the initial one. The exact location of earthquake foci was determined after a local network of 14 permanent seismic stations has been constructed. It turned out that the jolt epicenters have been positioned along the southwestern strike fault established by drilling in the upper part of the reservoir sandstones. The earthquake foci have been positioned at depths 1,830 to 3,660 m, i.e., within the water injection zone or deeper.

Beside these known cases of earthquakes emergence due to the water injection in the oilfields, a similar phenomenon has been observed in 1972 in Dale (New-York, USA). Lamont-Doherty geological observatory has constructed in the state of New-York and in adjacent areas a network of 18 seismic stations for studying the placement of weak earthquakes. Eight of these stations have been positioned in the western portion of the state and provided a possibility to observe any manifestations of seismic activity associated with located not far from the Buffalo deep wells used for the injection of used water. In Dale, water was being injected at high pressure in one well for supporting a hydraulic salt production method from a few closely located wells.

The wells have been drilled in the Clarendon-Linden fault zone and close to it. Prior to the water injection, seismic activity in the area was one earthquake in a few months. After the injection started, a station 1.5 km from the well began recording about 100 jolts a day. When the water injection was stopped, the seismic activity in two days declined approximately to the initial level. Also, some local jolts have been felt, with no earthquake of the magnitude stronger that 1.5 have been recorded. Epicenters of most accurately determined jolts recorded using portable equipment have been within a 5-km-wide zone extended along that fault.

The experiment has been repeated in a year. A new well-positioned 300 m north of the previous one has been selected for the water injection. Hydraulic salt production was now conducted from different wells. After the injection has begun, a number of registered seismic jolts turned out even greater than at the previous time.

Exactly these data enabled the aforementioned authors to substantiate and, for the first time in the world practice, to develop a technology of the vibroseismic action on the waterflooded reservoirs for improving the oil yield.

5.2 A Study of the Residual Formation Pressure in the Vibration Field

The first experimental studies of the residual formation pressure in the vibration field have been conducted in a low-frequency range. The selection of low frequency has been selected by the energy considerations—a possibility to create vibration offsets commensurate with a typical length of pore channels at minimum vibration field intensity. For instance, for ultrasound frequency vibrations (10 to 30 kHz), achieving a vibratory offset on the order of average pore diameter (10 to 17 mkm) is associated with a huge intensity. At low-frequency vibrations (1 to 50 Hz), similar offsets in a porous medium are observed at much lower intensity (on the order of 1 kw/m^2 and less).

The vibration effect on the pressure flow has been studied on a lab installation which included:

- A vibro-bench with equipment for mounting reservoir models;
- Hydrodynamic vibro-bench of shear vibrations;
- A system of saturating the reservoir models with fluids;
- A system of vacuuming the reservoir models and fluids.

A porous medium has been modeled by the artificially cemented sandstone. The models have been made by stuffing cylindrical viniplast tubes, 1 m long and with the internal diameter of 30 mm, with a mix of a quartz sand and marshallite having particle diameter less than 0.08 mm. After the stuffing and compacting, nitrogen has been pumped through the poured medium under pressure of 0.1 to 0.5 MPa and an alcohol solution of a synthetic glue in a volume of 5 to 6 pore volumes of the model. Then, the poured medium has been dried up blowing through it nitrogen under the pressure of 0.3 to 0.5 MPa during 5 to 6 days. For this reason, the models have been placed into a solenoid wound from a polyvinyl chloride tube, 10 mm in diameter. During the model dry up, water from a thermostat has run through the solenoid supporting the porous medium at a temperature of about 70°C. This cemented medium was anisotropic, its permeability in the flow direction has been higher by one order of the magnitude than its permeability in the cross direction, which is typical in natural cores. The artificially cemented porous medium had an intermediate wettability as part of sand grains surface has been covered with a synthetic glue, which was preferentially wetted by hydrocarbon fluids. Intermediate wettability has been ascertained by the motion of a distilled water droplet placed on the surface covered with a dried up glue film. The droplet was not spreading on the surface and remained localized. Besides, preliminary experiments of capillary soaking have shown that dry models of artificial porous medium equally well soaked in water and kerosene. This indicated that this artificial medium has been close to the natural media in which active oil components, adsorbed on the matrix, partially hydrophobized the rock.

The vibration effect on the pressure flow has been studied on the reservoir models in the shape of quadrangle parallelepipeds made out of 20-mm-thick organic glass, 735 by 195 by 10 mm in size. The models have been stuffed with a tightly compacted various quartz sand fractions. The sand has been then cemented by a synthetic glue and dried up by blowing through a hot nitrogen at 90°C to 95°C. This has made the reservoir models having porosity of 40% to 45% and absolute permeability of 10 to 20 D (for nitrogen).

The study of shear (crosswise) vibrations on the pressure three-phase flow, a vibrobench has been designed and made (shown in Figure 5.1). It included a model of porous medium and connected with it low-frequency hydrodynamic generator of shear vibrations.

Two schemes have been created of the reservoir models' saturation with fluids. The first scheme included a pump repumping lubricating oil

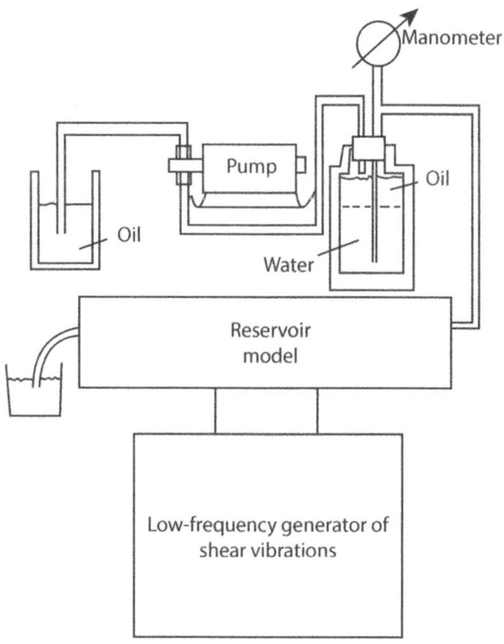

Figure 5.1 Schematic of the lab installation for a study of thc three-phase flow if a vibration field at a constant fluid flow rate.

from an open vessel in a water-filled vessel. The oil pushed water into the model (a case of water pumping into a reservoir) or in a different vessel filled up with an oil model which was pushed into the reservoir model.

The second scheme provided a possibility of saturation at constant pressure gradient. For this purpose, nitrogen has been pumped at a constant pressure in a special container (Figure 5.2).

Under either saturation schemes, sufficiently high pressures have been assured (under the first scheme, 50 MPa, and under the second, 10 MPa).

Productive reservoirs' reservoir properties have been determined experimentally. Among the most convenient one is a method under which a specific electric resistivity of water-saturated rocks is determined. Using this, a redistribution has been observed of the residual model oil and water in individual areas and along the entire length of the model. This method is based on a fact that a porous medium saturated with the formation (mineralized) water has certain electric resistivity. With a decrease in water saturation (increase of oil saturation), the value of the specific electric resistivity was increasing. Thus, between the resistance R (or in other words saturation parameter) and water saturation, there is a correlation:

$$R = cs^{-n},$$

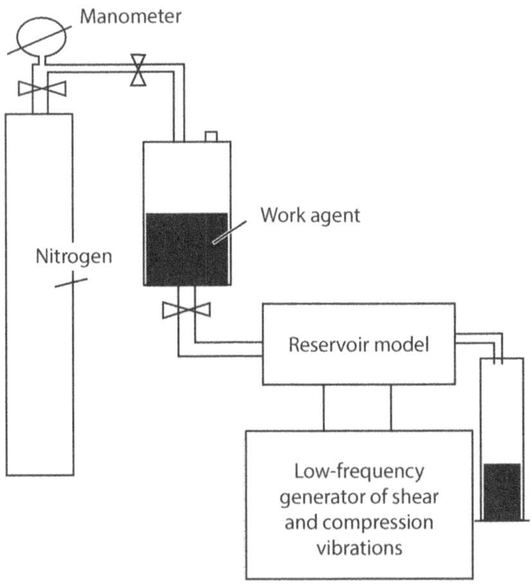

Figure 5.2 The schematics of the lab installation for a study of three-phase flow in a vibration field at constant pressure gradient in the fluid.

where c is proportionality factor, and n is an exponent varying with the reservoir type, salinity, and other reasons. The n value may vary between 2 and 2.7. If a logarithmic scale is used (logarithm of resistivity on the ordinate axis and logarithm of s on the abscissa axis), then the correlation is a straight line.

Experimental points needed for the construction of this straight line were determined by way of measuring electric resistivity of a part of the model at known water saturation values. For instance, drawing straight line through the point of maximum (100%) water saturation and water saturation measured after displacement of the oil by water, one may have determined water saturation values of a part of the model changing in time. For measuring electric resistivity, the models were equipped with special electrodes made of pieces of the orthodontal wire, 1 mm in diameter. Before the installation, the electrodes have been subjected to training by industrial frequency alternate current at electric voltage of 5 to 6 V in the solution of electrolyte which was used as a formation water model. The electrode training was needed for decreasing a distorting effect of near-electrode processes at the measuring time. The reservoir models have been carefully vacuumed, filled up with the vacuumed distilled water which then has been displaced by oil. This has been the way to model a saturated porous medium.

Flat models of the porous medium manufactured as described above have been used for conducting experiments of studying vibration effect on the pressure flow. First, the kerosene and water displacement from the model by distilled water has been conducted until a complete termination of kerosene flow from the model. Then, the displacement has been continued after the vibration field has been turned on. Results of the experiment are illustrated in Figure 5.3.

A number experiments have been conducted of the three-phase pressure flow in a reservoir model placed horizontally. In these experiments, the action has been implemented after the conduct in a usual regime of model oil displacement by gassed water until the total watering of the production (post-displacement regime).

In the next series of experiments, the model oil displacement has been conducted from a horizontally positioned flat reservoir model by an alkali solution (Soda Ash) of various concentrations. Upon reaching 100% water content of the fluid at the reservoir exit, the post-displacement has been conducted at the vibration with the same pressure gradients as in the previous cases. The correlation indicated that the vibration effect has been manifested not immediately but with some lag. Analysis of the displacement curves of this experimental series indicated that the action has been most active at frequency about 20 Hz. Thus, the vibration action resulted in a case of the two-phase flow in an increase of the final oil yield by 10%,

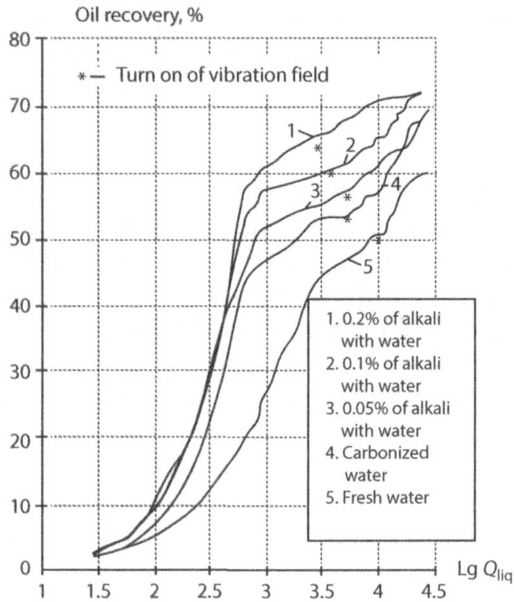

Figure 5.3 Oil displacement by water with various parameters in a vibration field. Q_{liq} is the volume of pumped-through fluid (cm^3).

i.e., in additional recovery of more than 20% of the reservoir model oil reserves residual after waterflooding.

In a case of the model oil displacement by water with dissolved carbon dioxide, the post-displacement after achieving 100% water content of the production enabled, as a result of an intense water-dissolved carbon dioxide release in the model, additional recovery of more than 35% of the residual oil reserves, i.e., improvement of the final oil recovery by 12.6%.

The great interest has been displacement parameters in the case of modeling the alkali flooding as shown in Figure 5.3. For instance, when the alkali weight concentration in distilled water has been 0.1, the post-displacement at the action enabled an increase of the final oil recovery by 10.2% and an additional recovery of about 26% of the model oil residual reserves from the reservoir model. The displacement character for a case of alkali weight concentration in water of 0.2% enabled an increase of the final oil recovery by 6.3% and an additional recovery of 18% of the oil residual reserves from the reservoir. Also is shown the displacement description for the alkali weight concentration in water of 0.05%. One obtains the post-displacement under action resulted in an increase of the final oil recovery by 12% and an additional recovery of 27% of the oil residual reserves from the model. It has been demonstrated that the alkali concentration (within the studied range) weakly affected the final oil recovery in vibration field.

Dyblenko and Sharifullin [5], studying the oil by water displacement in a vibration field, used core samples of low-permeability silty sandstones and oil models prepared out of water-free degassed reservoir oil viscosity of which has been brought to the oil viscosity under formation conditions. For displacing the oil, a formation water model obtained by way of dissolving table salt in water has been used. Bonded water has been made through a capillary extraction. After the cores have been saturated with oil, experiments were conducted of oil displacement at a constant flow rate of 0.2 to 1.0 m^3/day with a continuous recording of the pressure gradient.

Oil displacement prior to the vibration has been conducted until the total termination of its release from the porous medium. The 7 to 8 pore volumes of water have been pumped through until a total stabilization of the pressure gradient, and the vibration has been turned on at a gradually increasing intensity. The vibratory offset value has been maintained in the experiment within an average pore channel radius. When, by the change of the pressure gradient, the vibration effect has been identified, the vibration parameters (the vibratory acceleration and frequency) have been recorded and maintained constant to the total oil displacement from the porous medium. Figure 5.4 displays oil displacement and pressure gradient

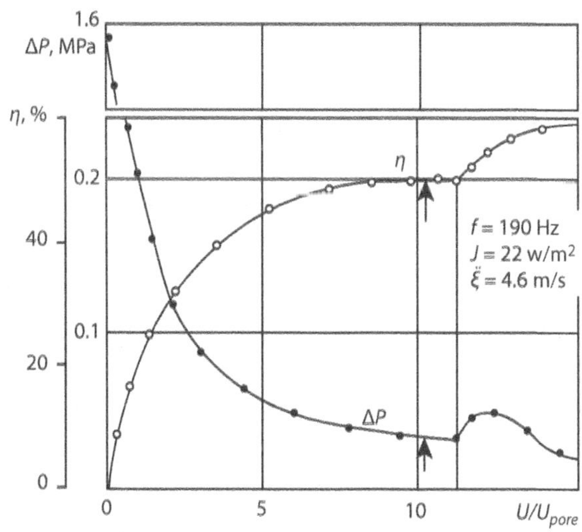

Figure 5.4 The displacement nature and variation of pressure gradient ΔP in the vibration field (after V.P. Dyblenko and R.Ya. Sharifullin [5]). The moment of the vibration turn-on is indicated by an arrow. η is the oil recovery factor; U is the water throughput; U_{pore} is the reservoir pore volume; f, J, and $\ddot{\xi}$ are, respectively, frequency, intensity, and vibration acceleration amplitude.

parameters in time depending on the pumped-through volume of water vs. the reservoir pore volume. As the derived results show, after the end of the oil displacement process and the achievement of the maximum oil recovery rate of 50.5%, the process turn-on has resulted in the oil post-displacement by water. The oil recovery rate has been increasing up to 59%. The pressure gradient changes have indicated that the vibration had involved the residual immobile oil in the flow. The phase permeability for water has been declining initially and, after displacement of the residual oil, has been rising above the previous level.

One can now review a single-dimensional option of the non-wetting fluid displacement by a wetting one in a vibration field.

On assuming that:

- The fluids do not mix and do not compress;
- Any phase transitions are absent;
- Viscosity factors are constant;
- Phase permeabilities and capillary pressures are known and monosemous saturation functions;
- Hysteresis phenomena are not taken into consideration;
- The porous medium's matrix and the wetting phase vibrate synchronously, with known offset amplitudes;
- Reynolds' acoustic numbers are small compared with the unit.

The motion equations in this case as follows:

$$u_{\mathrm{wet}} = -k/\mu_{wet}\, k_{wet}(s)((\mathrm{d}P_{wet}/\mathrm{d}x) - \omega^2 A\rho_{wet} + g\rho_{wet}\sin\partial), \quad (5.1)$$

$$u_{\mathrm{non}} = -\, k/\mu_{non} k_{non}(s)((\mathrm{d}P_{non}/\mathrm{d}x) - \omega^2 A\rho_{non} + g\rho_{non}\sin\partial) \quad (5.2)$$

Here, u_{wet} and u_{non} are flow rates of wetting and nonwetting phases; μ_{wet} and μ_{non} are the phase viscosities; k is the absolute permeability; $k_{\mathrm{wet}}(s)$ and $k_{\mathrm{non}}(s)$ are the relative phase permeabilities; s is the water saturation by the wetting phase; P_{wet} and P_{non} are the pressures in each phase without consideration of the action; ∂ is the reservoir dip angle; A is the matrix offset amplitude; and ω is vibration frequency.

The connection between pressures in each phase is regular:

$$P_{\mathrm{non}} - P_{\mathrm{wet}} + \int_0^z \omega^2 A\rho_{wet} \frac{1-\Delta}{1+2\Delta} dx = \sigma\cos\theta\sqrt{m/k}\,J(s), \quad (5.3)$$

where σ is the interfacial tension, θ is the wetting angle, $J(s)$ is the Leverett function, m is the porosity, and $\Delta = \rho_{wet}/\rho_{non}$.

If one adds continuity equation for each phase to these equations:

$$-m\frac{\partial s}{\partial T}=\frac{\partial u_{wet}}{\partial x}; \qquad m\frac{\partial s}{\partial t}=\frac{\partial u_{non}}{\partial x}, \tag{5.4}$$

One obtains a closed equation system of flow in the vibration field.

The solution of the above equations for a fraction of wetting fluid in a flow is as follows:

$$f_{wet}=\frac{\left[1+\frac{k_{non}(s)}{u\mu_{non}}\left[\frac{\partial P_{non}}{\partial x}+\omega^2 A\rho_{wet}\frac{1-\Delta}{1+2\Delta}-g\left(\rho_{wet}-\rho_{non}\right)\sin\partial\right]\right]}{\left[1+\frac{\mu_{wet}}{\mu_{non}}\frac{k_{non}(s)}{k_{wet}(s)}\right]} \tag{5.5}$$

where $u = u_{wet} + u_{non}$ is the total fluid flow rate.

Just as important was a study of vibration effect on the oil post-production from shaly reservoirs.

Shaly clastic reservoirs contain up to 15% to 20% clay minerals cement [4]. A specific of such reservoirs is clay minerals swelling on contact with water which had salinity lower than the formation water salinity. A decrease of porosity and permeability in the contact zone occurred.

Under the influence of vibration, micro-flows in a clay-containing reservoir may change. Besides, if active clay minerals are present, then their position relative to the reservoir matrix must change under influence of vibrations [3].

Experiments have to either support or deny these assumptions. The experiments have been conducted with the stuffed reservoir models 40 cm long with the internal diameter of 2.8 cm within a viniplast tube. On the tube end, a steel cap with a filter and pass through hole has been screwed-on. A sand layer (0.1- to 0.2-mm fraction) has been placed over a synthetic glue or a screw thread for preventing a fluid slippage on the internal tube surface on the internal tube surface. The experiments have been conducted at a pressure gradient of 0.07 to 0.1 MPa. A mixture of the quartz sand (0.05- to 0.2-mm fraction) as a porous medium has been used with the bentonite clay in needed proportions. The model permeability for nitrogen has been 0.3 to 0.4 D; with porosity of 25%. Solutions of 30 to 200 g/l of NaCl in the distilled water and kerosene with transformer oil mix as formation fluids have been used with viscosity of 5.0 cP. Before the experiments, a residual water saturation has been created in the reservoir models.

For this purpose, they have been saturated with formation water (150 g of NaCl in distilled water), which then was displaced by the oil.

The first series of experiments has been conducted on reservoir models containing no clay and also with clay content 5%. Figure 5.5 is a display of experimental correlations of oil recovery vs. the volume of pumped-through fluids. Curve 3 reflects the model oil displacement by water of 150 g NaCl salinity from a no-clay reservoir model without the application of a vibration field. After the beginning of water pumping-through, the oil recovery increases, first sharply and then slowly, reaching 67.6% with the water pumping-through volume equal to 1.5 of the pore volume. Curve 1 reflects the oil displacement from a reservoir model with the clay content of 5%. After pumping-through of the same amount of the same fluid, the oil recovery increased to 86.8%. This is totally compliant with ideas of waterflooding specifics of clay-containing reservoirs. The oil recovery vs. accumulated fluid volume correlation in the vibration field has been studied also using displacing water salinity of 50 (curve 5), 100 (curve 2), 150 (curve 3), and 200 g/l (curve 4). The vibration field parameters have been the same at various salinity of displacing water.

Same correlations are presented in Figure 5.6, i.e., the oil recovery factor vs. accumulated fluid volume. This represents the action efficiency more

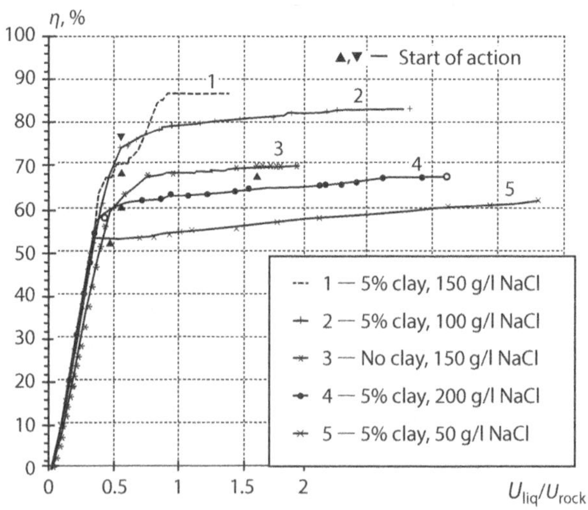

Figure 5.5 Oil by water (of various salinities) displacement from a clayey reservoir in a vibration field. U_{liq} is accumulated fluid volume; U_{pore} is the reservoir pore volume; η is the oil recovery factor.

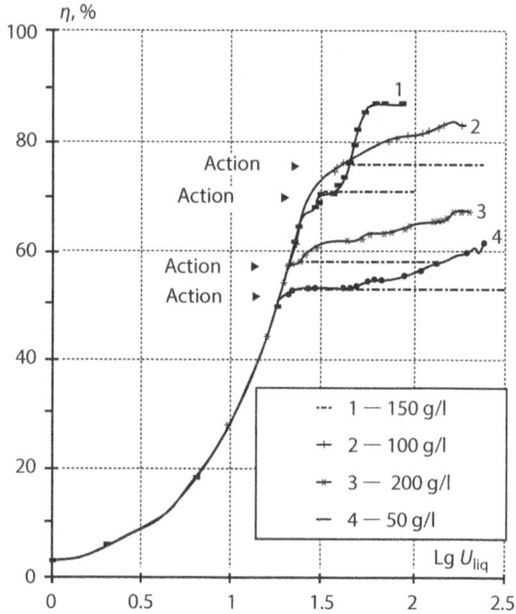

Figure 5.6 Oil by mineralized water displacement from a clayey reservoir in a vibration field. U_{liq} is accumulated fluid volume in cm^3.

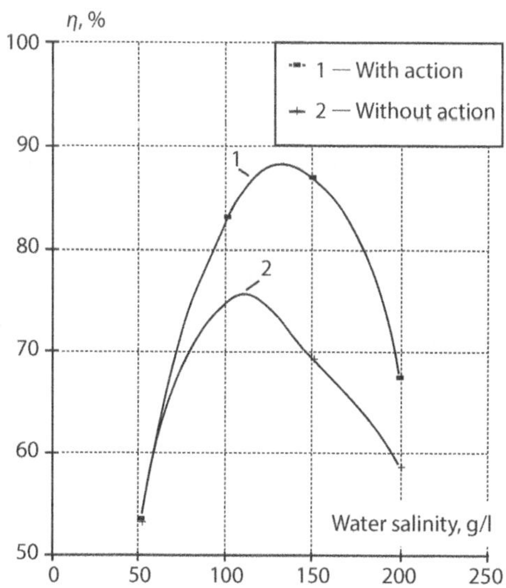

Figure 5.7 Effect of water salinity on the oil recovery from a shaly reservoir in the vibration field.

clearly. Vibrations substantially change the nature of pumped-in water salinity effect on the oil recovery. Thus, the maximum oil recovery from a uniform reservoir model used in experiments should have been expected at pumping-in a 50 g/liter salinity water. However, under vibration, the maximum oil recovery has been occurring at pumping in a slightly less saline water (150 g/liter) than formation water (200 g/l).

In the next experimental series, oil with a mineralized water in a vibration field was displaced. The reservoir models had the shaliness of 5% to 15%, permeability of 0.3 to 0.4 D and porosity of 25%. The pressure gradient varied from 0.07 to 0.1 MPa. The displacing water salinity has been 50 to 200 g/l (shown in Figure 5.7). A mix of kerosene and transformer oil, having 5.0 cP viscosity, was used as the model oil. At constant pressure gradient of 0.07 MPa, the injected water salinity decrease from 200 to 50 g/l resulted in the reservoir water intake decline from 1 to 0.19.

In the vibration field, the reservoir water intake ratio had been increasing to 1.45 to 1.64 at the salinity 150 to 200 g/l. Water salinity decrease to 50 g/l enabled the clay component to react, and at the same pressure, gradient of 0.07 MPa the flow stopped. The vibration has not restored the flow (in this case, the intake ratio has been equal to zero). Doubling the pressure gradient resulted in a sharp increase of the reservoir model intake in the vibration field at water salinity of 50 to 100 g/l. The water intake ratio in vibration field and without increased by a factor of 3.12 to 4.3. The reservoir model water intake at salinity of 50 to 100 g/l remained the same after switching the vibration off: the water intake ratio after vibration and without the field has been 3.0 to 4.1. The maximum water intake ratio has been reached at the injected water salinity of 50 g/l.

Figure 5.8 shows the oil recovery rate from the reservoir model with a 5% clay content vs. the injected water salinity in the vibratioin field (curve 1) and without vibration (curve 2). The oil recovery rate accrual in the vibration field at 200 g/l salinity water injection has been 8.8%. This figure also shows that there is the optimal water salinity value resulting in the highest oil recovery rate.

Experiments with the oil displacement from shaly reservoirs by an alkaline water in the vibratory field have been conducted with the same stuffed reservoir models at pressure gradient of 0.0085 MPa. The porous reservoir model has been a quartz sand (0.05- to 0.2-mm fraction) with 5% of bentonite clay. The reservoir permeability for nitrogen has been 0.28 to 0.4 D: porosity was 25% to 26%. Distilled water solutions of 50 to 150 g/l of NaCl and a kerosene with transformer oil mix (5.0 cP viscosity) have been used as the formation fluids. Based on the results of experiments with the oil displacement by the alkaline water in the vibration field, a 0.1%

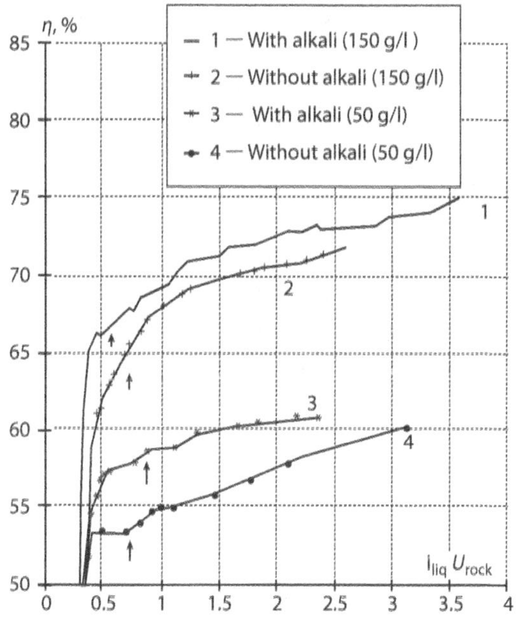

Figure 5.8 The displacement at the flow of alkaline water in a shaly reservoir in a vibration field.

NaCl solution in water has been used. Before experiments, the residual water saturation has been created in the reservoir models. The reservoir has been saturated with a NaCl solution in distilled water with salinity of 150 g/l. The formation water had been displaced by the model oil until the total termination of water coming from the reservoir. The first series of experiments was conducted on reservoir models at the oil displacement with 50 and 150 g/l salinity water with and without alkali. The results are displayed in Figure 5.8 which shows experimental correlations of the oil recovery rate vs. the cumulative volume of fluid pumped through the reservoir model. The arrows show moments of turning on the vibration corresponding to the production water content of over 85%. The vibration parameters were the same in all cases.

Water salinity decrease to 50 g/l had been causing an increased swelling of the reservoir clay component compared to swelling at 150 g/l. The result has been a decrease in the effective reservoir permeability to the oil and water. A result of the vibration has been that at salinity of 150 g/l the oil recovery factor was first sharply and then slowly (see curve 2) rising and reaching 72.5% at the pumped-through fluid volume was equal to two pore volumes of the reservoir. When the salinity has been decreased to 50 g/l, the vibration caused a gradual increase of the recovery rate to 60% at the same volume of the pumped-through fluid. The water salinity increase from 50 to 150 g/l caused the increase of the oil recovery rate from 60 to 72.5% due to vibration.

Adding 0.1% of an alkali to a mineralized solution had been causing additional increase in the oil recovery factor (at the 50 g/l water salinity, 0.1% alkali in the solution has caused an increase of oil recovery factor in the vibration field to 61%, at the 150 g/l salinity, to 75%).

The oil recovery factor correlation vs. cumulative fluid volume in the vibration field had also been studied at salinity of 100 g/l and 0.1% alkali of the displacing solution (curve 1 in Figure 5.9). The same figure shows also correlation factors at 150 and 50 g/l salinity (curves 2 and 3). Comparison of these correlations indicates that there is optimum salinity for the maximum oil recovery factor and the maximum development tempo at a constant alkali concentration of 0.1% in the displacing solution. According to the available data, such optimum concentration should be considered 100 g/l. At this salinity, the recovery factor has been reaching 75%.

The next series of experiments has been conducted for a water intake study of the model reservoir at the oil displacement by alkaline water in the vibration field. The model reservoir and vibration field parameters have been maintained the same as in the previous experiments. The experiments have been conducted in model with the clay content of 5% at a

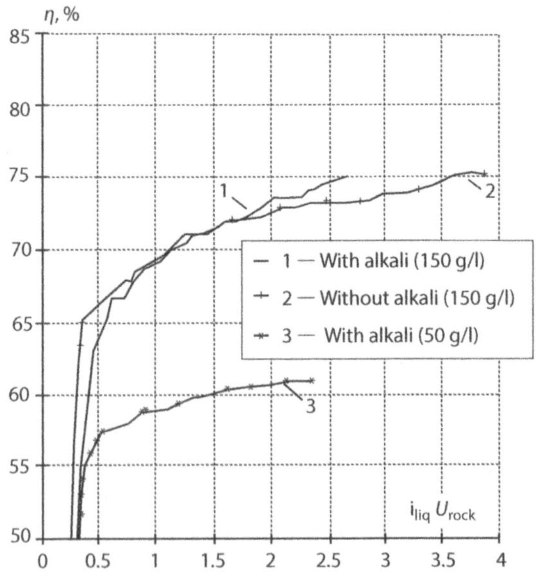

Figure 5.9 The displacement at the flow of mineralized water in a shaly reservoir in a vibration field.

constant pressure gradient of 0.0085 MPa. It was shown that a decrease in the water salinity from 150 to 50 g/l in water with 0.1% alkali has resulted in the water intake by models decline from 1.0 to 0.49. In the vibration field, the intake for models with a 0.1% alkali has been increasing to 1.58 to 2.03 at the 50 to 100 g/l salinity. This intake has been on average 1.5 times greater than the intake of the equal salinity water without the alkali. The reservoir model intake at water salinity of 50 to 150 g/l with 0.1% alkali in the solution remained equally high after the vibration field switch off—the ratio of water intake after vibration to water intake before vibration has been 1.25 to 2.03. The maximum intake of water with alkali (up to 2.03) has been reached at the injected water salinity of 100 g/l.

If the porous medium saturation has been conducted by two immiscible fluids (oil and water), then when one of them reaches by maximum residual saturation, phase permeability to this fluid becomes zero. In fact, there is cutting of the residual fluid into separate, unconnected droplets and accumulations surrounded on all sides by the opposite phase. The volume content of the immobile residual fluid may be quite high. This phenomenon is defining to a great extent the maximum residual oil saturation of the watered reservoirs.

5.3 A Study of the Oil Capillary Displacement in the Vibration Field

Capillary forces define completeness of oil displacement by water within the watered-out volume of an oil reservoir. The efficiency of these forces is usually evaluated by several processes modeled in lab conditions. Among such processes are the direct-flow and counterflow capillary soaking, which describe the capillary forces' efficacy at the front of displacing water and in watered-out reservoir volume behind the displacement front. Lab modeling of these processes is well worked out in practice.

The effect of vibration on the intensity of direct flow capillary soaking has been studied on various models of porous medium and cores used in a research of the oil recovery processes. Before conducting experiments, a residual oil saturation has been created in the porous medium. For this purpose, a reservoir model has been saturated with water under vacuum, and then water has been displaced by the oil until reaching the irreducible water saturation. Kerosene has been used as the model oil. The installation for studying the direct-flow capillary soaking in the vibration field has included a reservoir model, a water tank, a measuring cylinder, connecting tubing, and a vibro-bench. The reservoir model was fixed horizontally on the vibro-bench table.

The water level in the tank connected to one end of the reservoir model has been held at the level of horizontal upper surface of the model. Having exited from another end of the model, the fluid supplied through a connecting tube into the measurement cylinder and accumulated at this location.

The reservoir model has been represented by a compacted uncemented quartz sand placed in a cylindrical tube. Permeability of such reservoir for nitrogen has been 9.3 D, with a porosity of 0.38. The experimental results are displayed in Figure 5.10 as the oil recovery factor vs. square root of time. In the first experiment (curve 1), the capillary soaking has been implemented without vibration. After the experiment, initial conditions have been restored in the reservoir model, hydrocarbons had been washed off the sand by an alcohol and petroleum-ether. Then, the model has been dried with air and saturated in vacuum with water, which has been displaced by kerosene until reaching irreducible water saturation. After this, the capillary soaking process has been repeated in the described physical model which has been from the very start subjected to vibration. The results are displayed the curve 2. Within time intervals *AB* and *CD*, the process has been implemented without the vibration.

The vibration resulted in a substantial intensification of the direct-flow soaking (at a decline of the soaking duration by the factor of five, the oil

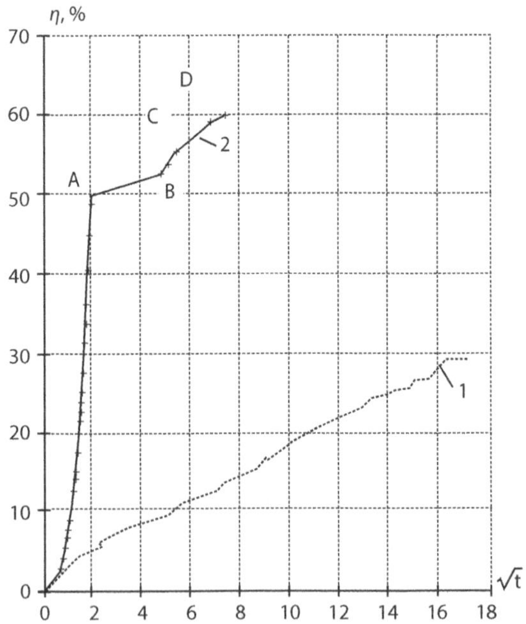

Figure 5.10 The oil displacement at the direct-flow capillary soaking of reservoir models in the vibration field. The time *t* in hours.

recovery doubled). Therefore, vibration renewal after an interruption has resulted in an acceleration of the direct-flow capillary soaking.

The vibration effect on the direct-flow capillary soaking has also been studied in natural cores. As a porous medium, cores from the Elabuga Oilfield, 26.8 cm long and 3.0 cm in diameter model has been used. Permeability of a composite sample for nitrogen has been 1.0 D. Experimental results on natural cores are displayed in Figure 5.11 (curves *1* and *2*). At the first stage, the direct-flow capillary soaking has been conducted without vibration (curve 1). After experiments, the initial conditions in cores have been restored. Then, the direct-flow capillary soaking has been implemented on the same model under vibrations from the very start of the soaking process (curve 2).

The rate of the direct-flow soaking vibration has increased compared with that without vibration. Vibrations increased the no-water oil recovery factor.

Experiments have also been conducted in an artificially cemented medium of a quartz sand prepared by the method described in 5.2. The reservoir model has been 400 mm long and 30 mm in diameter. The sample permeability to nitrogen has been 7.0 D. At the first stage, the direct-flow capillary soaking has been conducted without vibration. It has been

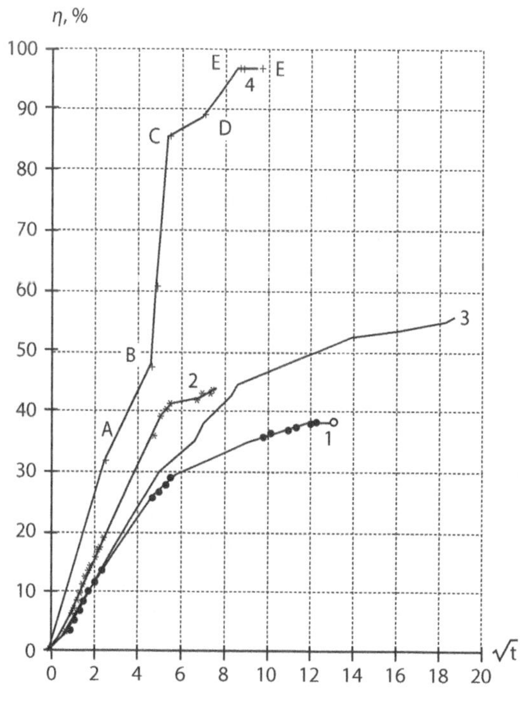

Figure 5.11 Direct-flow (curves 1 and 2) and counterflow (curves 3 and 4) natural cores' capillary soaking in the vibration field. *t* is in hours.

found that in such a case the capillary soaking was absent. At the next stage of the experiment, the soaking has been implemented under the vibration. As the vibration frequency has been increased, the vibration intensity has increased, which allowed to activate the process.

The installation for the study of the counterflow capillary flow in the vibration field was a kerosene-saturated rock sample placed in a vessel with water rigidly attached to the vibro-bench table. A pycnometer has been used for collecting and measuring the volume of kerosene displaced from the sample. The broadened end of the pycnometer was placed in an open water vessel above the sample and, after opening the valve at the op, has been filled with water to the upper mark of the measuring scale.

The kerosene-saturated rock sample from the Elabuga Field has been placed in a special clip positioned in the same water vessel under the pycnometer funnel. Thus, the entire surface of the sample has been in contact with water. The vibration was delivered to the sample through water which vibrated together with the vessel rigidly attached to the vibro-bench table. This way, conditions have been achieved of only counterflow capillary soaking when the displaced kerosene volume has been equal to the volume of water that invaded the sample. The vibrations have been delivered through the saturating fluid. Kerosene ejected from the sample has become buoyant and accumulated near the water surface in a graduated part of the pycnometer.

At first, the curve of kerosene-from-sample displacement dynamics (Figure 5.11, curve 3) has been recorded. The experiment has lasted around 272 hours. After restoring initial conditions in the sample (as described), the sample has been again saturated with kerosene and placed in the vessel with water. The kerosene from the sample displacement behavior (curve 4) has now been studied under vibration with the same parameters as in experiment with the direct flow soaking. Vibration has been applied from the start of the experiment and till the moment marked *A* on the curve 4 and also in time intervals *AB* and *CD*. As the curves 3 and 4 show, the counterflow capillary soaking of a natural core (a rock sample) has been going on much more intensively under vibration than without. The recovery fluid tempo doubled.

In connection with the discovered vibration effect on the direct-flow and counterflow capillary soaking, an attempt has been undertaken to estimate the vibration effect on the speed of establishing balance between the model oil and water in porous medium. A porous medium sample has been placed in a tube and saturated by the model oil. After that, waterflooding has been conducted at a constant injection rate until the moment the oil has stopped flowing. By the end of waterflooding period, the porous medium contained water and residual oil, i.e., modeled the trans-frontal area of a waterflooded reservoir. Next, the sample has been placed in the equipment in which the vibration effect has been studied. Each end of the sample has been connected through valves with piezometer tubes. One of these tubes has been filled with water and another, with oil. The design was for the pressure excess in the tube with the oil to have been greater than the capillary pressure leap at the oil-water separation boundary.

After connecting the sample with the piezometers and another with atmosphere, the oil level was beginning to do down and water, to get up. In a hydrophilic sample, this process has been going slowly and has been stopping at compensation of a gradient of the applied pressure by the capillary pressure at the oil by water replacement front. This compensation method has been used for evaluating the capillary pressure leap at the oil-water separation boundary. Figure 5.12 displays diagrams of applied pressure gradient vs. logarithm of time (in minutes). A sample has been used with porosity at 0.387 and permeability for nitrogen of 7 D. In two experiments (curves 1 and 2), it turned out impossible to stabilize the levels on both sides of the sample. Curve 2 has practically coincided with curve 1 before the beginning of vibration. At the moment of vibration, curve 2 is getting away from curve 1 and the difference at the end of experiment has been about 10%.

The final conclusion is that the vibration accelerates the oil and water redistribution in the samples.

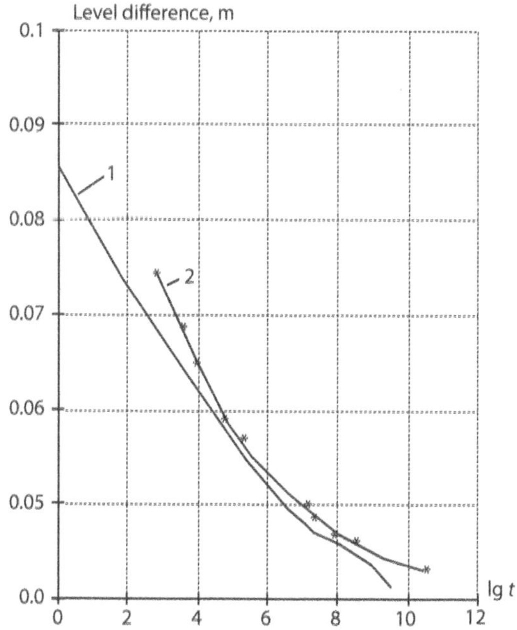

Figure 5.12 Pressure gradient change at capillary soaking with vibration. 1. No vibration. 2. In the vibration field.

5.4 Studies of the Oil and Water Gravity Flow in the Vibration Field

In conditions of a multiphase flow, the surface, viscous, gravity, and elastic forces are major forces acting on each phase.

The surface (or capillary) forces create on the boundary of liquid phases in a porous medium (on meniscuses) the pressure of about 0.01 to 0.3 MPa. The surface forces' direction of action and value are defined by the rock dominating wettability by one of the saturating phases and by the porous medium heterogeneity (micro-nonuniformities). That is why wettability and micro-nonuniformity of the medium are the most important parameters affecting the number, distribution, and motion conditions of phases.

Viscous forces, or forces of the hydrostatic resistance, are proportionate with viscosity and flow rate. Their effect in very slow flow processes is insignificant. However, possible deviations from the Darcy's law at low flow velocities, i.e., the existence of initial pressure gradient at flow, may significantly increase resistance forces.

Gravity forces create permanently acting pressure gradient numerically equal to the phases' density difference. The value of this gradient is 0.1 to 10 MPa/m. Its action results in buoyancy, for instance, of oil in water or gas in oil. Phase redistribution due to the elastic forces also may play some role in the residual oil motion. However, major in this process are interactions of the surface and gravity forces.

Experimental studies in the gravity flow of hydrocarbon fluid droplets in capillaries of a constant and variable cross-section indicate a possibility of such buoyancy at the availability of water solutions' wetting film with a low value of the interfacial tension. Gravity flow of oil droplets is observed also in the natural environment. There are numerous, often contradictory observation results of the oil saturation redistribution in the domal areas of watered-up fields, development of which has been stopped for various reasons.

One can now review an ideal porous medium made up various radi vertical capillaries of which one value is distributed as the probability density. We will consider the oil droplets' motion speed v in capillaries composing an ideal porous medium as sustained and depending on a droplet volume U_k and on the radius of the capillaries where droplets are positioned. Based on this porous medium model and oil particles' motion in pore channels model, statistical descriptions have been derived of the oil gravity motion process in a watered-up reservoir of the thickness H limited at the top and bottom by the impermeable top and base [2]. The oil saturation initially had the same value s_0 in all reservoir points. For solution of this problem, an algorithm and the programs allowing conducting of statistical calculations in various cases have been created. A case has been considered when all pore channels had the same radius and the correlation between a droplet motion speed and its volume has been governed by a linear law $v = U_k$. It has been assumed in calculation that $U_k = 10^{-15}$ m^3, $a = 10^9$ m/s^2, $s_0 = 0.5$, and H = 10 m.

The obtained results indicated that initially the oil and water separation process has been more intense. For instance, in the first three assessment months, 80% of the oil moved in the upper layer, the remaining 20% were surfacing more slowly (theoretically, infinitely long time). The explanation has been that the speeds of large and small droplets have been different. Large droplets have been surfacing faster than the small ones.

It has been shown that theoretical duration of a natural gravity separation process after a reservoir waterflooding has been many dozens and even hundreds of years. For instance, assuming the reservoir thickness equal to 20 m, permeability for the oil 0.1 D, the oil density 0.850 g/cm^3, then the natural oil buoyancy due to the gravity forces will be $8*10^{-9}$ m/s. Thus, the duration of a gravity oil separation from water in a 20-m-thick reservoir will be 79 years.

At the same time, the reviewed calculation examples have shown that it is not possible to talk about a total separation of the oil and water in the reservoir. It makes sense to talk about the duration of a certain fraction of the oil separation from water and the isolation of it as a pure oil layer. This fraction must be determined based on different conditions (for instance, economic). As estimation showed, the duration of initial stage in the oil and water separation in isotropic reservoirs may be short, shorter in the active development period, whereas in anisotropic reservoir the process may procrastinate for dozens and even hundreds of years depending on reservoir vertical permeability and thickness, oil viscosity and the distribution of initial oil saturation. In a low-angle reservoir with the thickness which is small compared with their horizontal expanse, the duration of the next stage is by 2 or 3 orders of the magnitude longer than of the first one.

Capillary forces acting on dispersed oil droplets are large compared to the gravity forces. For this reason, the dispersed oil is practically a low-mobility one. However, if it is possible to eliminate the capillary forces or strongly decrease than by lowering the interfacial tension or by way of disrupting equilibrium on the meniscus, then individual droplets become mobile under the gravity action and coalesce relatively easily.

Vibration usually causes increase in the absolute permeability, decrease of phase tension on the oil-water boundary and of the capillary pressure, in an increase of phase permeability to the oil and a decrease in its viscosity.

Based on this, computer modeling has been conducted of the effect of these parameters on the dynamic and static gravity filtration of the oil and water. The study results are shown in Figures 5.13 to 5.15. The following symbolic is used: $\xi = x/H$ is the relative reservoir thickness, x is the vertical coordinate, $f(s)=\frac{P_k}{\sigma}\sqrt{k/m}$ is the capillary pressure (Leverett) function, s is the water saturation, p_k is the capillary pression, σ is the interfacial tension of the oil/water boundary, k is the absolute permeability, and m is the reservoir porosity.

5.4.1 Absolute Permeability Effect

It has been mentioned in Section 4.1 that the reservoir absolute permeability increases in a vibration field almost 100-fold. For computer modeling of the absolute permeability effect on the oil gravity flow, these reservoir parameters have been varied according to the data in Table 5.1 (experiment 1).

Relative permeability and Leverett function vs. water saturation have been showen by curves in Figure 5.14. As shown in curves 1 and 2, vibrations increase gravity segregation of the oil and water compared to the natural process (curve 3).

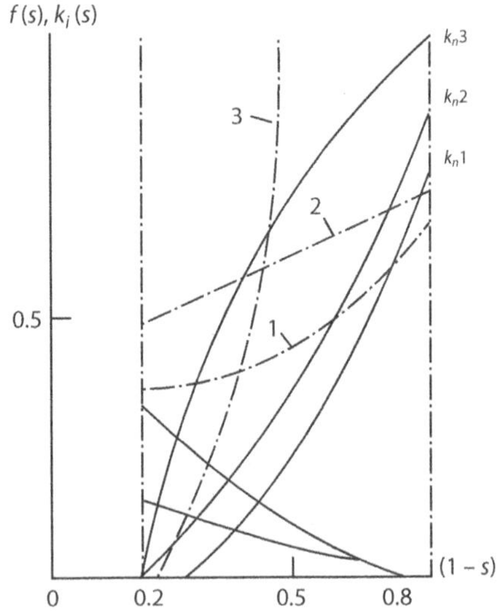

Figure 5.13 Capillary pression function (Leverett function) f(s) and relative phase permeability $k_i(s)$ vs. oil saturation (1 – s). 1. $f(s) = 0.9(0.8 - s)^2 + 0.35$; 2. $f(s) = 0.4(2 - s)$; 3. $f(s) = 0.56(s - 0.3)^2 - 0.24$; $k_{n1}(s)$, $k_{n2}(s)$, and $k_{n3}(s)$ are phase permeabilities for oil.

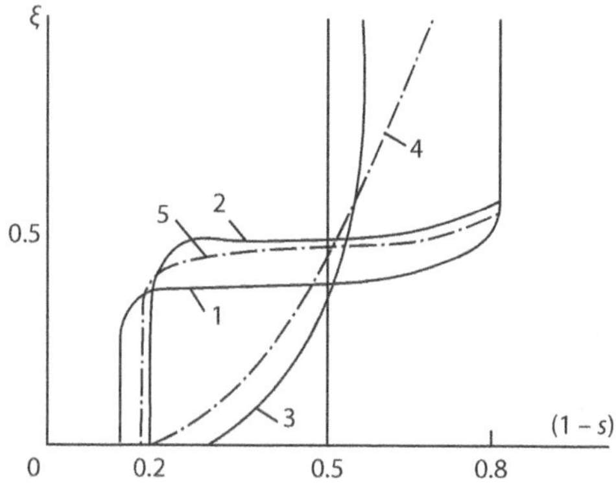

Figure 5.14 Gravity flow against the absolute permeability and function of absolute pressure. 1. $f(s) = 0.9(0.8 - s)^2+0.35$; 2. $f(s) = 0.4(2 - s)^2$; 3. $f(s) = 0.56(s - 0.3)^2 - 0.24$; 4. k = 0.01 D; 5. In the vibration field (increase in absolute permeability to k = 1 D).

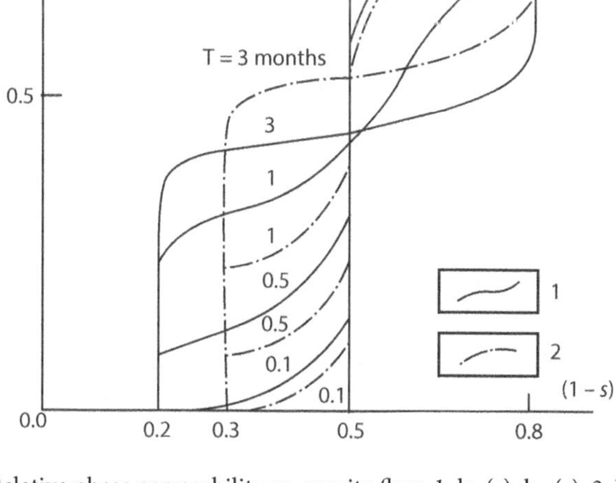

Figure 5.15 Relative phase permeability vs. gravity flow. 1. $k_{n1}(s)$, $k_{w1}(s)$; 2. $k_{n2}(s)$, $k_{w2}(s)$; $k_{n1}(s)$, $k_{n1} < k_{n2}$, $k_{w1} < k_{w2}$.

Table 5.1 The absolute permeability effect on the oil gravity flow on reservoir parameters.

Parameter	Number of experiment 1	2	3
Absolute permeability k, D	Varied	1	1
Initial water saturation s_0	0.5	0.5	0.5
Interfacial tension σ, 10^2 J/m^2	2	2	2
Oil viscosity μ_n, cP	2	Varied	2
Water-oil density difference $\Delta\rho$, kg/cm^3	340	340	340
Reservoir thickness H, m	15	15	15
Porosity m	0.25	0.25	0.25

Some of the results obtained showing the oil saturation variations in the reservoir thickness at the state of a static equilibrium are also presented in Figure 5.14. Curve 4 shows stationary distribution of the oil saturation in the reservoir (permeability 0.01 D) 100 months after the flow had begun. Curve 5 describes stationary distribution of the oil saturation after vibration, which caused an increase in absolute permeability by the factor of 100 (to 1 D) 3 months after the vibration. This curve has been compared with the natural gravity process demonstrated by curve 4. As a comparison of these two curves show, under vibration, the gravity flow may be quite efficient. For instance, in the lower portion of the reservoir ($\xi = 0.2$) under the natural gravity separation, the water saturation in 100 months was 0.62 and in 3 months under vibration, 0.82. Similarly, in the upper reservoir cross-section ($\xi = 0.7$) under the natural gravity separation, the water saturation in 100 months was 0.58 and in 3 months under vibration, 0.8. The static balance of the system as a result of the vibration has been achieved 33 times faster than under the natural regime (100 months vs 3 months). Thus, an increase in the reservoir absolute permeability under the vibration resulted in an increased oil yield.

5.4.2 An Effect of Oil Viscosity

As mentioned in Chapter 4, under vibration the oil viscosity multiply decline. In this connection, parameters included in Table 5.1 have been assigned (experiment 2). Computations indicated that the qualitative side of the gravity segregation process has been more or less preserved, i.e., the water saturation curves $s(\xi k)$ in the static balance by the moment of the phase redistribution coincide regardless of the oil viscosity. However, quantitative difference on the real time scale is significant. For instance, as may be seen from Table 5.2, at $\mu_n = 100$ cP, the static balance is reached

Table 5.2 The duration of gravity separation for the static balance of oil viscosity.

Oil viscosity μ_n (cP)	Duration of gravity separation (months)
100	50
50	10
5	3

in 50 months, whereas at the viscosity 20 times lower, the phase redistribution occurs already in 3 months, i.e., 16 times faster.

Therefore, the oil viscosity decline at vibratory action does not affect the stationary water saturation distribution but results in a substantial acceleration of the phase redistribution process.

5.4.3 The Capillary Pressure Effect

Next, the authors look at the initial computation data in Table 5.1 (experiment 3). The correlation of relative phase permeability vs. water saturation is shown in Figure 5.13. As interfacial tension, reservoir absolute permeability and porosity are fixed; the change in the Leverett function actually means a change in the capillary pressure. Stationary distributions of the water saturation are shown in Figure 5.14 (curves 1 and 2). As shown in the Figure, an increase in the capillary pressure results in a strengthening of the gravity flow process. Indeed, at the relative reservoir thickness, for instance, $\xi = -0.5$, the oil saturation corresponding with a higher capillary pressure (curve 2) is equal to 0.55 and corresponds to a lower capillary pressure (curve 1), which is 0.79. As vibration lowers the capillary pressure, one may conclude that vibration of the reservoir results in an intensified gravity separation of oil and water.

5.4.4 The Oil and Water Phase Permeability Effect

The source data for computations are included in Table 5.1 (experiment 3). The Leverett function has been assigned in a format $f(s) = -0.9(0.8 - s)^2 + 0.35$ and phase permeability changes presented as $k_{n1}(s)$, $k_{n2}(s)$, and k_{n3} functions displayed in Figure 5.13. Computation results for the oil saturation stationary distribution after 3-month-long vibration and for three values $k_{n1}(s) < k_{n2}(s) < k_{n3}$ are included in Table 5.3.

The table shows the higher relative permeabilities for oil correspond to higher oil saturation values in the upper portion of the reservoir.

Table 5.3 The computation results for the oil saturation stationary distribution after 3-month-long vibration for the three values: $k_{n1}(s)$ and $k_{n2}(s)$ and k_{n3}.

ξ	Oil saturation		
	$k_{n1}(s)$	$k_{n2}(s)$	$k_{n3}(s)$
0	0	0	0
0.2	0.2	0.3	0.4
0.4	0.25	0.35	0.55
0.6	0.66	0.7	0.8
0.8	0.7	0.75	0.85
1.0	0.75	0.8	0.9

Under the vibration, the ratios for oil the and water increase. So, in the next computer experiment, relative permeabilities to water and oil have been changed. Let one consider the Leverett function as $f(s) = 0.9(0.8 - s)^2 + 0.35$ leaving all other parameters the same as in Table 5.1 (experiment 3).

Figure 5.1 shows the dynamics of the gravity flow at various phase permeability values for water and oil. A stationary state in the immiscible phasers' distribution in both cases has been reached in three months. Analysis of the curves has shown that in the lower reservoir portion the increase in permeability ratios for the oil and water has resulted in an increased water saturation and, appropriately, in a decreased oil saturation.

For instance, at a relative thickness $\xi = 0.2$, in 0.5 of a month, the oil saturation corresponding to the values of oil to water permeability ratios (Figures 5.13 and 5.15) has been equal to 0.47 and with higher, 0.41. Whereas in the upper portion of the reservoir an increase in the oil to water permeability ratios has resulted in an increased oil saturation. For instance, at $\xi = 0.7$, in a month the oil saturation for lower permeability ratios has been 0.56 and for higher, 0.68. Therefore, increased oil and water ratio values has created conditions for more intensive phase redistribution: water has rushed down, oil up.

Beside computer modeling, physical experiments have been conducted for the oil and water gravity flow. Cemented models of the porous medium have been used. The models have been 1 m long, with 0.03 m in diameter, permeability of 5.4 D, porosity of 0.33, and have been saturated by a mix of saline water (5% NaCl solution in water) and kerosene. The saturation procedure has been as follows: the model has been saturated with vacuumed water, which has been displaced by kerosene and then again by water. Linear reservoir models in which the residual oil-water saturation has been modeled have been set vertically, and after equal time intervals, the water saturation of individual model areas has been determined [6].

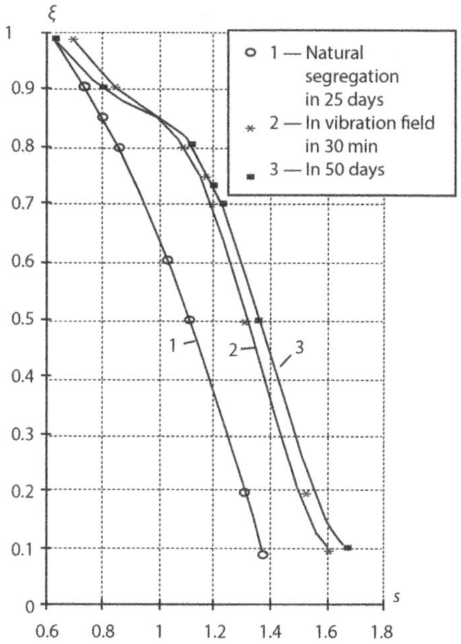

Figure 5.16 Oil and water gravity flow in the vibration field.

Figure 5.16 shows the water saturation distribution for the model areas after 25 days under the natural regime of gravity segregation (curve 1). The next series of experiments has been intended for a study of vibration effect on the gravity flow process.

The vibration has been implemented using a hydrodynamic vibrator. It has been rigidly attached to the reservoir model and created harmonic shear vibrations with the frequency of 120 Hz and amplitude of 10^{-4} m. Accelerations created in the model have been 2.5 *g*. The water saturation distribution for inter-electrode areas of the reservoir model by height in the same figure is shown: Curve 2, 30 minutes after the vibration beginning; and Curve 3, 50 minutes. The gravity segregation in the vibration field has been 800 times faster compared to the natural gravity flow regime. Such significant acceleration of the gravity flow in the vibration field may be explained by a change in the wettability nature of the porous medium model matrix by kerosene and water in the vibration field. The pore surface in the model has been covered with a layer of a solidified synthetic glue. Preliminary experiments have shown that on contact of kerosene and 5% water solution of NaCl with such surface, the contact angle for kerosene is positive (although less than 90°). However, under a vibration action kerosene forms a continuous film on the surface, and a return to the equilibrium state of partial wetting occurs only a few minutes upon the stop-of the vibration [7]. Obviously, the film formation has been due to the nonequilibrium of the wetting process.

Based on these experiments, one can estimate the model's relative permeability to kerosene. Assuming a linear distribution of the saturation along the model's length at any moment in time, one obtains flow rate for kerosene:

$$u_k = 8 * 10^{-3} \text{ m/s}.$$

Relative permeability for kerosene is

$$k_k = u_k \, \mu_k \text{grad} P,$$

where $\mu_k = 1.65 * 10^{-3}$ cP is the kerosene viscosity, grad$P$ $2.5 * 10^{-3}$ Pa/m is the pressure gradient emerging due to the density difference of kerosene and water. Thus,

$$k_k = 5.5 * 10^{-15} \text{m}^2.$$

Based on the model's absolute permeability, its relative permeability for kerosene is

$$k_k / k \approx 10^{-3}.$$

Accepting the assumption that the vibration has caused the formation of a continuous kerosene film on the surface of the model pores, one can estimate its effect on the rate of phase separation.

The matrix was composed of glued together sand grains with radius $r = 10^{-4}$ m. The sand grains' contact areas with each other are assumed to be taken by the glued marshallite grains. At porosity $m = 0.33$, the specific surface of the poured medium $S = 3(1 - m)/r$. However, the sand grain contact areas are excluded, so we assume $S = 1.5(1 - m)$. The film thickness $h = 9.2 * 10^{-6}$ m. The speed of kerosene motion within the film is $4.3 * 10^{-5}$ m/s. Thus, the flow speed $u_k = 2.5 * 10^{-6}$ m/s. Consequently, the flow speed on the application of shear vibration increases approximately 250-fold, satisfactorily complies in the value order with experimental results.

Another possible mechanism speeding up the gravity separation process may be the directional motion of kerosene droplet along the porous medium model under applied vibration. Before evaluating the time needed for gravity separation of water and kerosene, one can review the mix as a resonance vibratory system. Various kinds of vibration may cause the emergence of resonance frequencies under which the system is unstable, and qualitative changes may occur in the process flow. In this case, one of such characteristic frequencies is the eigenmode frequency of a kerosene

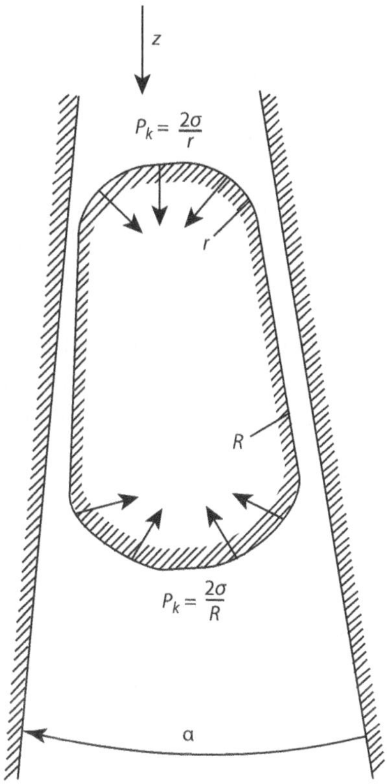

Figure 5.17 Scheme of movement of droplet in pore channel in the filed of vibration.

droplet squeezed in a narrowing channel. Capillary pressure is the function of radius r of curvature upper meniscus. In such a case, as follows from Figure 5.17, the droplet motions up the channel gradually slows down, and the force of the interfacial tension is

$$\delta F(z) = -2\pi\sigma r'(z)\delta z,$$

where σ is interfacial tension factor at the phase boundary, δz is the z variation, and z is the vertical coordinate.

The droplet eigenmode frequency is $\sqrt{2\pi r'\sigma/m}$.

For a capillary with a constant narrowing angle α $r'(z) = \sqrt{2\pi\alpha \mathrm{tg}\alpha/m}$.

We will now estimate the frequency of eigenmode vibration of a squeezed kerosene droplet and compare it with the frequency of the coercive force. For the estimate, we will assume $\sigma = 0.03$H/m and $m = \pi\rho_k r_k^2 l_k$ where r_k is the pore channel radius, l_k is the length of the droplet, and ρ_k is the kerosene density. From Kozeny-Carman formula for the connection between the pore radius and permeability of a fictitious ground follows that the pore radius is

$r_k = \sqrt{8k/m}$ and is 10^{-5} m for the model. May be $l_k = 10r_k$, then tg $\alpha = 1$ and $\omega = 10^4$ Hz. The length of the connected kerosene phase is about 0.1 m which is quite possible at the length of the model 1 m. Thus, in the performed experiments, in consideration of a selective wettability of the matrix walls, the resonance excitation regime appears quite probable.

The mechanism of a directional motion of a kerosene droplet from the lower to the upper end of a vertically positioned viniplast tube. Analysis of the forces affecting the droplet indicates that preferential droplet's motion will be rather up than down. This follows from irreversibility of the buoyant lift direction. A simplified model of this method of a leap-like motion is, for instance, that in the first half of a period the droplet more a greater distance compared to the reverse move, that is, in the first half of the period the droplet is moving up together with the matrix and in the second half is remaining in place. This is caused by the act that the total force affecting the droplet at its motion up is greater, and for this reason, the emerging adhesive force is decreasing as a result of molecular interaction. In our estimate, the time needed to the droplet for moving from the tube bottom to its top is about 3.5 min. It is clear that the estimate is overblown. In experiments, the gravity separation time was approximately 60 min. Thus, the gravity separation rate in the vibration field is different by the factor of 17 from the experimentally observed, i.e., is lower by one order of the magnitude. If we take into account that the adhesion forces' establishment time in our model has been simplified, then it is close enough.

We will now estimate the energetics of the gravity flow acceleration process. In our scheme of measuring the water saturation, the model was separated by electrodes into nine parts. The entire model contained 61 cm^3 of kerosene, thus each part contained about 7 cm^3 of kerosene. In experiments, the oil saturation by the segment changed by 10% and then moved from the lower part of the model into the upper one (0.7 cm^3 of kerosene, i.e., $56*10^{-4}$ kg).

One can assume that the work $A_{complete}$ expended for moving this amount of kerosene is equivalent to the work of overcoming the capillary forces. The capillary pressure is $P_k = \sigma/r$. Assuming the droplets' form a sphere, the work expended for moving one droplet is $A_1 = 2\pi\sigma rL = 2*10^{-6}$ J. The number of droplets is

$$N = (21*10^{-7})/(4\pi*10^{-15}) \approx 2*10^8.$$

Therefore, $A_{complete} = NA_1 \approx 4*10^2$ J. Energy released at operation of a hydrodynamic vibrator used for creating vibrations in the reservoir model is $A_{expended} = 27$ J, which is approximately 7% of the work moving 0.7 cm^3

of kerosene from the model's lower part to the upper. This fact is another indication of the possibility of a resonance excitation [2].

One can evaluate the energy which may be obtained at burning of 0.7 cm^3 of kerosene. Kerosene calorific value is $4.61*10^4$ kJ/kg. Thus, the amount of heat released at burning of $5.6*10^{-4}$ kg of kerosene is 25.8 kJ, which is about 1,000-fold the energy expended for moving that amount of kerosene in the model.

References

1. Simkin, E.M. *et al.*, The effect of elastic vibrations on the capillary soaking by water of oil-saturated porous media. *DAN USSR*, 317, 6, 334–336, 1981.
2. Ganzha, V.L., Zhuravsky, G.I., Simkin, E.M., *Heat-mass transfer in multi-phase systems*, p. 285, Science and Technology, Minsk, 1990.
3. Kawahara, T., Formation of saturated solutions in nonlinear dispersive system with instability and dissipation. *Phys. Rev.*, 51, 5, 381–383, 1983.
4. Kuznetsov, O.L. and Simkin, E.M., *Transformation and interaction of geophysical fields in lithosphere*, p. 265, Nedra, Moscow, 1990.
5. Dyblenko, V.P. *et al.*, *Raising productivity and well reanimation with the application of a vibro-wave action*, p. 380, Nedra, Moscow, 2000.
6. Simkin, E.M. *et al.*, Segregation of a hydrocarbon fluid and water in a porous medium in the elastic wave field. *DAN USSR*, 307, 3, 574–578, 1989.
7. Surguchev, M.L., Kuznetsov, O.L., Simkin, E.M., *Hydrodynamic, acoustic and thermal action on oil reservoirs*, p. 320, Nedra, Moscow, 1975.
8. Surguchev, M.L., Yu.V., Zheltov, Simkin, E.M., *Physicochemical processes in oil-gas reservoirs*, p. 330, Nedra, Moscow, 1984.
9. Simkin, E.M. and Surguchev, M.L., Advanced vibroseismic technique for waterflooded reservoir stimulation, mechanism and field test results. *Proc. 6th European Symp. On Improved Oil Recovery*, vol. 1, Stavanger, Norway, pp. 213–241, 1981.

6

Vibration Effect on Properties of Saturating Phases in a Reservoir

6.1 Changes in Interfacial Tensions and Rheological Parameters

There is only one experimental work [1] devoted to a direct measurement of the interfacial tensions in hydrocarbon fluids (transformer oil) at the boundary with the water in the field of vibrations at frequency of 200 to 800 Hz. It was shown that, at a long-term action, interfacial tension under vibration declined approximately by a factor of three. At the same time, indirect data in this publication indicated that the interfacial tension at the separation boundary oil/water in a vibration field may decline almost by the factor of 1,000. The information has been quoted in our publication about the effects observed in two samples of a sandstone with porosity of 6% and 12% [3]. The samples have been saturated with kerosene and then kerosene has been displaced with water. After such preparation of the samples, phase permeability for kerosene became equal to zero. The sandstone so prepared had been subjected to vibration at 20-Hz frequency and displacement amplitude of 500 mkm. Eight hours after the beginning of experiment, approximately 85% of kerosene contained in the sample was released. Thus, the interfacial tension at the kerosene/water separation boundaries declined almost by the factor of 100.

In experiments conducted by Zarev [13], a viscous oil had been being displaced from water-saturated sand samples. In a chamber installed on a vibrobench, a sample of the sand containing 5% (by mass) of viscous oil and 40% of sea water has been placed. The sample had been under the static load of 0.5 MPa and had been subjected to the vibration at frequency of 1.2 Hz and offset amplitude of 4 mkm. Fifteen days prior to the vibration, there had not been any oil release. Forty-eight hours after the vibration has begun, the oil started to be released from the sample. In experiments, changes have been estimated of the interfacial tension at the

George V. Chilingar, Kazem Majid Sadeghi, and Oleg Leonidovich Kuznetsov. Acoustic and Vibrational Enhanced Oil Recovery, (181–214) © 2022 Scrivener Publishing LLC

oil-water boundary in the vibration field. Prior to the vibration, this factor has been (2 to 5)*10^{-2} H/m and greater. After the action, it declined to 10^{-4} to 10^{-5} H/m.

Perhaps, the so-called "Konovalov effect" may be explained by a decrease of the interfacial tension of water (and other liquids) at the boundary with the air. The effect is a drastic increase of the liquids' uplifting speed in capillaries under the action of high-frequency vibrations.

In order to estimate the vibration effect on rheological parameters of saturating phases, the writers will review rheological behavior of various liquids in the vibration field.

6.1.1 A Newtonian Liquid

The rheological behavior of a Newtonian liquid is described by an expression as follows:

$$\frac{\partial v}{\partial z} = -\frac{\tau}{\mu},$$

where v is the flow rate, τ is the tangential shearing stress, μ is the dynamic viscosity, and z is the linear coordinate in the direction perpendicular to the motion direction.

It follows from this expression that tangential stresses describing a medium resistance to shear do not depend on pressure gradients emerging in the vibration field in the plane of motion direction. Therefore, under a laminary regime, they cannot result in a viscosity change. This is supported by the experimental studies of Surguchev *et al.* [12].

6.1.2 A Viscoplastic Liquid

Most reservoir fluids that behave as Newtonian, on motion in the real reservoirs display viscoplastic properties. For instance, in the case of a steady laminar flow within a round cross-section cylindrical capillary, their rheological behavior is described by an expression:

$$\frac{\partial v}{\partial r} = (\tau - \tau_0) \,/\, \mu' \qquad (\tau \geq \tau_0)$$

$$\frac{\partial v}{\partial r} = 0 \qquad (\tau < \tau_0),$$

where τ_0 is the maximum shear stress in a considered point of the capillary at the distance r from its axis; μ' is the plastic viscosity.

The major parameters of a viscoplastic medium are plastic viscosity and structural strength as measured by the maximum shear stress. When it is exceeded, the state of fluidity emerges described by the differential viscosity. The plastic viscosity is describing Newtonian flow of a medium with maximum destroyed structure.

For each pressure gradient, there are capillaries having small radius in which a viscoplastic fluid remains immobile. This means that, at the flow of such fluid in a porous medium, its part positioned in small pores is not displaced. In order to involve this part of fluid in the flow process, it is necessary to increase the shear stress in this fluid. This may be achieved in a case when the porous medium is vibrating. In such a case, the periodical inertia force $\rho \frac{\partial u}{\partial x}$ is added to the action of the constant pressure gradient creating shear stress. Directed against the pressure gradient, it will cause an increase in the shear stress at the opposite direction decreasing it. This transmits an additional amount of fluid in the flow direction.

For instance, let us consider the rate of flow of a viscoplastic fluid in a single cylindrical capillary with the round cross-section, in a state of harmonic vibration, and compare this rate with its value in an immobile capillary. The differential equation of a viscoplastic fluid flow in a capillary without vibration is

$$\rho \frac{\partial v}{\partial t} = \frac{1}{r} \frac{\partial (rr)}{\partial r} - \frac{\partial p}{\partial x}.$$

The boundary condition on the cylinder's wall is

$$\tau = \tau_c \quad \text{at} \quad r = R,$$

where R is the capillary radius, and t is the time. It is assumed that the x axis is directed along the capillary axis in the direction of flow opposite to the pressure gradient. The velocity profile at the steady flow of a viscoplastic fluid in an immobile capillary is composed of two parts: a rotational paraboloid (from the capillary wall to the cylindrical surface) and a flat circle perpendicular to the capillary axis. In the capillary's central portion, the viscoplastic fluid is moving as a solid cylinder.

We next consider the motion on an incompressible viscoplastic fluid in the capillary walls of which there is longwise harmonic vibration.

For the motion's computation, it is necessary to solve the following differential equation:

$$\rho\frac{\partial v}{\partial t}=\frac{1}{r}\frac{\partial}{\partial r}\left(r\tau_0+r\mu'\frac{\partial v}{\partial r}\right)-\frac{\partial p}{\partial x}$$

with the edge condition $v|_{r=R} = v_0\cos\omega\tau$, where v_0 and ω are the amplitude and frequency of the speed change of capillary walls.

The results of a numerical solution of this equation, at various interrelations of the constant pressure gradient and inertia forces' amplitudes in the case of different initial shear tensions and fluid viscosity suggest the following conclusions. In all cases when the fluid is not flowing through the vibrating capillary, there is also no flow, in the immobile capillary (with all other conditions being equal). Under vibration, the maximum radius of capillaries with no flow of the viscoplastic fluid decreases compared with the maximum radius of an immobile capillary. For instance, for harmonic vibration it is equal to

$$R_{max}=\frac{2\tau_0}{\partial p/\partial x|-|\rho v_0\omega\ \sin\omega t|}$$

This means that in a porous medium under vibration, a greater number of pore channels begin letting through the fluid at the same constant pressure gradient.

In all cases, the average flow rate of a high-plasticity fluid in a vibrating capillary, with all other conditions being equal, turns out greater than its flow rate in an immobile capillary. The resonance frequencies depend on the capillary size. In a porous medium with channels of various radius, this phenomenon will be smudgy. Denoting the pressure gradient exceeding that which disrupts the mechanical equilibrium as ∇p_0, one obtains

$$\nabla p_0 = 2\ \tau_0/\rho g R.$$

Despite the significant differences in the quantitative evaluations, experimental data support these considerations.

The simplest rheological properties are those of Maxwell liquid:

$$\frac{\partial v}{\partial z}=\frac{\tau}{\mu}+\frac{1}{G}\frac{\partial \tau}{\partial t}$$

where G is the shear modulus of a viscoplastic medium.

As this equation shows, at a slow $\frac{\partial \tau}{\partial t}$ tension change, a Maxwell liquid is similar to a Newtonian liquid with viscosity μ. An increase of the tension application rate results in a decrease of the tangential shear and, therefore, of the resistance to shear. At a maximum case of a high shear rate, the medium resistance to shear declines to a value close to zero. This corresponds with a maximum destruction of structure. After the vibrations stop, the medium gradually returns in its initial state. These considerations have been experimentally supported by Simkin and Sokolov [8].

The dynamic viscosity of the oil has been studied in the vibration field at the frequency of 18 kHz. The oil was rich in resins (20%) and asphaltenes (11.2%) with a high viscosity of 48 cP. Viscosity has been determined using a rotational viscosimeter. The vibration has been applied for 30 to 60 min. Before and after the vibration, the oil has been placed in a rotation viscosimeter for the determination of the dynamic viscosity vs. the rotor revolution frequency correlation. The study results indicated that before the vibration, the rheological properties of the oil have been typical of structured liquids: high oil viscosity at a low rotor revolution frequency has been getting lower with the increased frequency. Immediately after the vibration, oil viscosity has been declined by 28% which has been manifested at a low deformation speed. An increase in the deformation speed resulted in the viscosity increase to its initial value. After stopping the vibration, with time the correlation of the viscosity with the deformation speed was restored; however, viscosity substantially (by 10% to 25%) increased.

The oil viscosity decline in the vibration field has been believed to be due to the fact that high viscosity oils containing paraffins and resins have periodical colloidal structures. The vibration has been destroying such structures due to the heating and sign-alternating pressure in a wave. This has been manifested in the viscosity decline immediately after the vibration. Initial oil structure has been formed the in conditions of high temperatures and pressures in the reservoir. A newly formed under the atmospheric conditions structure may be different and might have acquired different properties, including a higher viscosity. One may note that the relaxation time of large colloidal particles, i.e., after the vibration time, may take a few days and even months. The viscosity change of the USA Field oil may apparently serve another example of a system transition under action of the vibration field from a state close to thermodynamically unstable to stationary (stable) one.

6.2 Permeability Changes

6.2.1 A Single-Phase Flow

The effect of a vibration field on permeability of the reservoirs saturated with a single phase is in tearing off immobile surficial fluid layers from the pore channel walls. According to the present-day concepts, at the fluid filtration in a porous medium, the separation of electric charges is occurring [2]. Affected by the electrostatic forces and by chemical adsorption forces, a "dual electric layer" is formed on the pore channel surfaces. This layer is composed of an immobile adsorption and mobile diffusion sublayers. Two possible cases are possible based on a consideration of the model structure of dual electric layers [14]:

- The fluid contains ions with the sign opposite to the sign of the surface charge. In this case, under the action of the electrostatic forces and chemical adsorption forces an immobile adsorption sublayer is formed on the boundary with the potential ζ ("Stern potential"), which is sign-opposite to the surface potential ψ.
- The fluid contains ions of the same sign as the surface. In this case, the electrostatic forces repel chemical adsorption forces and are directed against them. Stern potentials are higher than the surface potential and have the same signs. A change in the adsorption sublayer thickness on one and the same surface results in a proportional change of Stern potential.

A qualitative evaluation of the surface potential ζ and Stern potential ψ is usually difficult. That is why indirect measurements are used: in particular, zeta potential is measured whose value corresponds to the diffuse relative adsorption sublayers sliding plane potential. This value approximately equals the Stern potential, the adsorption sublayer thickness change causes a proportional change in the zeta-potential. Within the surficial layer, which includes the adsorption and partially diffusion sublayers, the fluid in pore channels remains immobile down to the fluid slippage plane. An explanation is the emergence within the indicated layer of a substantially higher shear strength compared with its weakly expressed properties in the fluid volume.

Based on experimental data, Deryagin [2] has introduced a concept of the surface layer as of a special fluid phase with viscoplastic properties different from those of the main volume fluid properties. The boundary between these phases is sharp. Average thickness of the immobile surficial water layer in pore channels depends on its salt concentration and is $5*10^{-6}$ m (in a NaCl solution, 100 g/liter) to $50*10^{-6}$ m (in a NaCl solution, 1 g/liter). The maximum shear tension within such surficial water layer is 9.45 Pa. The formation of surficial layers causes the absolute (for gas) rock permeability to be substantially different from the phase permeability values for water and oil. At water flow, this difference depends on the salinity level and salt composition of the water solutions.

Enous [3] studied the effect of these factors on the phase permeability of fine- and medium-grained sandstones with shaly cement and gas permeability of 10^{-5} to 10^{-4} D. Cores, 10 to 15 cm in diameter and 20 to 30 cm long have been welded by the Wood's alloy within grass tubes. The fluid flow has been implemented at pressure gradient of 0.02 MPa and temperature of 18°C to 23°C. The study of 200 samples has shown that the speed of a continuous flow of water and salt solutions have declined with time and the rock permeability depended on the salt composition and salinity of filtrating fluids. Maximum permeability values have been recorded for a 1% solution of $Al_2(SO_4)_3$ and saturated $CaCl_2$ solution.

A permeability decrease in the fluid flow process is due to the formation of water films on the rock surface. It has been shown that the shear strength of a fluid in such films is greater than the volume strength. For the evaluation of the thickness of the films with an elevated shear strength, Kovach theoretical model has been used (an analog of the dual electric layer). The model suggests a gradual change in the fluid's properties from the free volume to the solid surface. Such changes are caused by the fact that on contact of the solid and liquid phases a redistribution of electricity carriers occurs. Solid particles receive an excessive surface charge compensated by the solution. Due to the electric attraction, counter-ions are next to the electrode as an immobile layer (representing one of a flat capacitor coatings) and as a mobile diffusion layer (another coating) which has, same as the first one, a higher shear strength compared with the main volume strength. The influence area size of the solid surface on the solution structure corresponds to the diffusion layer thickness and is determined from the following equation:

$$\chi = \sqrt{\varepsilon KT/4\pi Ce^2z^2},$$

where χ is the diffusion layer thickness, e is the electron charge, z is the valency of dissolved salts' cations, C is the salt concentration, K is the Boltzmann constant, ε is the solution's dielectric capacitivity, and T is the temperature.

For a NaCl solution (100 g/l, $T = 20°C$, and $\varepsilon = 81$), the diffusion layer thickness $\chi = 5$ mkm and, at NaCl content of 1 g/l, its thickness $\chi = 50$ mkm. Experiments give thickness χ value of 26 to 5 mkm for reservoirs with permeability ranging from 0.1 to 1 D. Therefore, even in high-permeability reservoirs at the fresh water flow, a substantial part of pore channel effective cross-sections will be blocked by liquid films with an elevated shear strength.

According to the aforementioned concept, vibration causes the destruction of the surface layer and converting it into a Newtonian liquid with the viscosity the same as the main volume liquid viscosity. In an extreme case, this changes the fluid flow nature in pore channels to the Poiseuille flow with the simultaneous increase of the effective cross-section due to the destruction of the surface film layer. If one assumes that (1) without vibration the fluid motion in constant cross-section capillaries complies with the Bingham-Shvedov equation and occurs outside of the surface layer, (2) complies in the vibration field with the Hagen-Poiseuille equation, and (3) occurs in any pore channel cross-section, then the permeability in the vibration field may increase almost 100-fold. The validity of these conclusions has been confirmed in numerous experiments.

For example, in experiments with the vibration on water-saturated sandstone sample: (1) First, the gas permeability of the dry sample has been measured and then a fresh water was flown through the core reading the permeability by the factor of dozens. (2) The cores have been subjected to vibration with a simultaneous permeability measurement. (3) Piezoceramic irradiators have been installed 10 to 40 mm from the sample end. Vibrations with 26.5-kHz frequency have been spreading toward the fluid flow. In a few seconds after the wave generation has begun, a drastic (approximately 82-fold) permeability increase has occurred. After the vibration field removal, permeability has dropped to the values prior to the vibration. It is important that the flow improvement has occurred also in samples whose permeability in the process of the fresh water flow had declined to the very low values of $0.185*10^{-3}$ D. Compared to a continuous regime, the impulse action has turned out more efficient. For instance, appropriate permeability values have been reached at the intensity 10 to 15 times lower than at a continuous regime.

The permeability recovery in a vibration field has been apparently due to the destruction of water film with elevated shear strength. One possible mechanism of such destruction is as follows. Intensive vibration fields cause the emergence of intense flows on the solid body–fluid separation boundary. In natural rocks, this process is implemented as an inter-pore water turbulization as experimentally determined fluid velocities have been 1 to 50 cm/s. As a result of vibration capillaries' walls and turbulent water motion (the water is the charge carrier), the interaction between the dual electrical level occurs at the fluid separation boundaries and the flow which creates a crosswise magneto-hydrodynamic pressure. The computations have indicated that at speed of 1 cm/s in a 10 mkm radius capillary the amount of magneto-hydrodynamic pressure may reach 10^{-5} to 10^{-3} MPa. Thus, water films with average shear strength of about 10^{-3} Pa may be in part or totally destroyed in the vibration field, resulting in the improvement of the rock permeability. It may be assumed that for these reasons natural seismic phenomena (earthquakes and volcanic eruptions) affect fluid migration and accumulation by increasing rock permeability.

6.2.2 Two-Phase Flow

The flow of gassy fluids, in which the gas phase is in the form of finely dispersed bubbles, facilitates emergence of the so-called nonlinear effects associated with their pulsation in the vibration field. But, affected by ponderomotive forces, a directional displacement of the bubbles relative the fluid occurs. The most significant ponderomotive force is Bjerknes force. A gas bubble displacement under the action of this force is most obvious when a bubble is next to the pore channel wall. In the extension phase, the bubble is expanding. The distance between its wall and the surface decreases, and a tangential flow directed away from the bubble due to the fluid displacement in the gap emerges. In the phase of a bubble collapse, due to the fluid inertia, the flow does not have time to change its direction, so the bubble wall closest to the pore channel surface remains in place. The bubble compression occurs as a result of a motion by the opposite wall. The bubble center is displaced to the pore channel surface. Within a pore channel or a narrow fracture, the bubble is interacting with its surrounding solid surfaces and directionally displaces in the pore channel or fracture under the action of the resulting force. This phenomenon results in a displacement of gas bubbles in the pore channels and narrow fractures or in their accumulation in dead-end cavities.

Beside the Bjerknes force, another ponderomotive force manifested in a porous medium emerges at interaction of a pulsating bubble with the generated vibrations. Acted upon by this force, gas bubbles change the motion speed and direction. It has been experimentally shown that exceeding some threshold in vibration intensity, progressively advancing gas bubbles begin performing disorderly crosswise oscillations and spasmodic motions. For this reason, they are sometimes called "dancing" or "mad" bubbles.

The described physical behavior of a uniform and gassed fluid corresponds with the low-amplitude medium particle oscillations. It, however, turns out incomplete for the high-intensity oscillations with large amplitude. In such a case, phenomena associated with the emergence of acoustic currents and cavitation are added to the indicated nonlinear effects.

Acoustic currents are stationary vortex flows of a fluid emerging under vibration both in a free volume (in a nonuniform field) and next to various obstacles. The cause of their emergence is irreversible energy and impulse loss in the medium. The greater the losses, the more intense are the currents and the higher their speed. The formed vortex current may be tentatively subdivided into three types as follows:

- A current in a free but limited volume. The scale of vortices emerging is defined by the extent of a limited volume and by a substantially large wave length. These are large-scale vortices, which are possible either in large caverns and fractures or inside a well.
- A current outside of the boundary layer. The scale of vortices emerging here is substantially smaller than in the first case. One kind of such a current was analyzed by Rayleigh and is named after him. This is a two-dimensional current between the two planes (or in a cylindrical tube) emerging under the action of stationary waves. Linear sizes of the vortices in this case are equal to the wave length. At spreading of low-frequency vibrations they are possible in caverns and fractures. At spreading of high-frequency vibrations, the emergence of vortices is quite probable within large pores of high-permeability reservoirs.
- Cavity currents accompanying a transition from the low-amplitude vibrations to high-amplitude vibrations. Theoretically, a fluid begins boiling at pressure dropping to the saturated steam pressure. Actually, however, a pressure of the beginning cavitation substantially depends on a physical state of the fluid. If the liquid contains a large amount of dissolved gas, then pressure lowering results in its separation from the fluid and, in the formation of gas voids (caverns), the pressure in which is higher than the saturated steam pressure. Two phases are distinguished in the life of the emerged cavitational bubble: an expansion and compression with a collapse, which together form a complete thermodynamic cycle.

The appearance of cavity bubbles is accompanied by a number of physical phenomena. For example, at the moment of collapse a slight luminescence of the bubbles is observed, the cause of which is not yet totally clear. In some studies, the cause is a gas heating in the bubble due to a high pressure at the collapse. The flash may last 1/20 to 1/1,000 of a second. The luminescence intensity depends on an amount of gas in the bubble. If there is no gas in the bubble, then there is no luminescence. At a bubble collapse, high temperature and pressure emerge within it. It is established in the experiments that that temperature next to a collapsing bubble rises to 800°C and the shock wave emerging within the fluid reaches 400 MPa. Some scientists believe that high pressures are caused by cumulative small streams emerging at the asymmetric collapse. High, rapidly changing pressures and heat shock waves cause the matrix destruction near the bubble. The conditions at which significant cavity destruction occurs is not known precisely, but there is no doubt, however, that it increases as the fluid motion speed relative to the matrix grows.

A possibility of the chemical and electrochemical erosion in the formation water needs to be considered. Cavity bubbles in a fluid result in qualitative changes of the medium natural parameters, which become nonlinear. After the collapse of a cavity bubble, another cavity bubble, as a rule, emerges in the same place. That is why the duration of its existence in some area of the fluid may be quite long. A bubble has enough time to interact with the other cavity bubbles and with the pore channel surface.

The foundation of the mechanism of vibration influence on a two-phase oil/water flow are the processes of oil droplets (in a water flow) becoming mobile in the vibration field [4]. This may happen either due to clustering (droplets merging into "streamlets") or due to the emergence of droplets of much smaller size than the pore channel size. Vibration of such water-oil system results in changes of the flow dynamics mostly in the dispersed phase. Maximum effect of vibration is experienced by the large droplets and conglomerates with the shape different from the spherical. The vibratory accelerations of the formation matrix, moving synchronously with the dispersion medium, and the dispersion phase in vibration field depend on the interrelation of phase densities:

$$(\rho_1/\rho_2)\ a_1 = a_2;$$

here, ρ_1 and ρ_2 are the densities of the dispersion medium and the dispersed phase; and a_1 and a_2 are the acceleration acquired by the formation matrix moving synchronously with the dispersion phase and by the dispersed phase in the vibration field.

The lower-density dispersed phase has a higher acceleration compared to the dispersion medium. Such dispersed phase behavior dynamics with mobile wetting film on the pore channel surface is obstructing its adhesion interaction with pore channel surface, the interaction working through the wetting film and in proportion of the contact duration. As a result, at the external pressure gradient, the efficiency of the dispersed phase displacement must be substantially higher compared to the case of no vibration.

As a results of vibration, the dispersed phase specific micro-currents emerge. These micro-currents cause the appearance of the so-called Bjorknes attraction force between the droplets and formation. The Bjorknes attraction force gives rise in the formation clusters. If, however, the droplets' size is smaller than the pore size, the attraction forces between them are weak, they may fragment into even smaller particles due to a strong decrease in the interfacial tension. Beside the indicated vibratory displacements of dispersion phase relative to one another, droplets and conglomerates' pulsations accompanied by hydrodynamic micro-currents and by the removal of diffusion sublayers around them. As a result, thinning of the adsorption layers around the dispersion occurs. The sources of such pulsations are micro-heterogeneity of the porous medium and of the phase separation surface [12].

When the immobile wetting film exists between the dispersion phase and the pore channel surface, the dispersion phase does not experience any motion in the channel until overcoming the capillary forces. The dispersion phase movements in the first half-period of the vibration correspond with the soaking regime, i.e., the vibratory and capillary pressure gradients coincide in the same direction. In the second half-period, the vibratory pressure gradient changes and becomes opposite to the capillary pressure gradient, i.e., the dispersion phase motion now corresponds with the draining regime. Due to the presence of capillary hysteresis, as capillary pressure gradient increases and either in part or totally suppresses the dispersed phase motions.

A change in wettability in the vibration field has been supported by physical experiments on reservoir models [10]. The studies have been conducted on cemented porous medium models. The pore surface of each model has been covered by a layer of the solidified synthetic glue. The models have been saturated with a mix of saline water (5% NaCl solution in water) and kerosene as follows: the vacuumed model has been saturated by the vacuumed water which was displaced by kerosene and the kerosene has been displaced by water. Experiments have shown that on contact with such surface of kerosene and 5% water solution of NaCl, the contact angle for kerosene has been positive (although less than 90°). However, under migration, kerosene has formed a continuous film on the surface. A return to the equilibrium state of a partial wetting has occurred only a few minutes after the migration stopped [11].

In a case of the porous medium saturation by two immiscible fluids (water and oil) and reaching the maximum residual saturation by one of them, the phase permeability for a fluid which has reached maximum saturation, has turned into zero. It means the separation of the residual fluid into separate droplets and accumulations surrounded on all sides by the other phase and immobile at a flow under equilibrium conditions. The volume content of the residual immobile fluid in the formation pores may be quite high. This phenomenon defines to a significant extent the maximum residual oil saturation of the totally watered-out reservoirs and the productivity decline in production wells due to blocking action of the water that had gotten in the reservoir. The vibratory action restores connectedness and the flow of the residual phase. Experimental data indicate a possible advance of isolated oil inclusions in natural materials under increased flow speed caused by the hydrodynamic head. This effect is explained based on, developed in the percolation theory considerations, the available in the rock medium distributions of isolated inclusions by the size. With increasing hydrodynamic head sufficiently large, inclusions of the isolated phase (ganglia) become mobile.

As vibration causes the emergence of inertia effects, it is possible to estimate the mobility of the residual phase isolated droplets in a pore channel, which may be viewed as a system of pore widening's connected by pore narrowings. Such a packing system is appropriate, for instance, for a chaotic packing of the balls and for the compacted quartz sand. The stability of interfacial boundary (between the oil and water) in the pore channel narrowing defines the position of the jammed phase and depends on the properties of fluids, interfacial tension of liquid phases, and liquids adhesion to the solid phase. As the oil hydrophobizes the solid phase surface, the wetting hysteresis for water exists. The static angle of selective wetting exceeds at that $\pi/2$ and is unbalanced. On the wetting contour of oil/water an adhesive tension force is created acting from the side of the solid body on the unit length of the wetting contour and preventing water advance in the narrowing. If the adhesion forces' energy barrier of the oil with the solid phase cohesion is overcome, the selective wetting angle becomes acute and, due to the work of capillary forces, water expels oil from the narrowings in the wider part of the pore channel.

On application of a vibration field to the system, additional inertia force (F) appears proportional to the phase density difference and to the medium vibration acceleration. If to assign the vibratory offsets' values commensurate with the effective pore channels radius in the medium, i.e., $\zeta \sim r_{ef}$, then the value of these additional forces may be evaluated as

$$F \sim 2\pi r_{ef}^2(\rho_w - \rho_o)\ddot{\xi}, \tag{6.1}$$

where ρ_w and ρ_o are the water and oil densities and $\ddot{\xi}$ is the porous medium vibratory acceleration.

Then, the condition of vibration overcoming the adhesion wetting barrier and of breaking the system's metastability may be written as follows:

$$2\pi r_{ef}^2(\rho_w - \rho_o)\ddot{\xi} \geq (\sigma_{tv} - \sigma_{tn})2\pi r_{ef} \tag{6.2}$$

where the σ_{tv} and σ_{tn} are the interfacial tensions on the boundaries of solid body/water and solid body/oil.

The adhesion tension may be written as follows:

$$\sigma_{tv} - \sigma_{tn} = W - \sigma_{nv}, \tag{6.3}$$

where $W = 2\Phi\sqrt{\sigma_{tv}\sigma_{tn}}$ is the work of adhesion forces, and Φ is the non-dimensional factor describing properties of the contacting phases. For oil/water contact, $\Phi \approx 0.5$ to 0.6.

Considering Equations (6.2) and (6.3), the condition for evaluating the threshold value of the vibration acceleration on condition that the vibratory displacements are commensurate with the medium's pore channels' radius is

$$\ddot{\xi} \geq \frac{\sigma_{nv} - 2\Phi(\sigma_{tv}\sigma_{nv})^{1/2}}{r_{ef}^2(\sigma_v - \sigma_n)}. \tag{6.4}$$

One can now compare the values found by way of estimating using the above equations with threshold values obtained in lab experiments of vibration action on the oil by water displacement. At computations, one may use the following parameter values: $r_{ef} = 2*10^{-5}$ m, $\Phi = 6.05$, $\sigma_{nv} = 0.29*10^{-3}$ H/m, $\sigma_{tv} = 0.197*10^{-3}$ H/m, $\rho_n = 870$ kg/m^3, and $\rho_v = 1190$ kg/m^3. Substituting these values in Equation (6.4), one obtains $\ddot{\xi} = 6.15$ m/s^2. The experimental value of the threshold vibratory acceleration is 4.6 m/s^2. The agreement may be considered quite satisfactory.

In the attempts to account for the physical effects of the vibration, the classical Muscat-Leverett theory of two-phase flow encounters conceptual difficulties associated with the uncertainty of phase permeability functions $f_i(s)$ and the capillary pressure $p_i(s)$ which conclude phase equations of the mass preservation and of the generalized Darcy's law. The phase permeabilities and the capillary pressure may be functions not only of saturation but also of some parameters of the external physical action. A task of determining even qualitative properties of these functions needed for theoretical computations is becoming difficult. In a situation like this, advantages are manifested of an unconventional direction—the two-phase flow micro-mechanics, which is amending the existing macroscopic theory by detailed motion mechanisms at the pore channels and porous medium ensembles' levels. The two-phase flow micro-mechanics includes physical and mathematical experiments on porous medium models and a percolation description of the multiphase flow. Similar studies bring quite informative results. The use of this methodology is promising for modeling the vibration field effect.

In order to come up with the condition of the residual phase mobilization in a porous medium, one may consider the porous medium as a regular 3D grid of capillaries with the known coordination number. Each capillary is connecting the adjacent i^{th} and j^{th} nodes and has its own radius r_{ij} and permeability $k_{ij} = r^2_{ij}/8$. The pore radius is considered a random value distributed according to the numerical distribution function among radiuses $\psi_r(r)$ satisfying the normalization condition as follows:

$$\int_0^\infty \psi_r(r)dr = 1.$$

Let us assume p_i to be the value of a local pressure field in the subject discrete medium. Then, the matter flow I_{ij} through an individual pore is proportional to the pressure gradient:

$$I_{ij} = -\pi r_{ij}^2 \frac{k_{ij}}{\eta} \frac{p_i - p_j}{l} \tag{6.5}$$

or in the form of Ohm's law:

$$I_{ij} = -q_{ij}(V_i - V_j), \tag{6.6}$$

where the conductivity is

$$q_{ij} = -\pi r_{ij}^2 \frac{k_{ij}}{\eta_l} = \frac{1}{8}(\pi r_{ij}^4 / \eta l)$$

and $V_i - V_j$ is the voltage difference (in this case, of local pressures).

For the local flows I_{ij}, Kirchhoff's equations are valid:

$$\Sigma_j r_{ij}(V_i - V_j) = 0, \tag{6.7}$$

where summation is conducted in all nodes adjacent with the ith.

Equations (6.5) to (6.7) are in substance finite-difference presentations of the Darcy's law and of the continuity equation $\nabla_x I = 0$.

One can assume that in a pore volume are dispersed two phases, oil and water and the oil saturation is below the percolation threshold, i.e., the oil phase is broken into separate unconnected clusters (droplets) and in the balanced conditions is an immobile phase and is not recovered. Oil-filled pores are impermeable to water due to the capillary and other forces and have zero conductivity. On assuming s to be the volume fraction of pores filled up with water, then $s_n = 1 - s$ is the volume fraction of oil-saturated. The pores filled up with water and pores filled up with oil are chaotically distributed in the capillary grid. The fill-up probability of arbitrarily selected pore of one of phases does not depend on the phase filling up the adjacent pores and is equal to fraction of the pores filled up with this fluid. Assuming that the oil fills up all pores with a radius $r'_f < r'' < r''_f$, then

$$s = \int_{r_{min}}^{r'_f} \psi(r)dr + \int_{r''_f}^{r_{max}} \psi(r)dr, \tag{6.8}$$

$$s_n = \int_{r'_f}^{r''_f} \psi(r)dr \tag{6.9}$$

For evaluating the effective transfer factor by the phases in the subject random capillary grid, one may use the self-synchronized field method. The advantage of this method compared with the other methods of determining factors is that it qualitatively considers the effect of phase coherence rapture at sufficiently low saturation, which stops the mass transfer. The self-synchronized field method is actually a replacement of the subject random medium by some effective medium of the grid of similar elements. The conductivity of elements in the effective grid is based on the fitting condition: difference of potentials of one element with a random conductivity, surrounded by the effective medium, must have average value typical for the initial grid. The self-synchronized field method produces the equation for a single-phase effective radius r_{ef} determination:

$$\int_0^{\infty} \frac{(r_{ef}^4 - r^4)\psi(r)dr}{r^2\left[r^4 + \left(\frac{Z}{2} - 1\right)r_{ef}^4\right]} = 0 \tag{6.10}$$

For the determination of effective radiuses r^*_v and r^*_n at a two-phase flow of water and oil within a subject random capillary grid in consideration of Equations (6.8) and (6.9), one obtains

$$\int_0^{r'_f} F(r_v^*, r)dr + \int_{r'_f}^{r''_f} F_0(r)dr + \int_{r''_f}^{\infty} F(r_v^*, r)dr = 0, \tag{6.11}$$

$$\int_0^{r'_f} F_0(r)dr + \int_{r'_f}^{r''_f} F(r_n^*, r)dr + \int_{r''_f}^{\infty} F_0(r)dr = 0, \tag{6.12}$$

where

$$F_0(r) = \frac{\psi(r)}{r^2\left(\frac{Z}{2} - 1\right)}, \qquad Fr_i' = \frac{(r_i^4 - r^4)\psi(r)}{r^2\left[r^4 + \left(\frac{Z}{2} - 1\right)r_i^4\right]},$$

Z is the grid's coordination number describing a number of the pore channels converging in one node and defining the structure of the porous medium. The pore medium relative permeability for the subject phases are determined as follows:

$$f_v = \frac{r_v^{*4}}{r_{ef}^4}, \tag{6.13}$$

$$f_n = \frac{r_n^{*4}}{r_{ef}^4}, \tag{6.14}$$

where r_{ef} is the effective radius of a uniphase medium determined from Equation (6.10). The absolute permeability is

$$k = mr_{ef}^2/24,$$

where m is the porosity.

Within the proposed model, the vibration is described by the increment $\Delta r'_{ef}$ in Equations (6.11) and (6.12). On the assumption of a constant saturation over a small-time interval, the exciting value $\Delta r''_{ef}$ is determined from $\Delta r''_{ef} = r''_{ef} + \Delta r$ Equation (6.9). Here, $\Delta r = r - r_0$, r_0 is the minimum radius of an oil-saturated pore equal to one.

Relative permeability curves for oil and water have been computed vs. water saturation without vibration and in a vibration field. The model function of capillaries' distribution by radius of the following kind have been used:

$$\Psi(r) = b\frac{(br)^k}{K!}\exp(-br), \tag{6.15}$$

where b and K are parameters of the distribution function. It has been assumed that the medium permeability $k = 0.25$ D and the parameters of the distribution function $b = 0.6$, $K = 5$, and coordination number $Z = 8$.

The computations have shown that the distinction for the model situation is more significant in the threshold area of the appropriate oil and water saturations. Increments of phase permeabilities for the oil and water change the sign with the saturation change, at that minimum of the relative permeabilities' change under the influence of vibration corresponds with the intersection point of the relative permeabilities' change for the oil and water. This intersection point defines some critical value of the porous medium oil and water saturations relative to the vibration. The medium's effective radius is r_{ef}; therefore, the absolute permeability in the constructed model is defined by the distribution function Equation (6.15). By changing parameters K between 3 to 5 and b between 0.4 and 0.8, it was possible to cover the permeability range of 0.05 to 0.3 D.

The computations have also shown that with the absolute permeability change has been changing the nature of effect vibrations. In the process of computations, critical oil and water saturation values have been determined for the porous media with different phase permeability as functions of the porous medium absolute permeability. These functions have been obtained for two values of the coordination number Z which defines the porous medium's nature. The clastic type reservoir structure (the medium composed of granules) is described by a cube 2D or 3D capillary grid with the coordination number $Z = 6$, whereas the fractured porosity medium (appropriate for carbonate reservoirs) has the coordination number $Z = 4$. With decreasing permeability, the oil saturation value declined, at that sharper for a medium with the fracture porosity.

Calculations of the correlation of water and oil relative permeabilities before the vibration with the porous medium absolute permeability conducted for a granular type medium ($Z = 6$) at the set oil saturation value $s_n = 0.25$ have shown that with a decline of the porous medium absolute permeability, the vibration efficiency has increased.

These computations have also enabled evaluating the characteristic features of the vibration effect on the two-phase flow processes. This effect is most significant in the area where the saturation by one fluid dominated the saturation by another fluid. In such a case, the flow of one phase is absent or small due to the accumulation of cohesion raptures in large pores and the formation of "small phase" stagnant areas in the medium small pores. Vibration mobilizes the fluid in small pores and facilitates the cohesion restoration of a "small phase" and its flow in the porous medium.

As Figure 6.1 shows, if the water saturation at the intersection point of the oil and water phase permeability curves is equal to s_0, then within the initial range of the water saturation change ($0 \le s < s_0$), when the oil phase is still a dispersed medium and the dispersion phase is water, the phase permeability for water in the vibration field is greater than without the vibration. Whereas the phase permeability for the oil within that range experiences no serious changes. At the $s = s_0$ point, both fluids are mutually dispersed, phase permeabilities for the oil and water are equal to one another. At further water saturation growth $s_0 \le s < 1$, when the oil phase becomes a dispersed phase and water becomes the dispersion phase, the phase permeability for the oil is higher than without the vibration. The phase permeability for water within this range remains the same

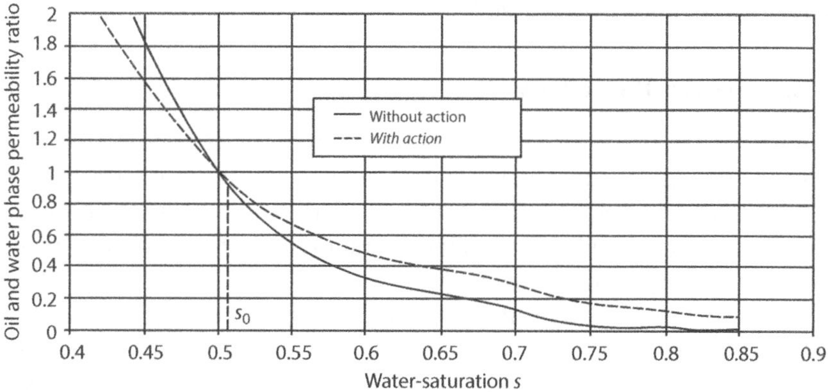

Figure 6.1 Computed interrelations between the oil and water phase permeabilities in the vibration field (two-phase flow).

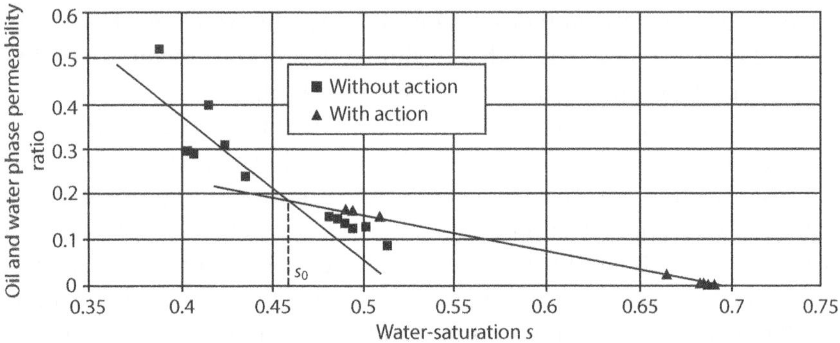

Figure 6.2 Experimental values of phase permeabilities interrelation between the oil and distilled water in a vibration field (two-phase flow).

as without the vibration, i.e., it does not experience the effect of vibration. It is necessary to indicate one more feature of the phase permeability picture associated with the dispersion phase behavior in a vibration field. The point where the dispersed oil phase permeability becomes zero defines the threshold water saturation above which the oil is immobile. At greater water saturation, the oil streamlets in the pore system tear into separate blocked droplets; therefore, the oil recovery of the watered-up reservoir has a limit. There is the threshold mobility for water as well when it is the dispersed phase. Under vibration, the threshold water saturation values for oil are moving right and for water, left.

Experimental data related to changes of the phase permeabilities and water saturation have been obtained based on study results of the double-phase flow under pressure in the vibration field and are shown in Figure 6.2. They indicate the conceptual rectitude of the assumed theoretical model.

6.2.3 Three-Phase Flow

If an oil-water mix flows in the reservoir, then the appearance of the third phase (a free gas) volume of which does not reach some limit (10% to 15%) results in lowering of the water saturation. Correlation nature of the oil and water phase permeabilities vs. water saturation in a vibration field remains the same as at two-phase flow [12].

Figure 6.3 displays typical changes of the oil and water phase permeabilities after vibroseismic action in the Samodurov Field. In the interval of low and medium water saturation ($0 \le s < s_0$), when the oil phase is still

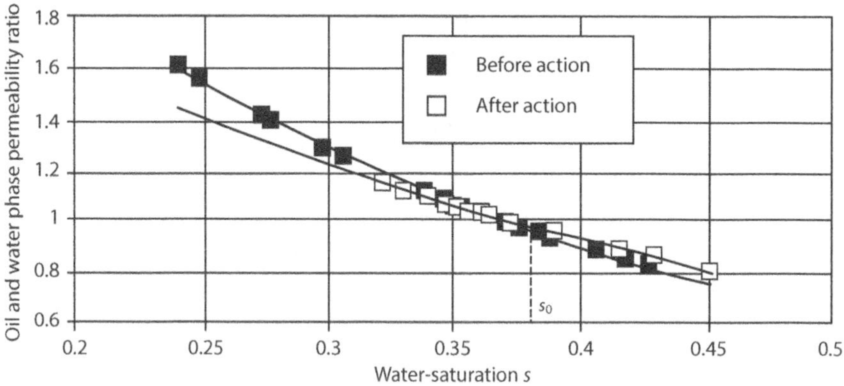

Figure 6.3 Changes in the oil and water phase permeability ratio vs. water saturation in Samodurov Field.

a dispersed medium, the phase permeability ratio after the vibration is lower than such correlation without the action. In the intersection point ($s = s_0$), phase permeability ratios before and after the vibration are equal to one another. At further increase in water saturation ($s_0 < s < 1$), when the oil phase becomes dispersed, the phase permeability ratios after the vibration are higher than before. These changes were also observed in other fields.

The authors have predicted this nature of phase permeability changes based on general theoretical concepts [10].

6.3 Capillary Pressure Changes

Accelerations emerging as a result of vibration of each saturating phase, under the second law of Newton, result in the emergence in them of inertia forces and of appropriate vibratory motions directed in the same direction as the vibratory accelerations. In as much as the wetting fluid (for instance, water) density is usually higher than that of the nonwetting one, a higher pressure develops in the wetting fluid.

Assuming that, during each vibration half-period, the vibratory acceleration remains constant and equal to maximum amplitude value, then according to the capillary pressure definition as the pressure difference in the nonwetting and wetting phases, then the emerging at that capillary pressure gradient is determined as follows:

$$\frac{\partial p_k}{\partial x} = \frac{\partial p_0}{\partial x} \pm \omega^2 A \rho_c \frac{1-\rho_n/\rho_v}{1+2\rho_n/\rho_v}.$$

Here, p_0 and p_k are the capillary pressures (respectively, initial, i.e., without vibration, and in a vibration field), x is the linear coordinate, ω is frequency, A is maximum displacement amplitude, ρ_n is the oil density, ρ_v is the water density, and ρ_c is the water-oil mix density.

The $\omega 2A\rho_c \frac{1-\rho_n/\rho_v}{1+2\rho_n/\rho_v}$ value may be called the acoustic capillary pressure gradient p_{ac}.

The oil phase motion in the first vibration half-period corresponds with the soaking regime, i.e., p_0 and p_{ac} coincide in the direction. Therefore, the total capillary pressure in the first half-period as a whole increases, and the acceleration obtained by the oil phase is directed toward the water phase. In the second half-period, the acoustic capillary pressure changes and becomes directionally opposite to the initial capillary pressure, i.e., the oil phase motion corresponds with the drainage regime. At that, due to a capillary hysteresis, the initial capillary pressure increases and kills a head-on motion of the oil phase. As a result, the capillary pressure in the second half-period of the vibration may be either lower or remain unchanged compared with

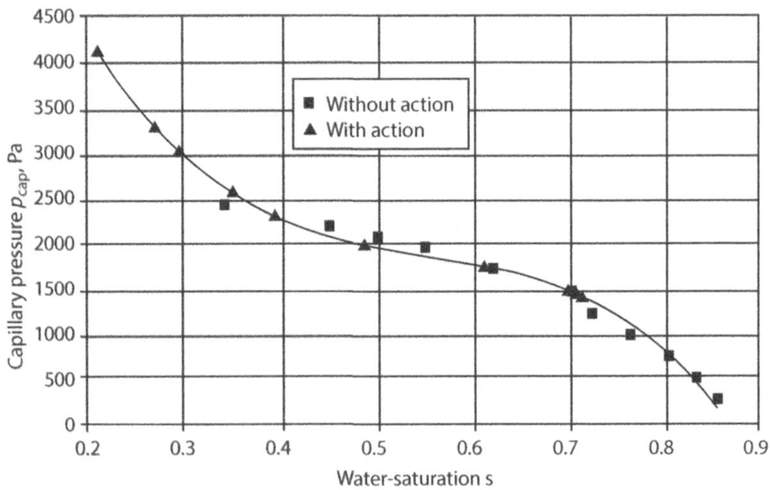

Figure 6.4 Capillary pressure vs. water saturation based on results of gravity flow studies in vibration field.

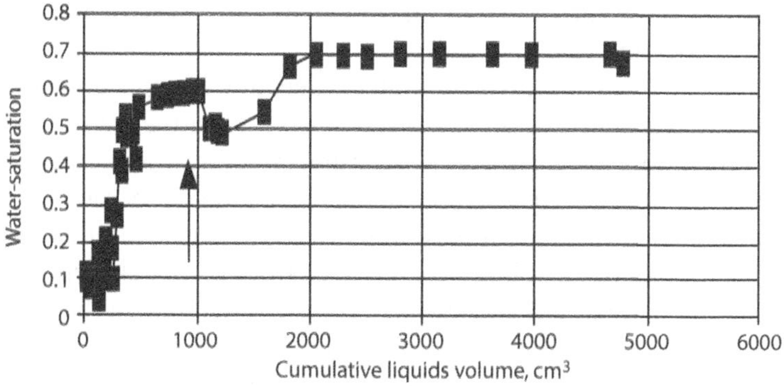

Figure 6.5 Water saturation change at the model oil displacement by distilled water in vibration field. The arrow indicates the beginning of vibration.

the first half-period. Overall, in one vibration period the capillary pressure will be higher than in the initial conditions (without vibration). As the capillary pressure is a single-value function of the water saturation, its growth in a vibration field causes the decline in water saturation [5, 11].

These considerations are well supported by experiments. For instance, Figure 6.4 shows capillary pressure vs. water saturation. It has been obtained in a study of the gravity flow in the oil-water-saturated sandstone in a vibration field [6]. The capillary pressure increase resulted in a decreased water saturation, and the vibration does not affect the nature of this correlation. Another example (Figure 6.5) indicates a decline in the water saturation at the capillary pressure increase under a pressure-caused increase in the model oil capillary pressure, when the oil is displaced by the distilled water in a vibration field.

6.4 Interformational Oil Degassing and a Decline in the Formation Water Saturation

The vibration effect on a fluid degassing is most visual in a case of opening a gassed water bottle. Immediately upon the bottle opening, plenty of large and small bubbles emerge in water. They are actively running up, to the bottleneck, in which water as if frothing. The "bubble mechanism" plays an important role in the process of gas release. The mechanism is in that the excessive gas molecules join and form bubbles. They surface according to the known laws and carry the gas out.

After surfacing of some gas the water quiets down, new bubbles do not appear. However, most of the dissolved gas is still in the water. It can be released now only through a slow diffusion process. Many excessive gas molecules are not removed by the "bubble mechanism," because after the first portions of surfacing bubbles, water becomes substantially degassed, microscopic embryonal bubbles are removed from it: they absorb excessive molecules, increase in size and surface. The gas remaining in the water has nowhere to run off, that is why no bubbles emerge and water, striving to get rid of unneeded gas may only use the diffusion mechanism. Now, in order to demonstrate vibration effect, it is sufficient to shake the bottle with the "quieted down" gassed water, to create in it ruptures and fractures which may become bubble embryos, and the bubble mechanism should work again. Similar actions maybe respectably any time filling at the dinner table the glass with a gassed water. It may remain quiet, without bubbles in the bottle; however, when getting in a glass, its integrity is disrupted the same way as at shaking, forming bubbles in the glass.

At the vibration on the oil reservoir, the nature of interformational appearance of gas bubbles depends on the interrelations between the current formation pressure, amplitude of the vibration pressure, and the oil with gas saturation pressure. A process is occurring at that illustrated by a scheme in Figure 6.6. In the decompression phase, when a variable vibratory pressure declines below the saturation pressure, the oil phase degassing occurs. But in the compression phase, when the variable vibratory pressure rises above the saturation pressure, no total inverse dissolution occurs of the gas phase in the liquid due the dissolution hysteresis phenomenon [12].

To an observer V who does not notice the "rapid motions", this process will look as an effect of interformational rise of the gas saturating/oil pressure in a vibration field [4]. In these studies, it has been for the first time shown experimentally that at vibration, an increase of oil with gas saturation pressure occurs [4]. These studies for the first time experimentally demonstrated that under vibration, the increase of oil with gas saturation pressure is observed, i.e., the emergence of gas bubbles at pressures higher than the saturation pressure without the vibration. Small gas bubbles, however, form with sizes much smaller than the pore size. The studies have been conducted on the installation schematically shown in Figure 6.7. The major part of the installation is a glass tube *1*, designed

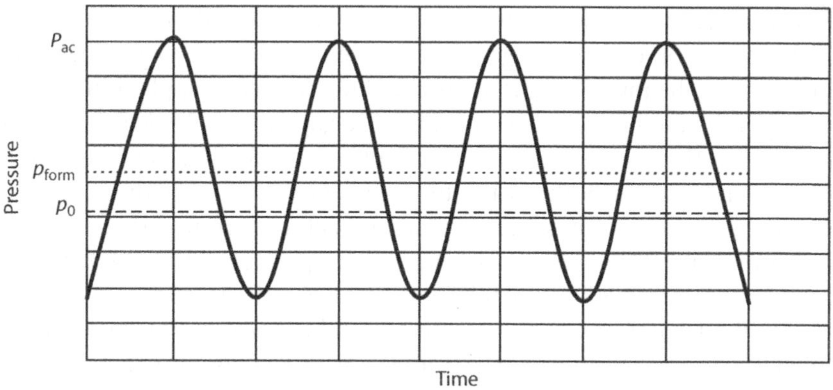

Figure 6.6 Schematic acoustic pressure change at the formation pressure higher than the saturation pressure. p_{form}, formation pressure; p_0, saturation pressure; p_{ac}, maximum acoustic pressure amplitude.

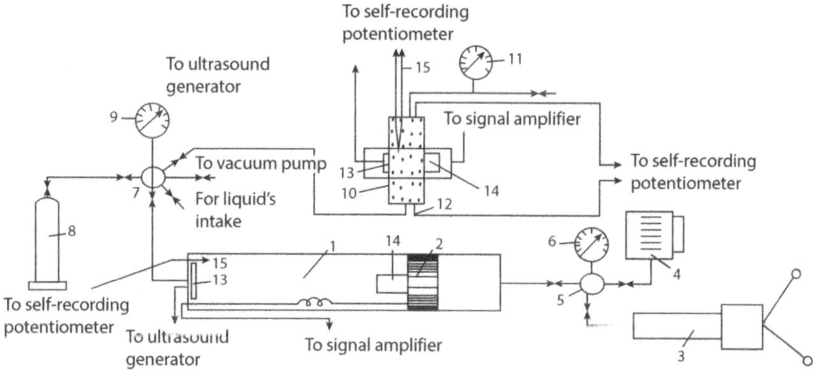

Figure 6.7 Schematic lab installation for a study of the effect of the intraformational oil pressure increase by the gas in vibration field.

for the work pressure of 0.6 MPa. In the tube, the saturation pressure has been determined in the volume phase. In order to be able to control the pressure in the work chamber (to lower, to raise or to hold constant), an organic glass piston 2 has been introduced in the tube with rubber ring gaskets. The pressure in the work chamber has been created using the press *3* supplying glycerin in behind-the-piston space from a tank *4* through the pipe *5*. It has been installed the test pressure gage *6* for 1 MPa in the pipe. A fluid has been supplied in the tube *1* through pipe *7* for which the tube *1* has been preliminarily vacuumed. Gas has been directed through this tube and through pipe *7* from a gas tank *8*. The test pressure gage *9* has been installed on the pipe *7*. A gas-saturated fluid has been supplied in a vessel with porous medium *10*. The test pressure gage *11* for 0.6 MPa has been installed at the porous medium exit. Probes *13* and vibration receivers *14* have been installed both on the tube and on the vessel with the porous medium. This vessel has been also manufactured from glass sustaining pressure of 0.5 MPa. For measuring the temperature, thermocouples *15* have been inbuilt both in the tube and in the vessel *10*.

The fluid has been pumped-in in the tube *1* using a vacuum-pump. Then, the gas has been supplied from the tank *8* at continuous stirring of the mix through shaking and rocking. At a certain amount of gas, the tank has been turned off and the pressure raised at continuous stirring of the mix until the entire gas has been dissolved in the liquid. After the equilibrium has been achieved, the pressure in the tube has been gradually lowered also at continuous stirring of the mix. The pressure at the moment the first gas bubble has appeared has been considered the saturation pressure. The saturation pressure has been recorded by changing the signal from a probe *12*.

The study has been conducted with a mix of propane and heptane. The saturation pressure has been preliminarily determined visually at intense stirring of the mix. After that, the pressure has been raised and gas has been again dissolved in the fluid. Then, the fluid has been subjected to a continuous vibration at various irradiation power and the saturation pressure has been determined. In most cases it has turned out higher that the initial (i.e., without the vibration). After the saturation pressure determination, the pressure has been raised again and vibration, stopped.

As Figure 6.8 shows, 18 to 20 minutes after vibration has stopped, the saturation pressure p_n has returned to the initial pressure p_0 before the one that has existed before the vibration. It should also be mentioned that in conditions of a continuous vibration, the appearance of the first gas bubbles has been clearly observed indicating a change not only of the signal amplitude but also of its shape (appearance of a low-frequency signal modulation). This indication has appeared in time ahead of the visual indication, i.e., it has marked the formation of invisible to the naked eye tiny bubbles. The correlation obtained as a result is shown in Figure 6.9. The saturation pressure under vibration may increase by 20% to 30% from the initial value. Under vibration, zones of elevated and lowered pressure have emerged and vanished in the fluid, the zones wherein gas bubbles have been formed (something like a fluid "fracturing" have been observed).

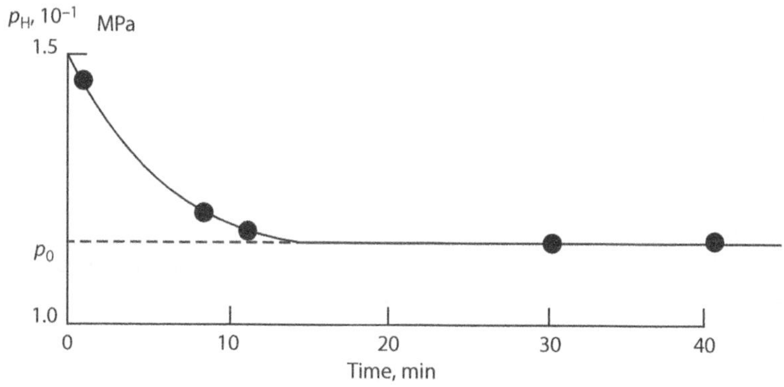

Figure 6.8 Saturation pressure vs. time after the vibration stops. p_0 is the saturation pressure before the vibration.

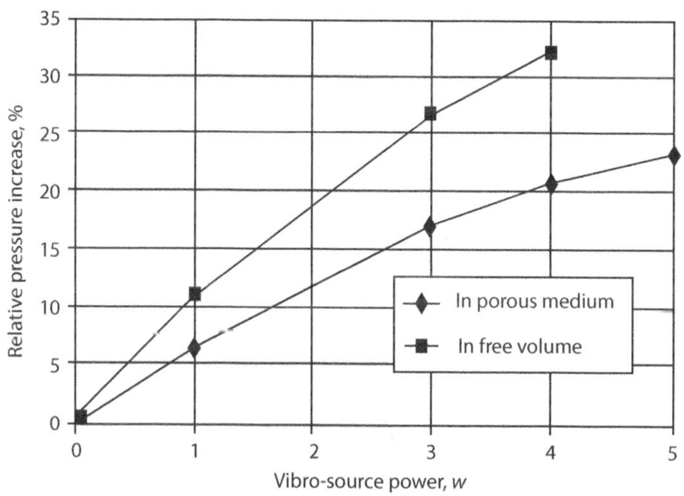

Figure 6.9 Heptane-propane mix saturation pressure vs. power of the vibration source.

A momentous raising of the pressure resulted in the disappearance of bubbles. However, at some pressure, equilibrium has occurred. The formed bubbles may have no longer been squashed and have continued growing. In the tube, it had been possible to observe visually as next to the vibration irradiator have formed barely noticeable tiniest gas bubbles, which have been raising and immediately vanishing, squashed by the external pressure. The oscillograph has clearly registered the appearance moment of the gas bubbles. It has been experimentally shown that some optimum power values of the vibration source have existed (characteristic for a given reservoir and for the saturating medium) at which the pressure of first gas bubbles' appearance has been maximal.

In real fluids, on the gas bubble surface, the adsorption of organic (surface-active) substances occurs, which results in lowering the interfacial tension on the fluid boundary and prevents their coalescence. Because of this, gas bubbles formed in real fluids in the vibration field are preserved stable even at very small size close to embryonic not increasing in size any more (Figure 6.10). The released gas bubbles are dispelled within the entire oil volume. If the pore volume fraction taken by the gas phase in the reservoir does not exceed a certain fraction of the pore volume, the oil phase is not degassed and the gas remains immobile relatively to the oil phase.

Analyzing a three-phase phase permeability diagram, Surguchev [12] has determined that the reservoir pore volume fraction taken in this case by the gas phase should not exceed 15% to 20%. At higher gas saturation, this phase has begun moving independently. The state of an oil-gas system when gas bubbles dispersed in the oil volume remain immobile relative to the oil phase has been described by Lysenko [7] as "oil ungassing without its degassing". Having assumed that the gas bubbles have been distributed uniformly and randomly in the oil volume, he has shown that such fraction of an oil reservoir was close to 33%. He assumed this value as the

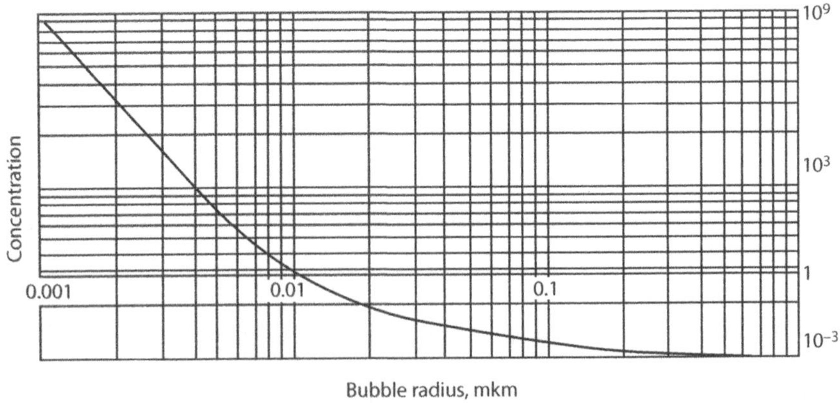

Figure 6.10 The gas bubble number in 1 cm^3 at water de-gassing in vibration field.

oil ungassing criterion without its degassing. In other words, a gassed oil phase that has emerged in the reservoir as a result of vibration may theoretically remain stable for years if the pore volume fraction in the reservoir taken by the gas has not exceeded 15% to 33%. The volume of such gassed oil phase will depend on the vibration pressure amplitude, the current formation pressure, and the oil by gas saturation pressure [9].

In order to prove this, we will denote the oil fraction in the volume of the gassed oil phase emerging under the vibratory action as follows:

$$\delta = U_0/(U_0+U_g)$$

where U_0 is the volume of purely oil phase, and U_g is the volume of the gas phase.

At dropping the vibration pressure below the saturation level by dp value, the following occurs:

- Oil with solution gas expansion by $(-\alpha\delta)dp$ value where $\alpha \approx \text{const}$;
- Free gas expansion by the $(1 - \delta)dp/p$ value (assuming isothermal regime);
- $(-C\delta/\rho)dp$ value gas release from the oil phase which is dissolved in it, under the Henry's law (ρ is the free gas density, C is the Henry's law constant).

Due to the isothermal equation of state, $p = \beta\rho$, and as a result, one obtains the local change of the oil fraction in the gas oil phase volume:

$$\frac{d\delta}{dp} = \left(\frac{\beta C - 1}{p} - \alpha\right)\delta + \frac{1}{p},$$

where β is a constant found from the equation of state. It is possible to assume that approximately $\alpha p = \eta = \text{const}$. Having introduced $1 - \beta C + \eta = z > 0$, one obtains $(1 - z\delta)p^z = \text{const}$, which may serve as a rough estimate of the oil fraction local change in the gassed oil phase volume depending on the change of the vibration pressure.

The oil fraction in the gassed oil phase is

$$d\delta = -\frac{U_0 dU_g}{(U_0 + U_g)^2},$$

and the result is

$$d\delta = \delta[1-\delta+\beta C\delta]\frac{dp}{p}.$$

Assuming $(1 - \beta C) = b_1$, one obtains

$$\frac{d\delta}{\delta(1-b_1\delta)} = \frac{dp}{p}.$$

The constant of integration is determined from a condition when $\delta = 1$ at $p = p_0$ (p_0 is the saturation pressure).

In this case, b_1 is a known function of the initial gas/oil ratio and of the oil formation volume factor. After integration, the following correlation is obtained between a relative volume change δ of the gassed oil phase and the current vibration pressure p:

$$\frac{p}{p_0} = \frac{\delta(1-b_1)}{(1-b_1\delta)}, \qquad \Delta p_{ac} \geq (p_{\text{form}} - p_0),$$

where p_{form} is the formation pressure and Δp_{ac} is the amplitude of the vibration pressure.

From here, the relationship is determined between the amplitude of the vibration pressure and an increase of the oil/gas mix volume as a result of the vibration:

$$\frac{p_{form} - \Delta p_{ac}}{p_0} = \frac{1-b_1}{\xi - b_1};$$

here, $\xi = 1/\delta = (U_0+\omega)/U_0$ is the relative volume of a gassed oil phase (the ratio of the gassed oil phase to the volume of the pure oil phase in the oil/gas flow).

The interrelation between the densities of pure oil, gassed oil, and gas phases is as follows:

$$\rho_{ng} = \frac{\rho_n}{\xi} + \frac{1-1/\xi}{\rho},$$

where ρ_{ng} is the density of the gassed oil phase and ρ_n is the density of the pure oil phase.

After opening the b_1 parameter, one has the following characteristic interrelation for the determination of the acoustic pressure, at which the state of oil ungassing without its degassing is still preserved in a reservoir:

$$\frac{\Delta p_{ac}}{p_0} \leq \frac{p_{form}}{p_0} - \frac{2G_0}{p_0 + 2G_0}, \qquad (p_{form} - \Delta p_{ac}) \geq p_0.$$

here, the pressures have dimension of kgf/cm², and G_0 is the gas/oil ratio before vibration beginning (in m³*kgf/cm²)/m³.

Figure 6.11 displays computer correlations of oil/gas mix volume increase vs. acoustic pressure amplitude, current formation pressure, and saturation pressure. Here, b is the oil formation volume factor, U_{ng} is the oil/gas mix volume in m³.

The oil volume gassed as a result of the vibration is substantially greater than the oils similar volume without the vibration.

Within a fixed amount of the formation pore space, an increase of gassed oil phase volume in the vibration field results in the appropriate displacement of the water phase from the formation and, therefore, in the lowering of the formation average water saturation. This is supported by the results after the vibration (Figure 6.12).

Immediately upon the beginning of the vibration, the average reservoir water saturation has been rapidly dropping. Overall, maximum formation water saturation decline exceeded 20%. Average (over the action period)

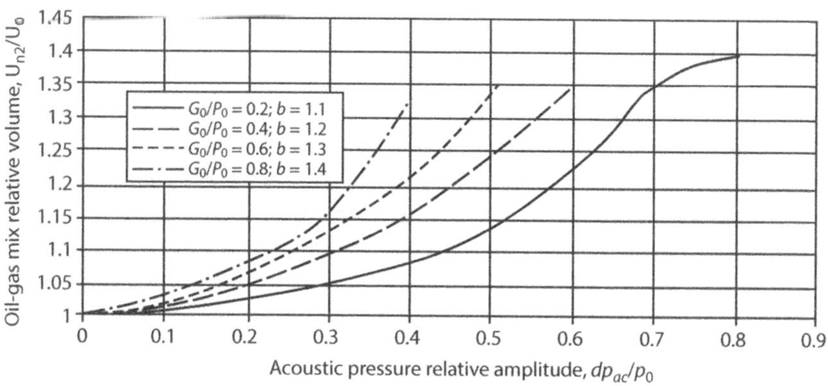

Figure 6.11 Increase in the oil-gas mix volume with the acoustic pressure increase in vibration field.

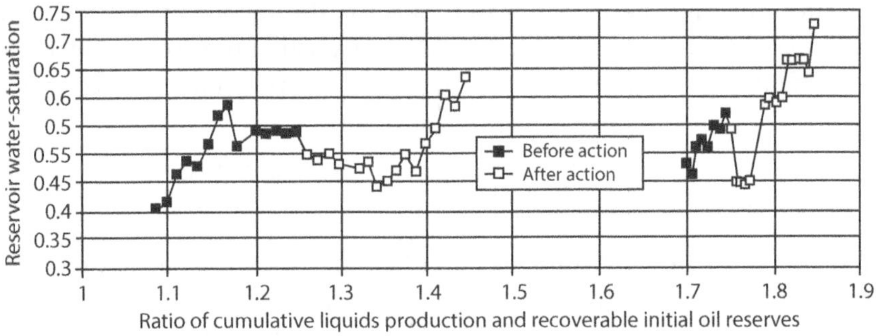

Figure 6.12 Formation water saturation change after the vibroseismic action in two fields.

the water saturation drop has been 40% relative to the water saturation, which would correspond to a computer forecast of its change without the vibration.

The formation water saturation decline dynamics after the vibration also testifies to the involvement in the development of new, not produced earlier oil reserves within the interbeds bypassed by the water during waterflooding.

Further on, as these reserves have been withdrawn from the formation, the displaced volume has been replaced by water and the water saturation has been beginning to grow again, gradually approaching the value corresponding to the computer forecast of the water saturation change without the vibration.

References

1. Vakhitov, G., Kuznetsov, O.L., Simkin, A.M., *Thermodynamics of Oil Reservoir Near-Bottomhole Zone*, p. 215, Nedra, Moscow, 1978.
2. Chersky, N.V. *et al.*, *The Tectono-Seismic Processes' Effect in the Hydrocarbon Formation and Accumulation*, p. 180, Nedra, Moscow, 1983.
3. Inous, K. *et al.*, *Capillary Chemistry*, p. 180, Mir, Moscow, 1983.
4. Kuznetsov., O.L. and Simkin, E.M., *Transformation and interaction of geophysical fields in the lithosphere*, p. 270, Oxford & IBH Publishing Co., New Dehli, 1994.
5. Kuznetsov., O.L., Mirzadjanzade, A.Kh., Simkin, E.M., Seismic and Rheochemical Technologies for Improving Productive Oil Capacity of Pay Zones and Monitoring of these Technologies. *Proc. 10th European Symp. EAGE on improved oil recovery*, pp. 241–243, 1999.

6. Kuznetsov., O.L., Simkin, E.M., Chilingar, G.V., Improved Oil Recovery by Application of Vibroenergy to Waterflooded Sandstones. *J. Pet. Sci. Eng.*, 19, 191–200, 1998.
7. Lysenko, V.D., *Theory of Oil-Field Development*, vol. 310, Nedra, Moscow, 1993.
8. Simkin, E.M. and Sokolov, A.V., Acoustic Field Effect on Rheologic Oil Properties. *Coll. Proc. VNIIYaGG*, pp. 60–63, 1981.
9. Simkin, E.M., Vibroseismic Method of Increasing Productivity of Water-flooded Oil and Gas Reservoirs. *Oil Gas Technol.*, 2, 24–25, 1998.
10. Simkin, E.M. and Surguchev, M.L., Advanced Vibroseismic Technique for Waterflooded Reservoir Stimulation, Mechanism and Field Test Results. *Proc. 6th European Symp. on Improved Oil Recovery*, vol. 1, pp. 213–241, 1991.
11. Simkin, E.M., Hydrodynamic Description of Oil Displacement by Water Process within the Elastic Vibration Field. *Proc. Internat. Conf., Flow Through Porous Media: Fundamentals and Reservoir Engineering Applications*, pp. 172–175, 1992.
12. Surguchev, M.L., Zheltov, Yu.V., Simkin, E.M., Physicochemical Micro-Processes in Oil-Gas Reservoirs. p. 330, Nedra, Moscow, 1984.
13. Zarev, V.P. and Kuznetsov, O.L., Experimental Study of Physicochemical Processes in Rocks at Small Elastic Deformations. *Proc. AN USSR, series Physics of the Earth*, vol. 6, pp. 94–101, 1978.
14. Chilingar, G.V. and Haroun, M., *Electrokinetics for Petroleum and Environmental Engineers*, p. 240, Scrivener Publishing (Wiley, New Jersey, 2014.

7
Energy Criteria

The results of numerous investigations confirm that elastic oscillations improves the fluid flow in porous media. The reason for this is that the elastic oscillations reduce the effective viscosity of the fluids in the porous media and improve the porous-medium wettability. This impact changes relative permeabilities, accelerates relaxation of mechanical stresses in the rocks, reduces their structural mechanical strength, and results in the thixotropic liquefaction of clay inclusions and disintegration of the bridging material [10, 14].

However, these effects are hard to conceive and investigate due to specific conditions that give rise to energy disturbances in the medium exposed to the elastic waves propagation.

Initially, many investigators used to consider the oscillatory impact frequency ω and intensity J as parameters that determine the initiation of various effects. However, the experience showed that such an approach is too narrow, scarcely informative and, therefore, is insufficiently productive in terms of practical application of the results. The known effects of the force impact (such as high-frequency and high-energy impact) do not at all imply any wave-frequency background as to the mechanism of their action. They are caused only by secondary, albeit intensive, wave radiation factors (i.e., medium heating along with potential gas liberation due to the shift of the thermodynamic state as well as direct dissipative medium structure breakdown). In addition, these effects occur within frequency ranges that are not of particular interest in terms of exerting a remote impact on the oil-saturated media due to heavy damping of elastic waves on these frequencies.

However, the impact of the low-frequency elastic oscillations at low intensity does not cause in most cases the direct force response of the medium; there is no direct relationship between the impact results and the frequency and intensity. Phenomena observed in the laboratory *in situ* show the selectivity of the low-frequency treatment. It is confirmed by the native metastable states existing in the nature of system attributed both to its internal properties and stress, and to superposition of continuous external forces or gradients of physical pressure and stress fields.

George V. Chilingar, Kazem Majid Sadeghi, and Oleg Leonidovich Kuznetsov. Acoustic and Vibrational Enhanced Oil Recovery, (215–260)

Rather low in intensity, but selective in its properties, the treatment can activate natural forces and gradients available in the Earth and result in manifestation of the effects that will correspond to already much greater total energy of the entire process of produced disturbances.

In such situation, the absence of an appropriate approach to select the governing parameters puts the investigator in a position of a sideliner who just marks the appearance of a solid result. This impedes the investigation of the manifestation mechanisms proper, related to the governing parameters of the oscillatory treatment and specific conditions, to the characteristics of the porous media structure and surface, to the saturating fluids and bridging materials, as well as thermodynamic conditions. Thus, for example, Westmark *et al.* [10] show the results obtained in the USA in the late 1990s when investigating the vibration influence on the waterflooding processes. The results show that excitation at 100-Hz frequency has less noticeable influence on the two-phase fluid flow in the *Berea* sandstone due to the wettability change of the oil-saturated core. Authors believe that this can become a dominant mechanism and will determine the oil production stimulation. The authors have also observed a strong destruction of shaly inclusions and subsequent increase of the absolute rock permeability exposed to elastic oscillations at this frequency range. These experiments were carried out later than the works by the other authors [14] which reproduce and competently confirm some of the investigation results shown in this book. However, they do not give any quantitative energy representations of the experimental data.

In the previous chapters of the book, the reader has already reviewed the results of experimental investigations where the authors have resolved the similar bottlenecks. When setting up the experiments, we put aside the specific source of impact and investigated the influence of the elastic oscillations proper within a rather wide amplitude and frequency range. The oscillatory medium acceleration $\ddot{\xi}$ and oscillatory displacement ξ were selected as elastic oscillation parameters controlled in the course of the experiments. It is not sufficient to select only the oscillatory acceleration $\ddot{\xi}$ as a criterion, as done by some authors. Oscillatory displacement ξ is an important structural parameter that characterizes "proportionality" of the treatment with regard to the internal space-time and mass interactions of a heterogeneous system. For example, in the case the reservoir is water-wet, inertial forces that arise at the interface between water and oil phases, when wetted, are proportional not only to the difference in the oil and water densities, or to oscillatory medium acceleration, but also to the square of the radius of "oil" porous channels in the reservoir. In addition, for example, coagulation structures do not break down when only oscillatory inertial forces are applied, that exceed the forces of short-range interaction between the particles. It requires sufficiently large oscillatory displacements commensurate with the size of a structural cell. Oscillatory displacement at the given oscillatory acceleration determines the breakdown velocity that should not exceed the structure recovery velocity. Equally important is the fact that the assigned oscillatory acceleration $\ddot{\xi}$ and displacement ξ at the assigned wave medium resistance determine both the density of the oscillatory-energy flow (oscillations intensity) and oscillations frequency.

Inertial forces that arise at the phase interface due to the difference in their densities, given that more than one phase saturates the medium, as well as the inertial forces that arise in the pore fluid structures due to the difference in densities of their structural components, determine to a great extent the mechanism of influence the elastic oscillations have on the flow processes in the saturated porous media. According to the physical mechanism of vibrational displacements, basis of which was laid by Mandelstam [16], a certain asymmetry in the system, under the influence of the "rapid" oscillatory inertial forces, causes a "slow" macroscopic movement that results in redistribution of phase saturations in the pores or in the breakdown of the intrapore structures. However, this asymmetry can be caused by wettability hysteresis, external pressure and stress gradients, kinematic features of the oscillatory process itself, competing near and far interactions between the particles in dispersed clogging structures, etc.

7.1 Parameters of Oscillatory Treatment and Conditions for Manifestation of Useful Effects in Saturated Geological Media

Selection of oscillatory acceleration and displacement parameters as determining ones allows not only cast in certain cases the frequency selectivity mechanism itself as a set of explicit physical regularities but also gain useful data on the manifestation of these effects in various conditions.

Based on the results of our investigations, the authors got a generalized energy threshold criterion to determine the onset of the effects in the reservoir exposed to elastic oscillations. This criterion determines the values of oscillatory acceleration $\ddot{\xi}$ and medium displacement ξ given the minimum value of

$$G=\left|\ddot{\xi}/g-a\right|\left|\xi/\bar{d}-b\right|, \tag{7.1}$$

where g is the gravity acceleration; $\bar{d}$ is the dimensional characteristic of the porous medium, e.g., the effective diameter of pore channels; and a, b ($0 \le a, b \le 1$) are dimensionless empirical coefficients that are functions of the properties of interacting liquid and solid phases in the porous medium and on the specificities of its structure.

This generalized criterion determines the required treatment energy and frequency selectivity reflected in its oscillatory parameters and applied with respect to the specific medium features.

The authors got threshold values of the parameters for the specific geological and physical conditions and could identify frequency ranges of the treatment the most efficient in terms of energy. Based on the experimental tests and the resulting criterion, we identified the optimum frequency range for the most representative geological and physical conditions of productive reservoirs. Treatment of a productive reservoir within the given frequency range of 10 to 500 Hz is the most efficient in terms of maximum manifestation of the observed effects the elastic oscillations have of the reservoir fluids flow and achievement of the greatest treatment depth.

To illustrate the above, Figure 7.1 shows the curves of the threshold values of the oscillatory displacement and acceleration, calculated for the specific type productive-reservoir conditions using known relationships of these parameters as a function of frequency f and elastic oscillation intensity J plotted in the coordinates of oscillations intensity vs. frequency.

Parameters of the saturated porous medium:

Elastic wave propagation velocity of 3000 m/s; density of 2650 kg/m^3; and effective pore diameter of $\bar{d} = 15\ \mu$m.

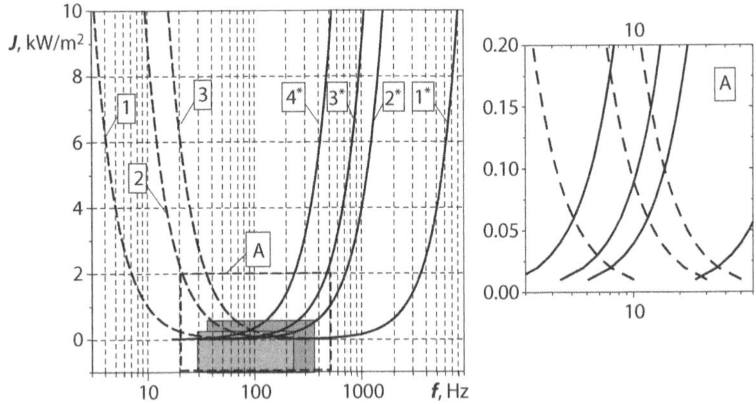

Figure 7.1 Elastic wave intensity corresponding to minimum threshold values of oscillatory acceleration and displacement of the medium particles.

Codes of the curves: 1, 2, and 3 are respective ***J*** estimates for the oscillatory accelerations of 0.1g, 0.3g, and 0.5g; 1*, 2*, 3*, and 4* are those for oscillatory displacements of $0.1\bar{d}, 0.3\bar{d}, 0.5\bar{d}$, , and $0.1\bar{d}$.

The following saturated-medium parameters were used in calculations: density of 2650 kg/m^3; elastic wave propagation velocity of 3000 m/s; and pore-channel average diameter of 15 mm. One can use these curves to identify the rational frequency ranges for this medium treated to improve the formation-damage elimination (Figure 7.1, area of 20 to 250 Hz) and flow processes in general (Figure 7.1, extended area of 30 to 400 Hz), where the required relations of the threshold values of ξ and $\ddot{\xi}$ are achieved at rather low oscillations intensity.

Here, the intersections of intensity curves that correspond to the certain acceleration and displacement threshold values determine the frequencies at which the given criteria parameters can be achieved at minimum energy of the oscillatory oscillations.

The elastic oscillation intensity within these ranges, required to get the wanted effects, is substantially small (see scale up A). This confirms the energo-informational purpose of this treatment mentioned above given the feasibility of its remote implementation in the reservoirs.

To achieve in reality, the flow and other effects within the confines of entire productive reservoirs, one has to use the reserves of the petroleum-reservoirs stress energy and the large energetics of the geological structures surrounding the reservoir. This issue was discussed in sufficient detail; however, no quantitative results have been published.

High-amplitude elastic oscillations and pulses have rather significant influence on the behavior of geomaterials exposed to the mechanical stresses, which impacts the elasticity, kinetics of accumulation of structural defects, micro- and macrofractures, and other processes. However, from the perspectives of the most efficient and deepest reservoir treatment, the most important thing is to implement the mechanism of trigger effect of the elastic oscillations influence that may occur when the elastic oscillations have rather low amplitudes, several orders of magnitude lower than the rock stress.

The mechanism of trigger effect is tied to the block fractal features of the Earth structure, while the influence itself of the small oscillations originates at the lowest micro-levels of the structural hierarchy, resulting in the change of probability function of accumulation of dilatons, dislocations and micro-fissure nuclei, subject to the load, and in the qualitative change of the micro-failures nature. The composite nature of the Earth suggests a potential explosive structure breakdown at each individual level of the hierarchy, while fractality brings about the rapid level-sensitive information exchange, when failure processes move stepwisely to the higher level of the structural hierarchy, and so on, until micro-fissures develop in the medium. Developments of micro- and macro-fissures are accompanied by block movements and earth-matrix transformation that occurs also at various hierarchal levels. Here, medium fluid saturation plays an important role. It does not only intensify the transformation processes but is, in addition, a component of the reservoir system self-organization. The latter results from the elastic oscillations that trigger impact and bring about appearance of a new long-lasting intrapore pressure and flow gradients.

7.2 Wavelike Nature of the Oil-Saturated Geological Media Stress-Energy Exchange. Elastic Oscillations as an Energy Exchange Indicator and Regulator

Constant energy and substance exchange with the environment is taking place in the rocks; various thermodynamic processes are under way related to the micro- and macroexogenous Earth transformations exposed to the various external natural and man-made impacts. These processes generate wave radiations in the form of natural acoustic emission (AE) and AE induced by external impacts, seismic acoustic emission (SAE), electromagnetic emission, etc. [8, 13, 20, 21–24, 35].

Propagation of elastic waves and other physical emissions in the Earth is linked with these processes. In this case, a special role belongs to the phenomena linked to the porous-medium fluid saturation, and interaction of the pore-filling fluid with the mineral rock "skeleton" given changes of its stress-strain state accompanied by anomalous secondary emissions with maximum energy in elastic waves of wide frequency range (1.0 to 1.0 × 10^3 Hz) [1, 5, 6, 31]. These phenomena differ from the ordinary SAE, which is linked with solid fracturing, specificities of the emission spectral characteristics and its power. They do not fit the description of the general seismicity and that caused by the external impacts. The scale, energetics, and spectral features of the radiations in the observed phenomena can be described using only specific flow processes, for example, by origination of "Hines" flow jumps [41], or using change of state and properties of the reservoir oil, such as degassing and viscosity reduction, due to the external acoustic impact. The nature of the registered phenomena has been scantily explored so far. Definition of the mechanism of this energy exchange and conditions of its manifestation will allow considering the saturated rocks in their natural occurrence as an active heterogeneous system, physical properties of which can be efficiently controlled through exposure to external waves.

Results of the experiments aimed to investigate the behavior of the stressed geological media exposed to external static and wave disturbances allowed the authors to identify a new property of the rock energy exchange. This new property presumes a receptive energy accumulation in the oil-saturated medium, which determines to a great extent parameters of the productive-reservoir wave-treatment optimization [28, 29, 40].

7.2.1 Manifestation of Seismoacoustic Radiation in Oil-Saturated Media Exposed to Internal Stress Disturbance and Elastic Oscillation Treatment

The authors investigated the acoustic radiation from the geological media in the laboratory under various internal stresses and small external disturbances. In addition to static mechanical stresses, we used small external wave-like disturbances such as low-amplitude elastic harmonic oscillations and elastic pulses of defined shape.

Figure 7.2 shows a laboratory layout. The authors used cores from the terrigenous and carbonate reservoirs as geological media samples where porosity and permeability varied over wide ranges. The authors also used specially prepared artificial cores that were "clones" with identical structural, mechanical, and flow properties. They were in accordance with the proven methods using quartz sand, marshallite and liquid glass that exhibit rather high structural and mechanical strength, and immunity to the change of the pore filling fluid. Core permeability varied from zero to 0.150 mm^2, and porosity from zero to 30%. Cores were saturated in compliance with a standard procedure with dead oil from the Yugomashevskoye Field and brine. Liquid properties were as follows: 8.0 MPa-s oil viscosity; 1.1 MPa-s water viscosity; 800 kg/m^3 oil density; and 1,010 kg/m^3 water density.

The tested samples were placed (see Figure 7.2) into special core holders [1] and exposed to an axial compression using a hydraulic press [2]. Special gauges [3, 4] recorded the time series of acoustic-radiation signals emanated from the samples along with the treatment parameters. A special measuring and computing system created by the authors was used to record and process the signals. The system included piezoelectrical vibration metering converters, compact measuring device VShV-003-M3 [5], designed to measure and analyze noise and vibrations, analog-to-digital converter (ADC) E-330 [6], a notebook [7], and software package

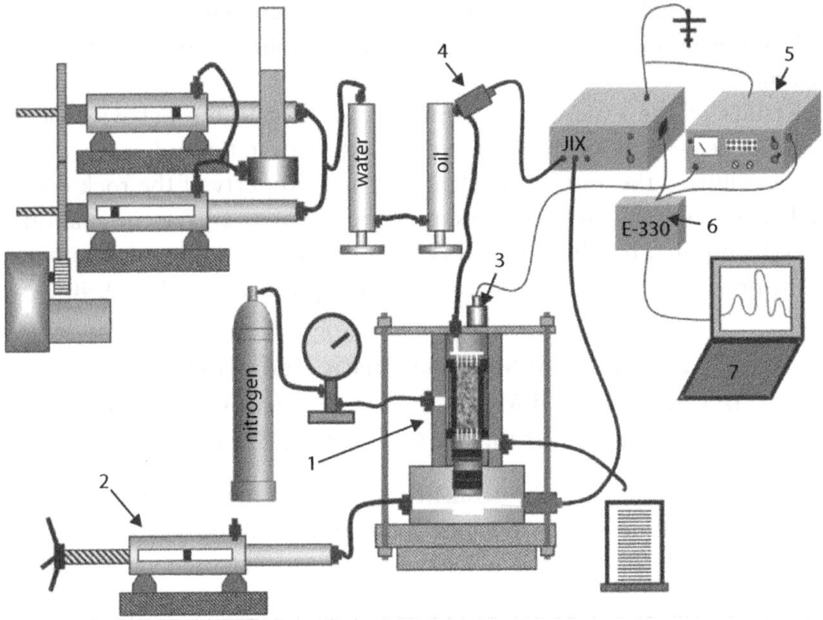

Figure 7.2 Laboratory setup layout to diagnose acoustic radiations of the stressed oil-saturated reservoir media.

to accumulate and process the results. Advanced methods were used to analyze the non-linear dynamic processes. The schematic diagram of the measuring systems is shown in Figure 7.3.

Geological media in their natural occurrence, especially fluid saturated ones, fit the description of the complex non-linear dynamic system. Elements of their structure, fissure faces and irregular failure surfaces, as well as distribution of various heterogeneities, faults, and fractured and microfractured zones in the lithosphere have a fractal or, more precisely, multifractal nature [35, 36, 38, 43]. These media consist of a hierarchy of fractal structures of widely differing scales. Their stress-strain state and its variation due to the emergence of new fissured fields and flow fields are subject to the interaction of micro- and macrostructures, as well as their state, and interaction of all the three phases, including solid rock matrix, liquid fluid, and gas.

Change in the medium state is subject to a great extent not only on the strength and strain properties of the matrix material but also to the surface interaction of all phases. In the case of structured oils, it is, in addition, subject to their volumetric, physicochemical, and mechanical interactions. In addition to the matrix deformation, development of the new fissures

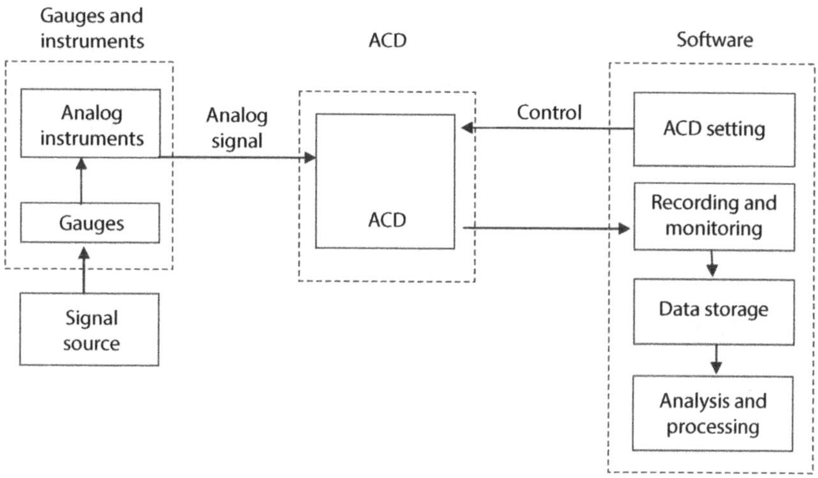

Figure 7.3 Schematic of the measuring system.

within the medium requires also that the new surface be completely wetted with liquid phases, providing liquid phase mobility and flow. In these circumstances, specificities of physical and chemical phase interaction can be observed, i.e., water and oil wettability of the solid phases and details of their viscosity variation that are different in the carbonate and terrigenous reservoirs. All the phenomena that occur here are interrelated and interdependent. The processes that occur in the geological medium exposed to external impact (development of micro-fissuring, fissuring and, in general, discontinuities in the medium at all the hierarchical levels, up to faulting) are accompanied by changes in the fractal structure. That is why the radiations that they emanate also have fractal characteristics.

To explore and describe the behavior of geological media that are non-linear dynamic systems, rather fruitful was the approach based on the theory of dynamic systems, fractal sets, and mathematical apparatus of fractal analysis.

The reservoir as a non-linear dynamic system exhibits a strange attractor, which is an attracting set in a phase space that contains trajectories $V(t) = \{V_1(t),\ldots,V_m(t)\}$, showing its behavior in a multidimensional phase space of dimension m. Numerical signal-series processing by Packard and Takens method restores the attractor properties, i.e., embedding dimension m and Hausdorff correlation dimension D_c from the time sequence of one of the constituents $V(t)$ [44]. A time sequence of the measured microseismic signal taken with a certain time lag $Vp(t) = \{V(t), V(t+\tau), V(t+2\tau),\ldots, V(t+(m-1)\tau)\}$ serves as vector components. As the vector components that describe the dynamic system are independent, the first value at which the autocorrelation function reaches a local extremum (becoming zero or reaching its minimum) is selected as value τ. The procedure reduces to the following: the vector $Vp(t)$ components are added in sequence, and at each $m = 2, 3\ldots.$, the correlation integral is calculated as follows:

$$C_m(r) = \frac{1}{N^2} \sum_{i,j=1}^{N} H(r - |V_i - V_j|), \tag{7.2}$$

where $H(z)$ is the Heaviside function:

$$H(z) = \begin{cases} 1, & z \geq 0, \\ 0, & z < 0. \end{cases} \tag{7.3}$$

for all couples of values of i and j, if $i \neq j$, $|V_i - V_j|$ is an absolute value of the distance between the set points in the m-dimensional space; N being the size of the point sample from the signal. The sum value depends on r; when r are small, $C_m(r) \approx r^{D_c}$; therefore, the correlation dimension is derived from the slope of the relationship ln C_m vs. ln r, the plot showing variation of $C_m(r)$ at various values of r, given rather large amount of N. When calculating data points above the mentioned relationship, one can identify a scaling area; the function is approximated by a straight line by the least-square method, whereas the slope of the straight line gives the D_c value.

The m dimension of the phase space, at which D_c ceases to vary notably, is the minimum embedding dimension, i.e., the smallest integer dimension of the space that contains the entire attractor.

In reality, the signal is recorded and processed on the computers discontinuously. A rather significant amount of the signal sample can be processed within a short time (about 20,000 points). The needed dynamic system parameters are calculated, and digital value and patterns of the resulting attractor are displayed to describe the current reservoir condition. Samples are analyzed automatically one after another. When the computer is fast enough, data on the screen follow in a discontinuous succession and reflect the processes that occur in the reservoir as a non-linear dynamic system.

The Takens procedures are used to analyze the radiation, assess the fractal correlation dimensions that determine the dimension of the phase space comprising the system state points, and describe the level of complexity and orderliness of its behavior. One can also calculate the Lyapunov indicators that measure the degree of determinateness and time interval depth of the system behavior predictability. Restoring the patterns of the strange attractors of the geological media dynamic state in real time, they allow judging the energy stability of the states and the nature of variation of those states in time. In general, we used an integrated approach to explore the physical radiations of the microseismic emission, where the above procedures of fractal mathematics are supplemented by wavelet spectrum analysis and energy accumulation dynamics analysis in the course of signal arrivals in real time.

Application of this advanced method to diagnose the state of real geological media using the acoustic-signal time sequences is shown on the examples below.

The method of investigations run on the testing setup (see Figure 7.2) was as follows. The tested core sample was secured in the setup; then, the core medium was exposed to stepwise increase of mechanical stress. The stress grew in the range of zero to 150 MPa. A computer recorded and processed the acoustic signals and other information arriving from the measuring gauges.

The monitor screen showed the fractal parameters of the current medium state along with the pattern of the restored attractor. They were analyzed, and results were reflected on the computer screen. The processing was done in real time; the lag was rather short. As another test was over, the recorded time sequences of AE were analyzed using fractal, spectral, and other methods.

The laboratory setup was used to break down cores by applying a gradually increasing load. Figure 7.4 shows the most distinctive screenshots of the Sutorminskoye core stress state that were selected from among the samples that are continuously reconstructed in two-dimensional phase space.

As a result of the tests, the following regularity was identified in the behavior of the medium samples exposed to stepwise loading.

As the internal stress grows, the level of the recorded AE gradually increased. Thus, the resulting patterns of the phase state that continuously follow one another, pulse that suggests, along with changing computed fractal dimensions, that the medium state goes through cyclic steady-state and transient phases, that correspond in terms of time to accumulation and release of the internal system energy. Pulsing cycle time was about 500 ms.

In the course of the testing experiment, the load applied to the core samples grew continuously, the attractor pattern varied (see Figure 7.4), and one could clearly recognize a stage of equilibrium state at a weak load. At later times, the stages followed one another and featured precursors of local microfissure initiation, and the final stage featured a precursor of macrofissure formation.

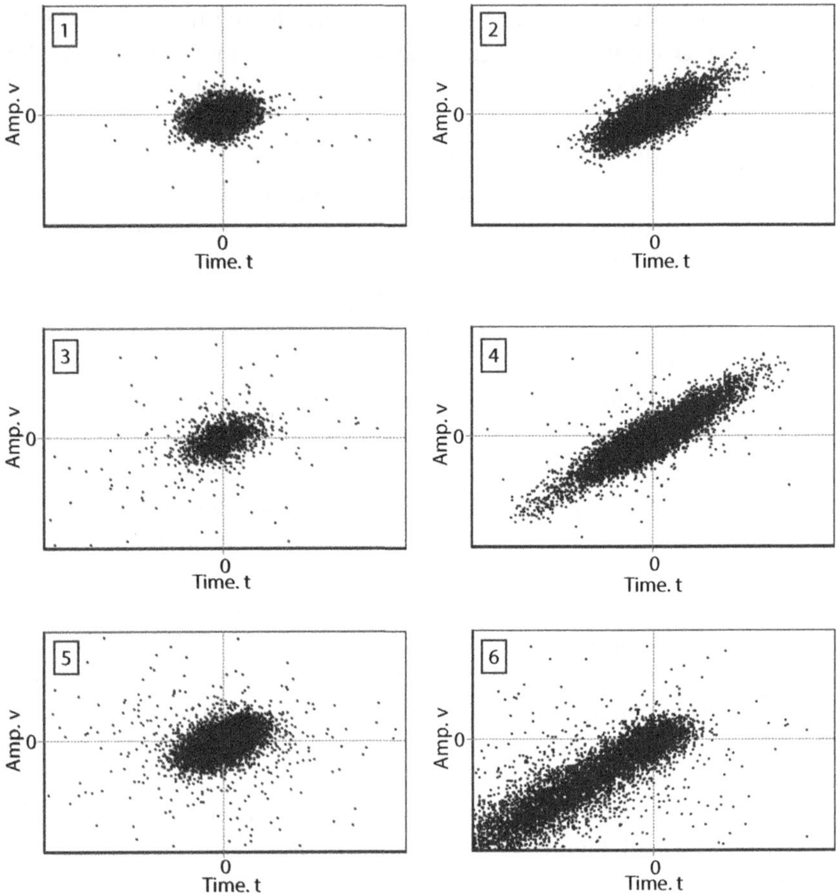

Figure 7.4 Screenshots of the restored core-medium state attractors at various loading steps from the weak loads to the moment of the microfissuring failure.
Sutorminskoye core parameters: coring depth 2,640 m; porosity 16%; and permeability 0.014 μm^2:

1. initial loading moment, weak stresses;
2. start of microfissuring in the medium;
3. precursor of local failure in the medium;
4. formation of local microfissure network shortly before failure;
5. precursor of core failure;
6. formation of a continuous network of local microfissures and formation of a core failure macrofissure.

At the lower steps of mechanical loads up to 3.0 to 4.0 MPa, the rocks showed a slow response to the applied wave disturbances (oscillations and pulses) in the form of elastic rectangular pulses. At a constant load,

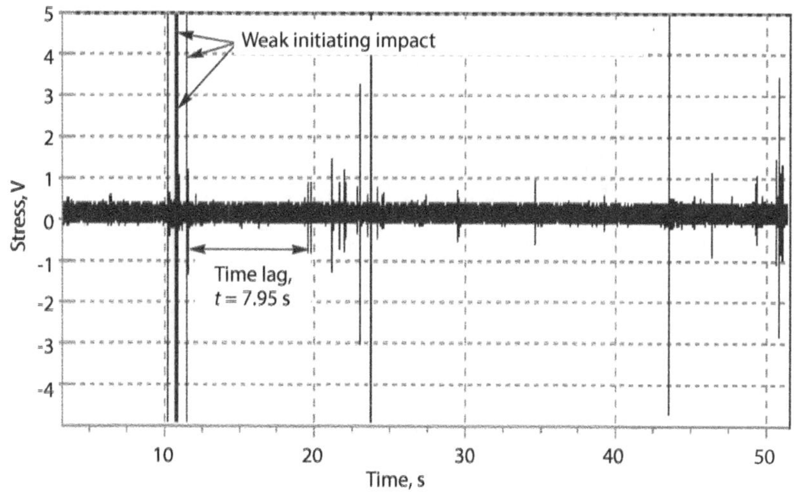

Figure 7.5 Initiation of AE signals from the oil-saturated core exposed to rectangular elastic pulses. Core porosity $m = 23\%$; permeability $k = 0.087\ \mu m^2$.

that is smaller than the failure load, the AE gradually fades out down to below the instrument discrimination threshold. Application of the external disturbance at 8.0 MPa results in a sharp AE signal increase (see Figure 7.5). The medium response to the external impact shows a rather long lag $\Delta t = 7.95$ s.

To assess the fluid saturation influence, the authors used specially prepared artificial clone cores manufactured in accordance with the above procedure. These clones are immune to the change of the fluid phase type.

Analysis of the acquired signal series allows a rather clear data separation according to varying sample saturation.

Figure 7.6 shows the time series (N) of energy releases (E, conventional units, c.u.) of the SE samples acquired immediately after the SE burst as the end load gradually increased to 6 MPa for both dry and water-saturated samples. The time series were processed on PC. To avoid the influence of various pickups, this series was recorded using a threshold discrimination of the recorded signals.

The fractal analysis results are as follows:

For the dry core, the correlation dimension is 6.25; the fractal dimension is 1.567; dimension of the embedding phase space being 10. For the water-saturated core, the parameters are, respectively, 4.805, 1.567, and 3.0. Thus, the water saturation improves significantly the degree of determinateness of the SE emissions (and the accompanying process of the core structure fissuring), improving self-organization of the system exposed to the external loading.

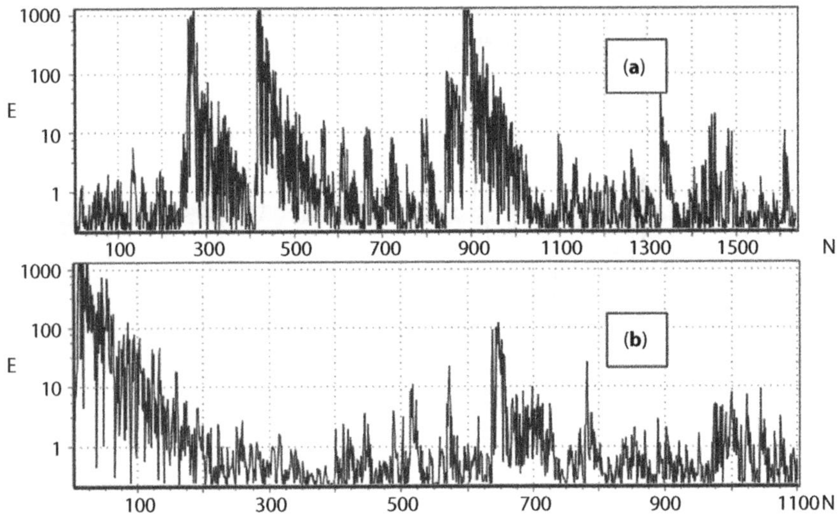

Figure 7.6 Variation of energy releases E along the time series N after the load increase to 6 MPa. Typical for a dry core (a) and water-saturated core (b).

Figure 7.7 shows the results of the SE signal wavelet analysis after the load was increased for the dry and water-saturated cores. The wavelet analysis diagrams present the progress of spectral variations of the signals over time. Appearance of the light strips suggests initiation of the energy release harmonics; variation of their thickness and the brightness on the ordinate axis shows the power distribution over the spectrum.

Comparison of the resulting patterns shows that, for the dry cores, stress relaxation and energy release, accompanied by the post-loading SE emission, runs rather smoothly; the spectrum frequency composition remains the same. As to the water-saturated cores, this process differs noticeably, its runs much faster, with impoverishment of the high-frequency harmonics in the spectrum over time.

Figure 7.8 shows the energy, frequency features of AE radiation, and the manner the mechanical stress energy accumulates in the rock samples with different saturations at loading stages from 6.0 to 7.0 MPa. Similar core samples were used in the experiments, with about 30% porosity and about 0.09 μm^2 permeability. The dispersion values of AE radiation for the oil- and water-saturated and dry cores are, respectively, 1.98, 1.23, and 1.07.

The results suggest that the properties of the pore-filling fluid have a significant influence on the microfissuring processes as well as the frequency spectrum of the radiation in the rocks that bring about the AE radiation. Results of the investigations on a series of oil-saturated cores showed a

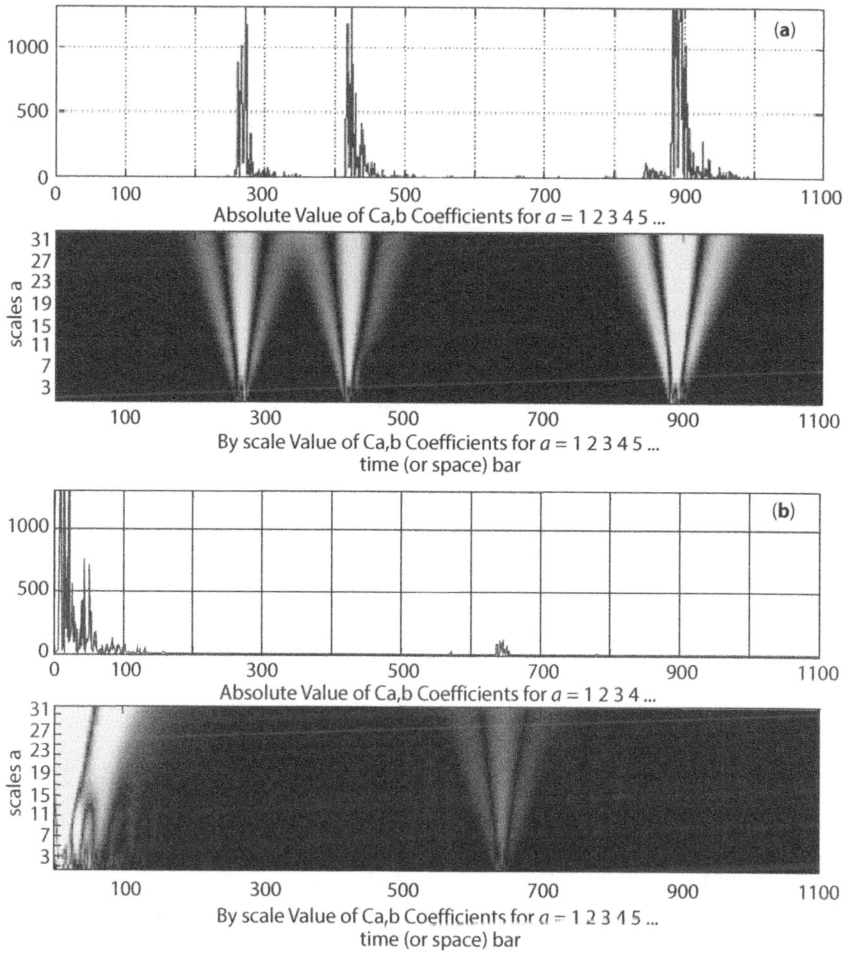

Figure 7.7 Results of AE signal wavelet analysis after the load increase, type results for dry cores (a) and water-saturated (b) cores.

significantly different nature of the SE radiation under the load as compared to the dry and water-saturated cores.

The water-saturated medium exposed to a mechanical load (see Figures 7.8a and b) releases energy regularly and monotonically as AE signal radiation.

For the oil-saturated medium, the energy exchange volatility increases noticeably. Here, one can observe already a noticeable stepwise elastic energy radiation exhibiting a higher dispersion and lower average energy. Summation of the AE energy at all the loading stages shows that the oil-saturated medium releases in the course of loading less energy in the form

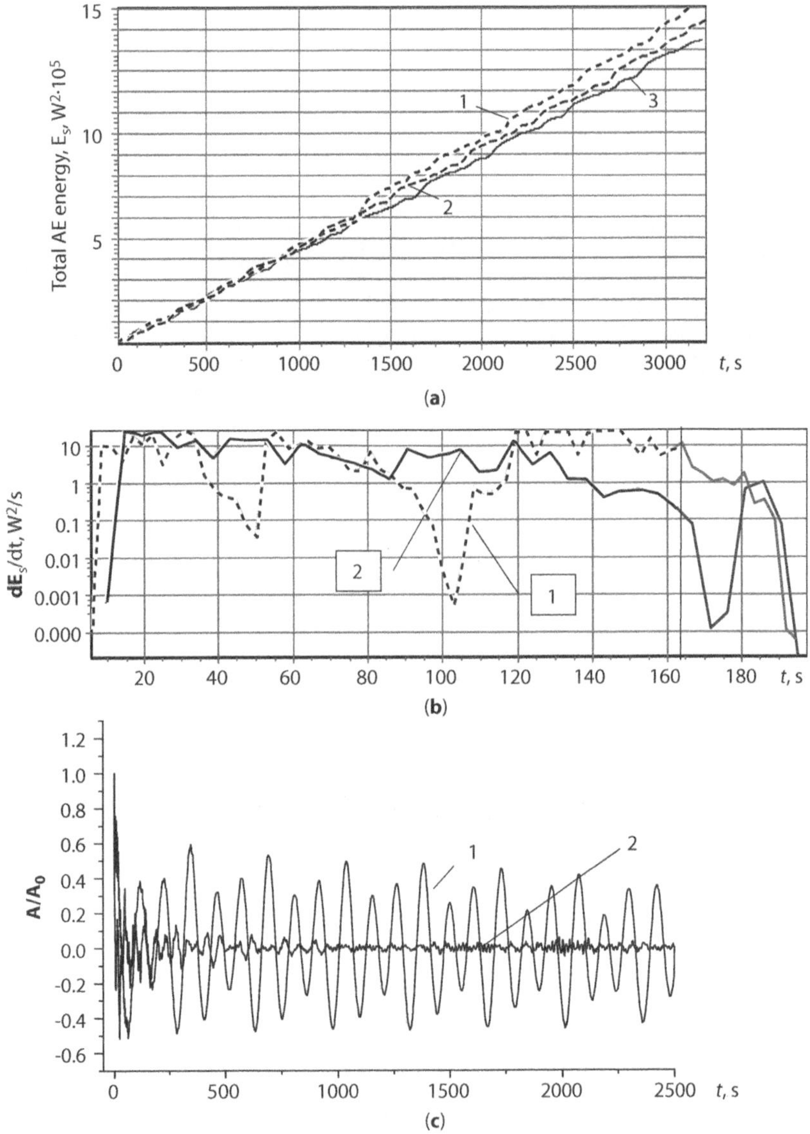

Figure 7.8 Specific AE energies and frequencies and mechanical stress energy accumulation at increasing loads for cores with different saturation: 1, oil-saturated; 2, water-saturated; 3, dry cores. (a) accumulation of AE energy Es over time; (b) variation of AE energy release rate dE/dt over time; (c) dynamics of auto-correlation function of the AE time series early section (A/A_0 is a normalized auto-correlation function; A is a function value at time *t*; A0 is a function value at time *t* = 0).

of AE as if accumulating it. However, the accumulation process is accompanied by discontinuous metastable states. Frequency features of the acoustic energy radiation released by the oil-saturated medium are important indicator of this process.

Dynamics of the autocorrelation function of the AE time series for the cores with different pore fillers (Figure 7.8c) common to all the experiments confirm this finding. Note that the oil-saturated medium generates an acoustic signal where the low-frequency harmonics prevail as compared to the signals from the water-saturated medium, especially at the early section of the time series where the response energy recorded by AE has its maximum value.

Figure 7.9 shows the wavelet analysis patterns for the SE signals radiated by a failing oil-saturated core. They suggest that the SE-release spectrum is impoverished in high frequencies and energy as compared to the radiation spectra from the failing water-saturated cores. Figure 7.10 shows relationship of the AE radiation energy initiated by a weak pulse wave impact for the cores with different saturations. Here, we also used cores with about 30% porosity and about 0.09 μm^2 permeability. AE radiation dispersions for the oil- and water-saturated and dry cores (see Figure 7.10a, section III) are, respectively, 2.78, 1.04, and 1.60.

As shown in Figure 7.10a, the energy accumulated in the medium, while under loading, is released with higher AE energy indicators. The observed

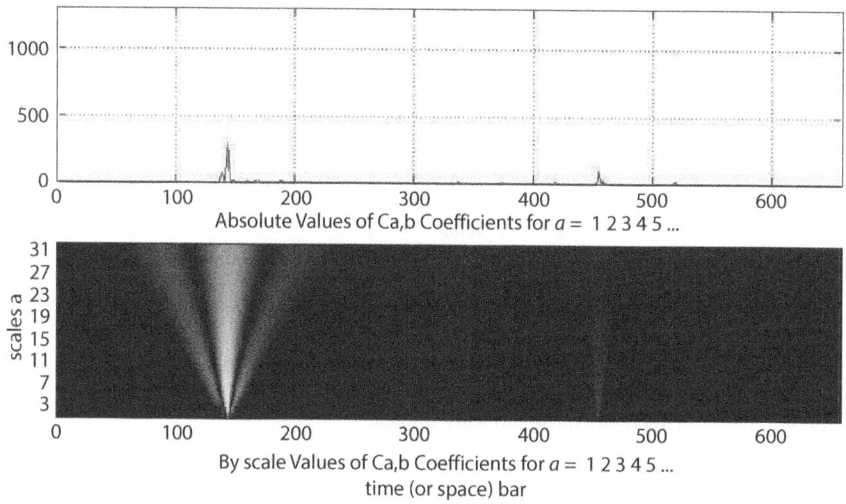

Figure 7.9 Results of the post-loading AE signal wavelet analysis typical for oil-saturated cores.

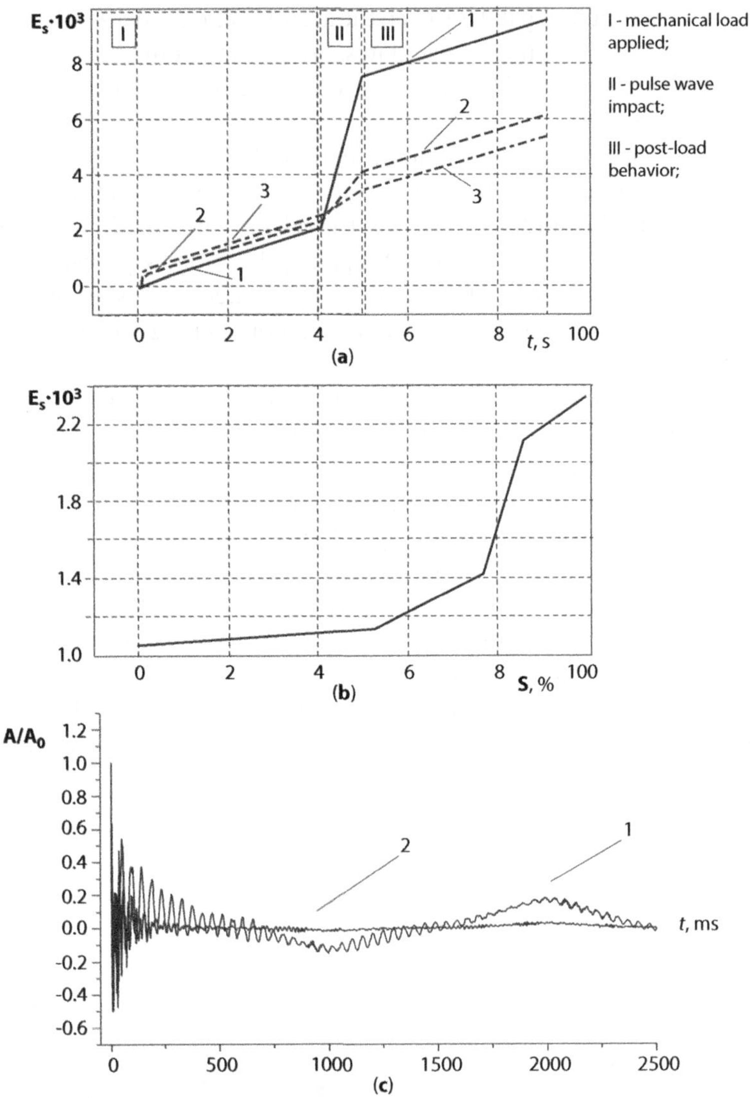

Figure 7.10 AE energy initiated by weak pulse wave impact, for cores with different saturations: 1, oil-saturated; 2, water-saturated; and 3, dry cores. (a) accumulated AE energy for oil-, water-saturated and dry cores; (b) relationship of AE signal energy vs. core saturation (S, %); (c) normalized autocorrelation function of the AE time series for oil- and water-saturated cores exposed to wave impact.

AE radiation anomaly is reflected in the fact that AE energy values that are smaller than those for the water-saturated cores and dry cores prior to the treatment increase significantly after the initiating the wave treatment.

The higher the medium oil saturation, the stronger is the manifestation of this anomaly (Figure 7.10b).

In addition, the autocorrelation function of the AE time series exhibits significantly different values for the oil-saturated and water-saturated cores (Figure 7.10c).

The fractal analysis results in the following parameters.

For the oil-saturated cores, the correlation dimension is 5.73; fractal dimension is 1.567; dimension of the embedding-phase space being 6.0. Hence, oil saturation brings about a substantial determinateness in the core structure changes, when under load, which appears to confirm the onset of a noticeable metastability.

If one compares efficiency of the regular harmonic and pulse wave external impacts, the can see that in all the experiments, the stressed reservoir medium responded to the weak impact earlier and more intensely when exposed to the rectangular (or triangular) pulses disturbance spectrum which has multiple frequencies in contrast to the single-frequency harmonic treatment.

We also assessed the influence of the core permeability on the intensity and nature of the detected phenomenon. We have not detected any noticeable permeability influence within the investigated range of the core reservoir properties.

7.2.2 Mechanism of Receptive Accumulation of Mechanical Stress Energy in Failing Oil-Saturated Media

As a result of investigating seismoacoustic radiations from the oil-saturated media exposed to the internal stress disturbances and external impacts, the authors identified a *new property of the rock energy exchange that implies a receptive energy accumulation* in the oil-saturated medium.

The saturated geological medium is represented by a hierarchy of fractal structures of all scales. The stress-strain state and its evolution due to the onset of a new fissured areas and new flow fields are subject to interaction of micro- and macrostructures as well as relation and interaction of all the three phases, including solid rock matrix, liquid fluid, and gas [9, 25].

General change of the medium state is subject to a great extent not only on the rock matrix strength and stress-strain behavior but also on the integrated surface interactions of all phases. In case of structured oils, volume interactions and physical, chemical, and mechanical interactions of all phases are also involved. Development of new fissures within the medium requires not only mechanical matrix deformation but also intrusion of the liquid oil, water, or gas phase into the volume voids. When no gas is present, the new surfaces must be wetted with liquid phases, their mobility ensuring the liquid phase flow. The phases exhibit their specific physical and chemical interaction with the solid phase that can be either oil-wet or water-wet. Liquid phase viscosity changes differently in carbonate and terrigenous reservoirs. All the phenomena that take place in these circumstances are interrelated and interdependent.

As a result of genesis and permanent internal thermodynamic process, productive fractured porous oil-, water-, and gas-saturated reservoirs featuring compound stresses possess their space, time, and function structure. One can, therefore, classify them as a self-organizing systems. Main fundamental properties of such systems that determine their functioning and mechanism of response to the external impacts, include *openness*, *non-linearity*, *dissipativity*, and *non-equilibrium* [37].

Actually, the reservoir medium *in situ* is open, and constant exchange in energy and matter with the environment brings about unstable states in it. Therefore, the non-linear dissipative reservoir rock medium is in a set of interrelated metastable equilibrium states that constantly interchange in various space and time scales.

Let us determine the mechanism responsible for the above experimental laboratory and field results, starting from the processes occurring at the microvolume level of the geological medium up to the generation of secondary radiations from the medium. Let us consider treatment with waves of harmonic elastic oscillations or pulse wave treatment, keeping in mind that the suggested consideration holds, to a certain extent, also for the electromagnetic as well as treatment of any other physical nature.

Mechanical stress fields always exist at the micro- and macrolevels of the geological medium organization, given their specific configuration of the solid phase surface and fluid saturations, physical and chemical interaction of the fluids with solid surfaces, and between themselves. When considering these stress fields, one should allow potential formation of a new surface due to fissuring and potential free gas liberation from the pore-filling fluids. Stress fields change constantly. This results in the formation and disappearance of non-metastable states distributed within the medium. Lifetimes of individual metastable states are subject to the distribution and local stability of the existing mechanical stresses, fields of fractures, and saturations.

Oil saturation is a special factor that has an influence on the duration and space scale of the medium metastable-state evolution.

Most of the rock minerals have water-wet surfaces; water saturation favors fracturing at stress disturbances. In water-saturated media, metastable states arise and disappear quickly (at low microlevels of the medium structure) and do not cause any noticeable disturbances at macrolevels. Oil phase that competes with water makes the surface of the solid phase oil-wet and slows down the spontaneous transition to the equilibrium through formation of new surfaces, i.e., fracture borders.

This can be tracked in noticeable change of the nature of acoustic radiations both in the results of the above laboratory experiments where, at the rise of mechanical stress, one can notice SAE bursts separated by rather long time intervals in oil-saturated cores and in the extensive field observations that capture SAE radiation in the oil-saturated areas of the medium featuring high dispersion. These features allow space delineation of the oil-saturated reservoir zones in the field and assess oil saturations of the productive reservoirs [1]. In addition to oil wettability, other "order parameters" are also possible. They can be linked to the specificity of the oil-saturated medium. They include, in the first place, gas saturation of the oil phase that can favor, in certain conditions, a high-intensity stress-energy radiation when a long metastable state is disturbed.

Sufficient static-stress gradients existing in the geological medium is required to cause such phenomena.

For this reason, a *receptive process of* mechanical stress *energy accumulation* occurs in the oil-saturated media, giving rise to the hysteretic phenomena. These media can accommodate sharp stress gradients and long-lived non-equilibrium states at all levels of the medium structure. Energy accumulation implies slowdown (as compared to the media saturated with other fluids) of the energy exchange between those zones of the structure and the external environment of acoustic and seismoacoustic radiations at constant or growing stress in the medium.

A rather weak though "informative" external impact that will be a receptor for the medium state, can upset the non-equilibrium state of the microvolume cell. Due to the *fractality and non-linearity* of the rock structure, the microprocesses *can become coherent, congruent, and synchronized.* They can give rise to a cooperative interaction and common rate of evolution. Fractality ensures maximum interaction of the structures and processes of various space dimension. The impact effect starts at microlevels and quickly moves to macrolevels, releasing large amount of accumulated energy, the medium having a self-oscillatory low-frequency response to the external impact. Coherence of the interrelated phenomena, such as pulses of mechanical stresses, dilatancy and flow processes, and gas-liberation processes, determines the intensity and scale of self-oscillatory processes occurring in the reservoirs. They give rise to hysteretic phenomena that manifest, for example, in low-frequency radiations from the oil-saturated reservoirs and large-scale oscillations of the fissuring fields.

State-change processes occurring in the non-linear reservoir media must feature a non-linear transient characteristic that can be described by a certain threshold, namely, the state changes abruptly while the external conditions change smoothly. The above results confirm that the elastic oscillation treatment has a threshold nature [15, 27, 30]. However, weak disturbances in the process of general changes can grow stronger. The external low-energy impact is a “trigger” that disturbs quasi-stationary states. The reservoir medium goes to excited state and, while moving to another state, it releases energy and generates damping oscillations. Elastic oscillations and pulsating that starts in the rock medium at the microlevel cause abrupt redistribution of the medium stress. This creates additional fissuring, initiates movement of the fluid interfaces that move faster when the new fracture voids get filled with fluids. Secondary pressure oscillations are informational nuclei of the reservoir *trigger response*. Reservoir-medium fractality and dissipativity give quick rise to secondary pulsating at all levels of the structural hierarchy, starting from the microlevel up to the largest fractures and blocks. At the macrolevel, the reservoir response takes form of low-frequency waves. The reservoir medium has a lag to respond to the treatment.

Thus, integration of the laboratory and field experimental investigations allows the authors to claim that they have discovered a new property of the energy exchange between the oil-saturated rocks that consists in the following.

Oil-saturated rocks are capable of receptively accumulating the mechanical stress energy. Such energy accumulation is characterized by the manifestation of metastable states along with hysteretic phenomena of wetting, gas liberation, and fracturing. In addition, this generates anomalous secondary wave radiation under wave impact (natural or man-made) when initiation and activation effects are achieved [28, 29].

The discovered new property describes and explains the class of phenomena and events observed in natural rocks and related to their oil saturation. Moreover, given large number of fields observed within the confines of a petroleum accumulation, the total response of the oil-saturated fields within the medium featured a low frequency [1]. In this case, we believe that the low-frequency Hines contrasts [41] caused by the impact of elastic waves on the oil-saturated zones with unsteady interface fronts, can serve as a microlevel response “information nuclei”. When observing strictly local medium microvolumes in the well logs [12], we can observe growing AE energy that achieve its maximum in the KHz spectrum component. We can capture the microfissure initiation processes where random fluctuations of the emission background, that permanently exists in the rock medium, can serve as "nuclei". Those fluctuations are subject to receptive selection.

Cumulative integration of synergy processes within the reservoir rock featuring the new property allows us to:

- manage a controlled wave treatment with the most informative amplitude-frequency mode resulting in the maximum reservoir response given its space-time state; whereas the expected efficiency is achieved at a minimum cost;
- develop efficient hydrocarbon-field production methods, and enhanced oil and gas recovery methods based on physical treatment;
- develop new tools and methods to identify the productive-reservoir intervals, and assess their oil saturations.

7.3 Justification of Rational Wave Treatment for the Near-Wellbore Zone and Entire Reservoir

Efficient wave treatments are available mostly for the near-wellbore zones. As to the individual areas of productive reservoirs and entire oil accumulations, their justification is yet at the initial stage.

Near-wellbore zone treatment, taking into its thermodynamic state and stress, and consequences of drilling operations and long-term production, on to face the following problems. Loose mud *cake; its adhesion to the surfaces of the rock pores and fractures diminishes; colloidal structures in the reservoir fluids break down; wedging effect occurs that forms and opens fractures; fluids start to flow in tight layers; damming effect of oil, water, and gas phases diminishes; chemicals and heat-transfer fluids are injected into tight layers; better vertical and horizontal treatment efficiency is needed.*

To justify rational regimes of the near-wellbore zone treatment to improve productivity, we must set up (1) threshold energy criteria; (2) efficient treatment depth; (3) conditions for efficient input of the elastic oscillations energy into the reservoir through the wellbore; (4) well-design stability factor in the case of exposure to elastic oscillations; and (5) choose operating amplitudes and frequencies, etc. These issues have been discussed in the literature [4, 11, 26, 32, etc].

Formation damage products are removed from the porous medium, as indicated above, at threshold values of $\ddot{\xi} \geq (0.1 \div 0.3)g$ and $\xi \geq (0.3 \div 1.0)\bar{d}$, and when pressure fluctuations exceed their threshold level of 0.02 MPa. Rational frequency range that allows a better removal at a smaller treatment energy varying from 20 to 300 Hz.

The efficient treatment depth varies from several meters to 10 and more meters. It is subject to space energy distribution of elastic oscillations and energy thresholds of the onset of operations.

These are the distances in the reservoir for the frequencies of the wavelength order. Near-wellbore zone treatment with elastic oscillations covers the medium-vibration area and wave formation area. In this case, the "vibrowave treatment" term is applicable. This is the term that the authors of the book decided to use in their investigation in 1990s. It describes the real depth of the near-wellbore zone covered with the treatment and brings the sense of optimality of its implementation in terms of frequency range.

Use of resonant excitation of elastic oscillations energy in the wellbore and in the reservoir improves the efficiency of elastic oscillations energy input into the reservoir through the wellbores as well as the efficient treatment depth.

Oscillation amplitudes about 3 to 8 MPa that the downhole oscillators must generate to threat the near-wellbore zones for about 8 to 15 hours not to cause any loss of cement sheath and wellbore integrity.

Combination of the elastic oscillation treatment with pressure drawdowns and overbalances as well as with injection of chemicals and heat-transfer fluids improves both vertical and horizontal productive-reservoir coverage and several-fold increase in the well productivity. Pressure-drawdown and overbalance treatments are used as components of integrated operations implemented in a certain sequence, including multiple cyclic alternation.

Cyclic alternation of pressure drawdowns and overbalances is one of the most important operations that improves well treatment efficiency. To justify operational cycle parameters (their duration and number) one must take into account specificities of the reservoir-pressure behavior not only at the bottomhole. This can be measured by a downhole gauge, and also at various distances through hydrodynamic computations.

One can use a sequence of five operation stages: production, injection, production, injection, and production. Thus, one can discuss the dynamics of pressure distribution within the near-wellbore zone with the following type parameters: 30-min production times (pressure drawdown); 10-min injections; ratio of the injectivity (Q_3) and phase transmissibility (k_3/μ_3) during injection (overbalance) to production rate (Q_o) and phase transmissibility (k_o/μ_o) during production (pressure drawdown) is equal to $(Q_3{*}\mu_3{*}k_o)/(Q_o{*}\mu_o{*}k_3) = 2$. Formation piezoconductivity is 0.011 m^2/s.

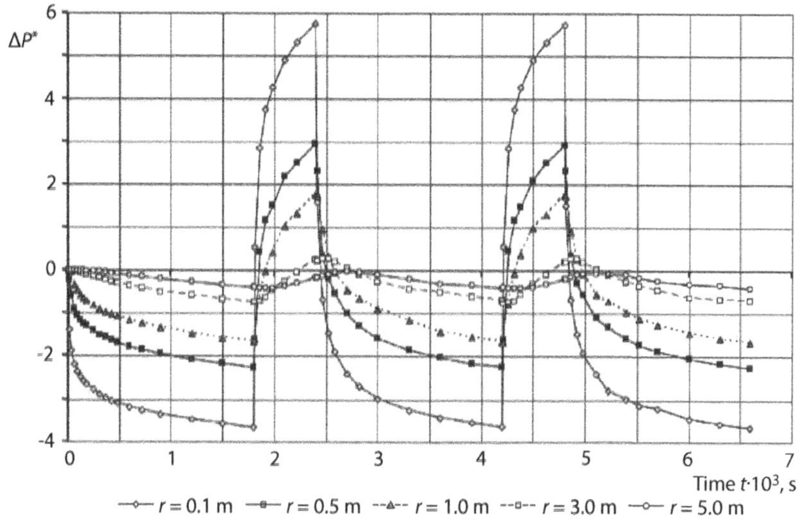

Figure 7.11 Dimensionless *in situ* ΔP^* pressure variation at various r distances from the wellbore axis at integrated wave and pressure-drawdown/overbalance treatment.

Total operation time is 30 × 3 + 10 × 2 = 110 min. Due to the short time, all this period corresponds to the transient flow. Computations were made by superposition method based on commonly used alternation of pressure-drawdown and overbalance stages.

Figure 7.11 shows dimensionless pressure. $\Delta P^* = \Delta P/(Q_o{}^*\mu_o/4^*\pi^*k_o{}^*h)$ behavior against steady-state pressure at the following distances: r = 0.1 m, 0.5 m, 1.0 m, 3.0 m, and 5.0 m from the axis of the well that goes through the terrigenous oil reservoir having parameters common for the Dachno-Repinskoye field, reservoir D_3fm (Kolganskaya Formation). As shown in the diagram, pressure variations are insignificant at the distance of 5 m from the wellbore axis, irrespective of the total process time. Hence, one can conclude that formation damage can be eliminated near the wellbore zone within 5 m, given the assumed initial conditions.

Figure 7.12 shows the curves of dimensionless pressure ΔP^* distribution in the reservoir that correspond to the various times of pressure-drawdown/overbalance alternations derived from the similar calculations. The calculated curves confirm that high- and low-pressure areas exist at some distance from the wellbore. At 5.0 m away from the wellbore, the pressure is always below the initial pressure (that corresponds to zero at the horizontal axis), i.e., the fluid does not move into the reservoir, and the mud cake does not dissipate in the near-wellbore zone.

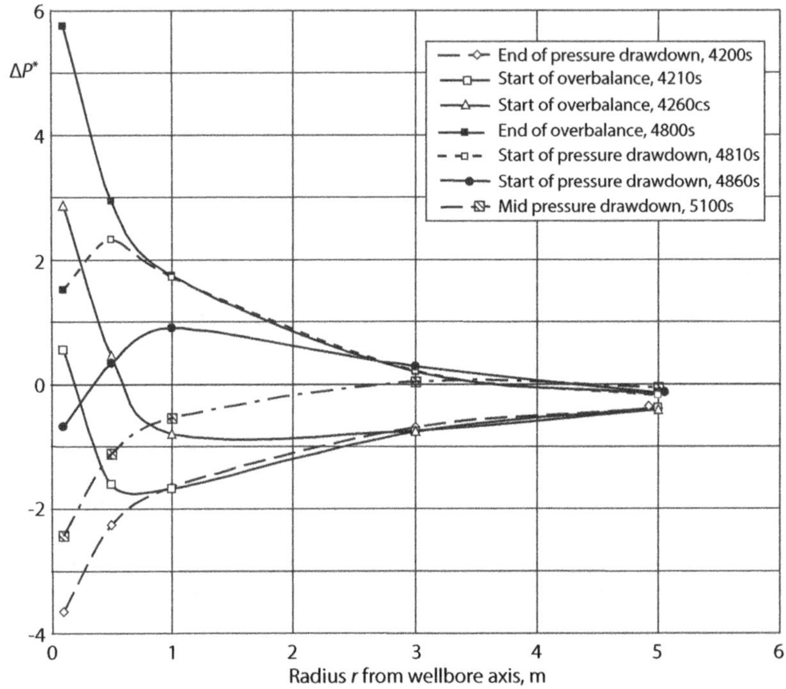

Figure 7.12 Distribution of dimensionless pressure ΔP^* in the reservoir under integrated wave and pressure-drawdown/overbalance treatment.

One can illustrate schematically the pressure variations that occur at the bottomhole and in the reservoir with respect to the reservoir pressure during pressure-drawdown and overbalance treatments as shown in Figure 7.13.

To simplify plotting the piezometric lines and to approximate the reality, it is assumed that the reservoir pressure amounts to ⅔ of the hydrostatic pressure in the well $P_{res} = ⅔\ P_{hs}$, pressure drawdown $\Delta P_{drd} = ½\ P_{res}$, and bottomhole pressure at overbalance $P_{ovb} = 1.5{*}P_{hs} = 2.25\ P_{res}$, i.e., overbalance is $\Delta P_{ovb} = P_{ovb} - P_{res} = 1.25\ P_{res}$. For example, in a well about 2,000-m deep filled with working fluid of 1000 kg/m^3 density, $P_{hs} \approx 20$ MPa, $P_{res} \approx 13.3$ MPa, $\Delta P_{drd} \approx 6.7$ MPa, and $P_{ovb} \approx 30$ MPa. The wellhead pressure at overbalance will be $P_{wh} = P_{ovb} - P_{hs} \approx 10$ MPa and usually corresponds to the casing-string pressure test, where ΔP_{drd} often does not exceed 6 or 7 MPa to prevent crossflows and sand production from poorly consolidated reservoirs.

In Figure 7.13, lines 1 and 2 correspond to the radial steady-state pressure distribution, when the bottomhole pressure is kept constant for a long time

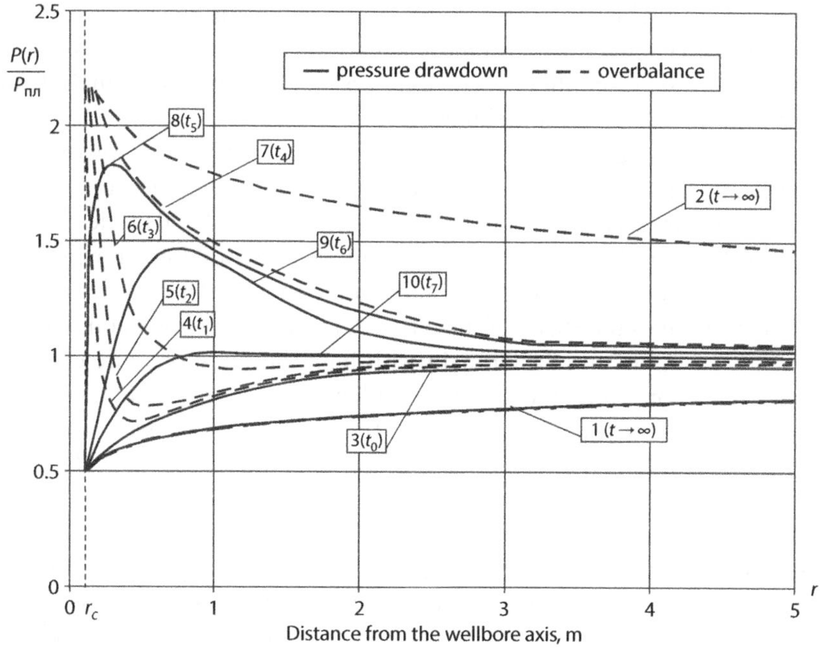

Figure 7.13 Pressure distribution in the reservoir under combined wave and pressure-drawdown/overbalance treatment.

at 0.5 P_{res} (lower line) or during long injection at P_{ovb} = 2.25 P_{res} (upper line), drainage area radius being 250 m. In the case pressure-drawdown or overbalance treatments are rather short, the piezometric lines shall lie between these lines; the shorter the treatments, the higher the angle of those lines.

At the first stage, combined wave and pressure-drawdown treatment is carried out to remove mud solids from the near-wellbore zone. The bottomhole pressure will drop to 0.5 P_{res}; reservoir fluids start to flow to the wellbore, and the pressure distribution in the near-wellbore zone takes the form of line 3 by the time t_0. The so-called "cone of depression" forms where the pressure-gradient vector is pointed from the reservoir toward the well.

At the second stage, the working fluid is injected into the well. The bottomhole pressure increases to P_{ovb} = 2.25 P_{res}, and the overbalance is maintained for the time needed to reach a quasi–steady-state pressure distribution in the near-wellbore zone (see Figure 7.13, curve 7). During this time, large potential amount of elastic energy is accumulated in the wellbore vicinity due to the liquid and rock compression. Given the finite porous-medium piezoconductivity, pressure redistribution lags behind, and by the time t_1, the piezometric line takes the shape of curve 4 that evolves into the curves 5 and 6, and then by the time t_4, it takes the shape of the curve 7, which is a cone of overbalance. During time $t_0 < t < t_3$, an annular domain with low intrapore pressure exists in the near-wellbore zone, and pressure-gradient vectors are pointed toward this domain from the reservoir and from the wellbore, respectively.

At the third stage, combined wave and pressure-drawdown treatment is carried out. The bottomhole pressure drops to hydrostatic P_{hs} and, by the time t_5, pressure distribution in the reservoir takes the shape of the curve 8. Then, as the fluid is produced, the bottomhole pressure drops further, and the piezometric line evolves into the curves 9 and 10. Afterward, as at the first stage, it takes the shape of the curve 3, i.e., cone of depression. During time $t_5 < t < t_7$, an annular domain with high intrapore pressure exists in the near-wellbore zone, and pressure-gradient vectors are pointed from this domain toward the reservoir and toward the wellbore, respectively. This phenomenon favors stress reduction in the working fluid and in the rocks.

The overbalance/pressure-drawdown treatment cycles are repeated several times. Large pressure changes occur during those cycles, in our case, pressure goes up to 1.5 P_{res} (see Figure 7.13, curves 4 and 8); this usually happens in the zone most damaged by mud solids. As compared to the hydroswabbing, when after injection the well is allowed to flow back, and the pressure drops to hydrostatic level, i.e., the pressure drop amounts to 0.75 P_{res}, in our case, the pressure drop is twice as large. Moreover, hydroswabbing, given the above conditions, covers only a very narrow zone adjacent to the wellbore. The fluid is continuously lost to the formation, entraining mud solids deeper to the near-wellbore zone. It is, therefore, reasonable to hydroswab mainly the wells drilled in the high-pressure reservoirs.

Increasing the injection pressure to $P_{ovb} = (2.5 \div 3)*P_{res}$, and pressure drawdown to $\Delta P_{dwd} = (0.75 \div 0.85)*P_{res}$, i.e., P_{dwd} be 0.15 to 0.25 P_{res}; then, the pressure changes will amount to 2 to 2.5 P_{res}. For a well about 2000-m deep, this corresponds to pressure drops up to 27 to 33 MPa. This drop occurs in the annular zone 0.2 to 0.5-m wide, the pressure gradient can reach up to $\Delta P/\Delta r$ = 90 to 110 MPa/m. To reach such gradient, the wellhead injection pressure should be $P_{wh} = P_{ovb} - P_{hs} \approx$ 13 to 20 MPa. The casing string should be in a good working condition, or a packer should be used.

Thus, along with alternating fluid-flow direction in the near-wellbore zone, from time to time, the high-pressure or low-pressure zones form at some distance away from the wellbore, and the fluid flows out of these zones in all directions or flows into these zones. These alternating flows move the particle aggregates in the reservoir pore throats, squeeze them out from the pore throats. Combined with the wave impact, the aggregates disintegrate, and when the pressure drawdown is high enough, the long quasi–steady-state flow removes bridging materials from the reservoir.

Moreover, the high-pressure gradients cause diverse rock deformations. Combined with regular elastic oscillations treatment, they reduce the rock strength, open natural micro- and macrofractures and initiate new ones. In some cases, in order to improve these effects, one can carry out also the pulse and wave treatment, using downhole tools, such as valves and percussive mechanisms.

Cyclic change of deformation directions transforms the near-wellbore zone into a source of rather strong infrasonic elastic oscillations that can exert dilation wave impact on the reservoir.

The above processes in the near-wellbore zone will also take place in the reservoirs with various geological and physical characteristics, including carbonate reservoirs.

To confirm that the wave treatment combined with the pressure-drawdown and overbalance treatment described above eliminates the formation damage, we carried out some tests. We continuously recorded bottomhole-pressure variations with self-contained pressure gauges and processed them later, plotting bottomhole pressure-drawdown curves for individual pressure-drawdown stages.

Figure 7.14 shows the bottomhole pressure variation for the Dachno-Repinskoye well 376 that covers three stages of the integrated wave and overbalance/pressure-drawdown treatment.

Bottomhole pressures measured at the three liquid-production stages at pressure-drawdown were used to plot corresponding pressure-drawdown

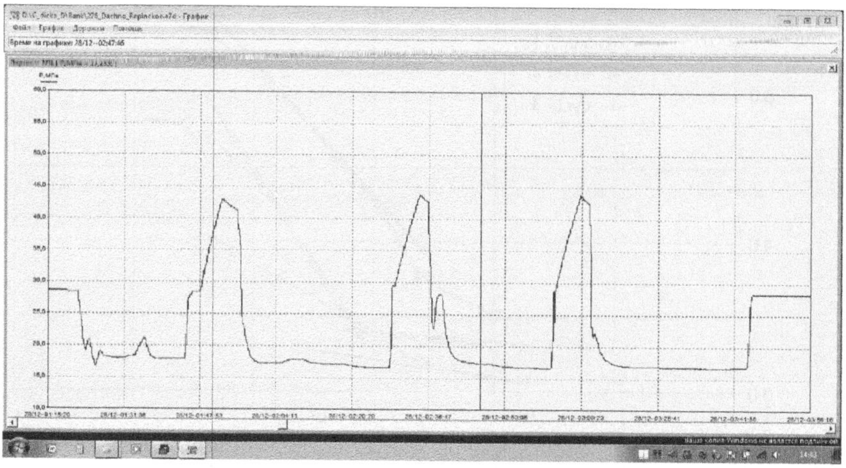

Figure 7.14 The bottomhole-pressure variation during the integrated wave and pressure drawdown/overbalance treatment of the Dachno-Repinskoye well 376.

curves in coordinates ΔP vs. log(t), that are shown in Figure 7.15. A short observation time is sufficient (within the production time at the previous rate) to ensure a reliable characterization of the near-wellbore zone state in these coordinates.

As shown in Figure 7.15 (curve 1), after the first stage of the overbalance/pressure-drawdown cycle, formation damage was eliminated in the close vicinity of the wellbore. Let this zone radius be R. Assuming the average reservoir piezoconductivity $\chi = 0.03$ m^2/s and curve inflection time $t_{infl} = 35$ s and $R = 1.5\sqrt{\chi t_n} \approx 1.5$ m. The zone located farther away performs much worse. After the second pressure drawdown/overbalance cycle, movement occurred only from the farther zone (see Figure 7.15, curve 2). As a result, the state of the near-wellbore zone worsened; however, after the third cycle, the entire near-wellbore zone of 5-m radius was free of bridging materials.

Thus, the above example shows that the near-wellbore zone state can get even worse at some treatment stage. A reasonable number of pressure-drawdown/overbalance cycles are required along with control in the course of treatment.

In practice, the number of cycles can go up to 10 and more. However, in order to improve the efficiency of the mud-particle removal from the reservoir, to increase the treatment zone radius and increase both vertical and horizontal productive-reservoir treatment coverage, one should control

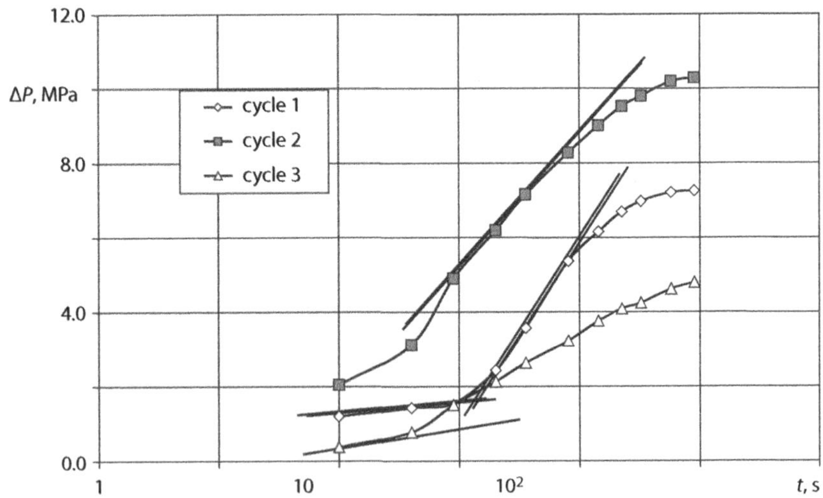

Figure 7.15 Bottomhole-pressure variation curves for well 376 in the Dachno-Repninskoye Field at overbalance stages during integrated wave treatment of its near-wellbore zone.

the balance between the pressure-drawdown and the overbalance stages, as well as parameters of the additional pulse and wave treatment.

Reasonable number of cycles, optimum timing of the stages, pressure-drawdown and overbalance values, as well as the parameters of the additional pulse and wave treatment are determined using special software.

The following scenarios are reasonable for the chemical injection in the field of elastic oscillations:

- Monocyclic treatment with injection of one slug or a continuous sequence of slugs;
- Multicyclic treatment;
- Spotting operations under subscenarios:
  - static chemical bath at hydrostatic pressure;
  - static chemical bath at overpessure;
  - dynamic chemical bath with cyclic pressure increase and decrease;
  - dynamic chemical bath with reciprocating flushing along the wellbore over a large interval (e.g., openhole or uncemented screen in the horizontal wells or sidetracks).

Wave treatment encourages not only deep chemical penetration into the streaks and low-permeability zones of the reservoir, and more uniform filling of the pore space with the chemicals. It also enhances their action and improves the efficiency of their implementation due to the synergy of impact elastic oscillations and chemicals. Timely implementation of pressure-drawdown and wave treatment facilitates removal of both liquid and solid or gel-like reaction products. This allows using chemicals or composites with much less additives up to refusal to introduce a number of expensive detergents, inhibitors, etc., that double or triple the cost of the chemical composites.

Among diverse generators of elastic oscillations the most reasonable one to treat the wells are the hydrodynamic oscillators that operate on the liquid-flow energy. They have a high efficiency of up to 60% and higher. Thus, the well treatments are combined with the scheduled well interventions and workovers as well most traditional bottomhole treatment methods.

7.3.1 Reservoir Treatment With Elastic Oscillations

Reservoir treatment with elastic oscillations has been implemented for several dozens of years, and methods of reservoir wave treatment with elastic oscillations have been described in the literature [2, 7, 22, 24, 33, 34, 36, 39, 45].

Results of our investigations [7, 14, 15, 30], along with those published by other authors [3, 11, 24, 32–34, etc.], including the derived energy criteria, allow pointing out the main effects that occur in the productive reservoirs exposed to the low-frequency oscillations at realistic amplitudes and frequencies:

- liberation of gas phase from the reservoir fluids in the form of microbubbles;
- improved flow mobility of residual oil along with the formation of mobile clusters and streams;
- improved oil relative permeability;
- initiation of oil displacement from the small pores and fissures;
- bringing bypassed oil to production; and
- improved reservoir properties (porosity and permeability) as a result of dilatation wave treatment;
- generation of new fractures in the dilatation zones.

Given complex geology of productive reservoirs and necessity to create elastic oscillations in the reservoir with the required energy parameters, justification of reasonable regimes of reservoir treatment with elastic oscillations is yet at the development stage. It is vital to find and justify new approaches to solve the above problems.

Optimization of the seismoacoustic reservoir treatment should include the acquisition and analysis of the geological information, directly or indirectly related to the variations a stress-strain state, fluid dynamics, and physical and chemical processes that take place in the reservoirs over time and within the confines of the accumulation exposed to the influence of the external and internal factors.

Informative parameters of the above processes include the signals of the emissions radiated by the Earth that reflect the processes that take place at all the scale levels of the medium fractal structure. These processes include deformations, initiation of fissures, faults of all sizes, generation and behavior of various heterogeneities, qualitative transformations of the structure states, etc. Time series of these signals are also fractal and carry important information on the current state of the system.

According to the main principle of synergetics [Haken], augmentation of complexity and orderliness (from chaotic components) have the same algorithm, irrespective of the nature of the systems where they are implemented.

Hence, the internal regularities of the Earth behavior are akin to the general regularities of the complex natural-systems behavior. The medium state goes through time activity cycles that repeat at various scales. When they start, during short periods of "calm", the system is at the nodal *bifurcation* points where the further evolution path is determined, whether it will improve its orderliness and accumulate energy, or it will get more chaotic and will release its internal energy. These bifurcation points are bound both to the time of the oil-saturated Earth evolution and to its space dimension.

As to the optimum treatment engineering, an important point is that one can efficiently control the self-development of the complex systems, stress-strain state of the medium, and its evolution by creating a rather low-energy wave and pulse disturbances at the bifurcation points.

We believe that, at the initial stages and in the course of the treatment implementation, time series of emission-radiation signals from certain medium volumes should be analyzed at various time scales. The purpose of such analysis is to detect internal hidden regularities in the behavior of saturated Earth exposed to combined stress, to determine various time dimensions of the hidden internal cycles of its current state, and to locate nodal bifurcation points of treatment which could be the most advantageous in terms of time and energy. These points are determined by time and are linked to a specific accumulation volume.

However, the most informative and efficient procedure is to analyze the time sequence of emission-radiation signals, while transforming the time scales that correspond to the energy and structure components of the hierarchical states of the matter. The time scale is sequentially reduced or increased in the ratio equal to number $\mathbf{\Phi} = (\sqrt{5} + 1)/2$.

This time transformation is based on the natural principle of self-sufficiency. It manifests in the evolution and structure of the complex non-linear systems, such as various natural formations, including the Earth. Application of this principle to the analysis ensures quick identification of hidden deep regularities and explicit relations.

Such analysis an determines the time points for the control action. In addition, this analysis allows identification of the frequencies that correspond to the distinctive time cycles of various time scales.

For this reason, in order to achieve a maximum medium response, it is preferable that the control energy action have a wave nature and several frequencies. The waves and/or periodical pulses must be emitted simultaneously at two or more frequencies that are derived from the emission analysis of radiations and correspond to the time dimensionalities of cycles of various scales.

Especial smallness of the treatment energy and the threshold treatment triggeredness allows estimation of the initial influence at microlevels on the molecular bonds in the fracture nuclei and interfacial tension. Thus, such medium characteristics as fractality and non-linearity bring about coherence, consistency, and synchronization of structural processes of all scale levels. Moreover, various mechanical, dilatation processes, fluid dynamics, phase exchange, etc., exhibit a cooperative interaction and common rate of development.

A trigger response occurs in the medium on the principle of resonance synchronization, i.e., the medium state changes at the main structure levels of various space scale at once. The Earth releases its proper energy, active fissuring takes place, and changes are under way in the saturation and flow fields. Thus, capillary imbibition processes start in the saturation discontinuities of the compacted zones, isolated clusters are mobilized, flow-pressure drops occur between zones with different permeabilities, and fluids start to flow faster. Under certain conditions, these changes contribute also to the hydrocarbon inflow from the mantle depths that replenish the accumulation reserves.

Oil production is supported by a more uniform and efficient oil displacement from the reservoirs, thus enhancing the oil flow to the wells.

One can discuss some real cases that can be explained by manifestation of the general regularities in the Earth behavior as a complex natural system, including cyclicity of its state, and going through the temporal cycles of activity and nodal points of *bifurcation* that repeats at various scales.

1. When monitoring the induced fracturing and hydraulic fracturing operations in the producing wells using seismic side scan sonar in the horizontal plane that goes through the wellbore, one could detect well-pronounced distinctive cyclic oscillations of the SAE intensity. They concentrate in certain areas in the near-wellbore zone when the stress state changes due to the fracturing fluid injection [30].
2. Cyclic antiphase variations of various frequency components of the subsurface seismic emission with respect to the solar component were observed when investigated non-linear effects of the seismoacoustic emission excitation in the Institute of Earth's Physics of the Russian Academy of Sciences but did not get any explanation [5, 6]. As applied to the above mechanism, this cyclicity and antiphaseness arise from the possibility to set up our own cyclic dynamic state of the subsurface medium at the observation point, its cyclicity parameters arising out of interaction with the solar component of the external disturbance.

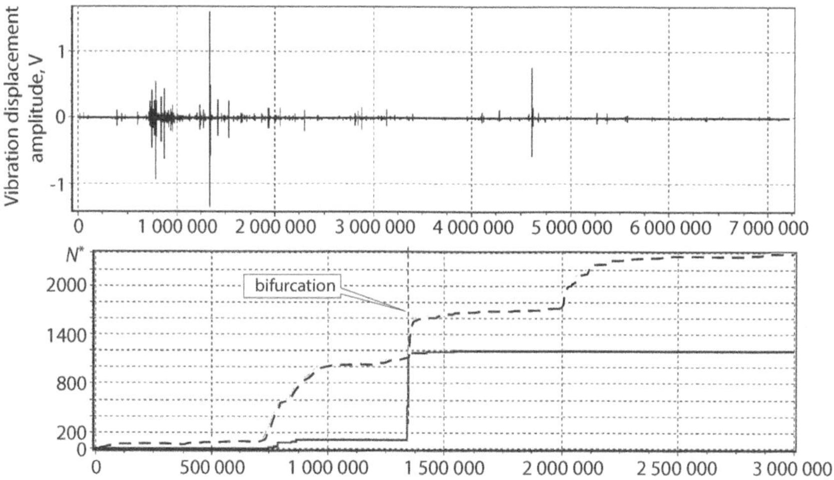

Figure 7.16 Variation of the AE signal count N* functional during loading at two levels of the structure scale (core: 100% oil saturation, initial loading stage).

The authors processed on computers the AE signals acquired in the course of the above experimental tests when we mechanically loaded the oil-saturated cores having different levels of saturation (see Section 7.2). When doing this, we assessed variation over time of the N* functional, which is the total count of the AE signals. Figures 7.16 to 7.19 show the results of the computer processing of the AE signals acquired in the process of mechanical loading of oil-saturated cores having different levels of saturation. Figures 7.20 and 7.21 show the processing results for 100% water-saturated and dry cores, respectively.

In order to improve the analysis informativeness using scale variation method on the principle of universal natural self-sufficiency, we extracted from the initial signal the components that corresponded to two structural levels of different scale, for which the count functionals were estimated separately over time. To control correctness, the Fourier transform with spectral analysis was applied to the extracted components. We did not detect any alias components and other anomalous spectral changes of the signals. The above curves show the particular stress dynamics of the Earth with different fluid saturations, specific processes of energy accumulation, and release while generating seismoacoustic emissions.

Institute of Advanced Petroleum Technologies (INNT) of the Russian Academy of Natural Sciences ran seismic surveys to establish particular features of the *in situ* seismic emissions (SE) [30, 31]. Those seismic surveys found out that the reservoir areas with higher oil saturations exhibit higher dispersion and lower average SE energy, which is in good agreement with the results of the above laboratory tests. Analysis of the autocorrelation functions (ACF) of the seismic emissions observed near the oil producers that are completely or partially watered out, also showed full compliance with the above results.

The applied AE signal analysis and the above data confirm and give a new insight into the experimental results obtained by the authors. They identify specifics of the Earth dynamics in terms of the new property of the oil-saturated rocks (article on Geophysics). Moreover, the detected time periodicity and other findings allow further exploratory analysis aimed to determine the frequency ranges where the external impact could be synchronized with the times and bifurcation-energy cycles of the medium state, and where an optimum treatment could be performed.

Thus, Figure 7.17 shows distinctly the cycles and stages of the energy accumulation and release processes in the course of loading. At periods of elevated AE activity alternate with longer quiet periods of energy accumulation. In addition, as shown in Figure 7.17, the count curves connect in a singular time point for different scale levels just at the moment of the maximum energy release simultaneously at both levels. This corresponds

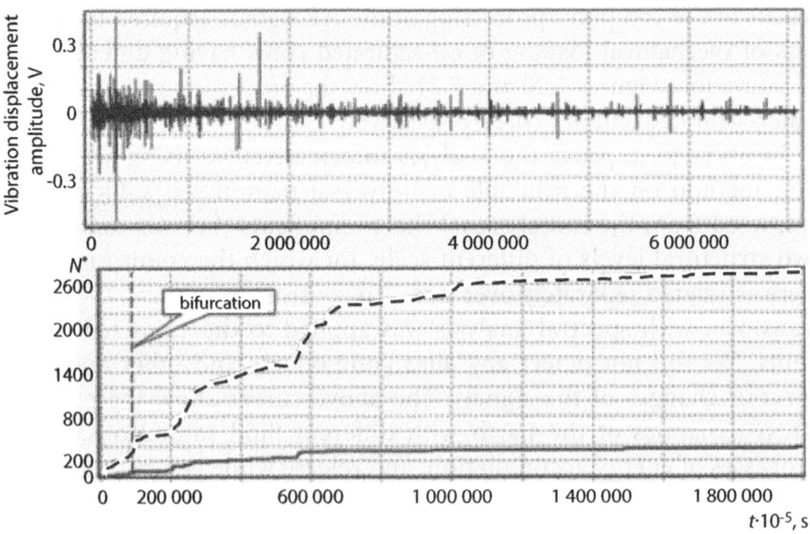

Figure 7.17 Variation of the AE signal count N* functional during loading at two levels of the structure scale (core: 100% oil saturation, intermediate second loading stage).

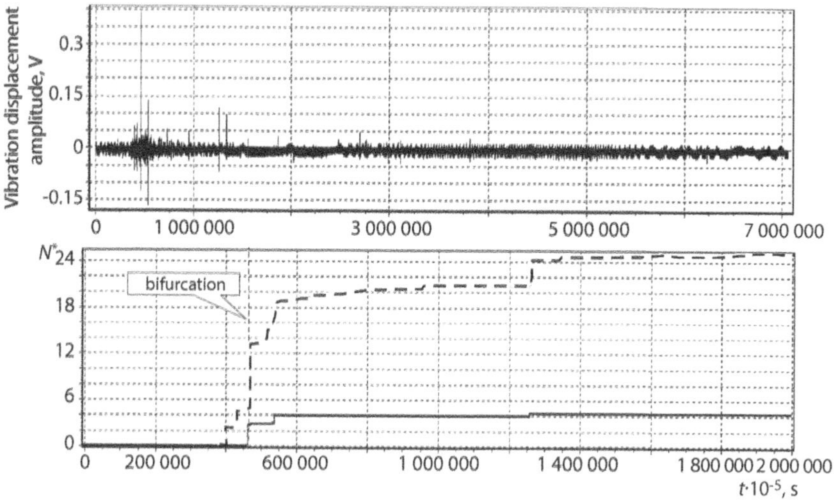

Figure 7.18 Variation of the AE signal count N* functional during loading at two levels of the structure scale (core: 45% oil saturation, initial loading stage).

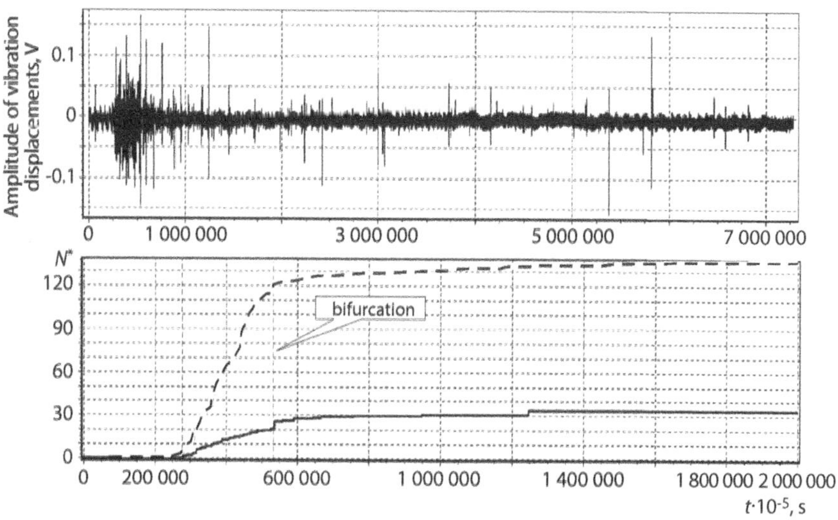

Figure 7.19 Variation of the AE signal count N* functional during loading at two levels of the structure scale (core: 45% oil saturation, intermediate second loading stage).

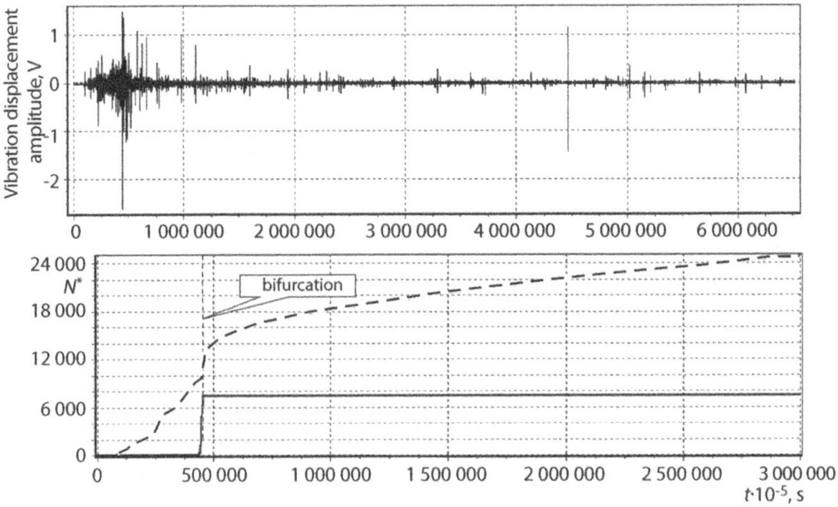

Figure 7.20 Variation of the AE signal count N* functional during loading at two levels of the structure scale (core: 100% oil saturation, initial loading stage).

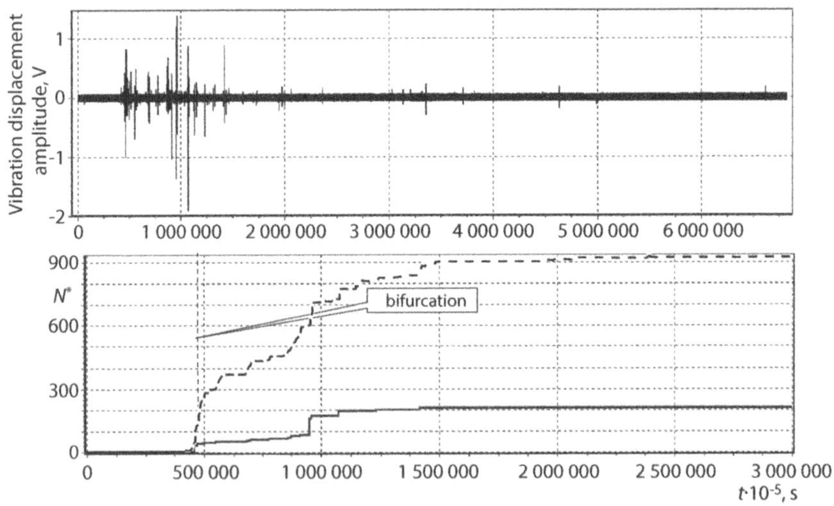

Figure 7.21 Variation of the AE signal count N* functional during loading at two levels of the structure scale (dry core, initial loading stage).

to the onset of coherence and consistency between the processes in these levels, maximum energy, and information exchange between them. This allows defining this time point as a point of the medium-state bifurcation. Similar results were obtained also for the threshold oil- and water-saturated cores, where oil saturation was 45%, though the curves did not connect in the bifurcation points. The 100% water-saturated core monotonously releases AE energy in the course of loading, without energy accumulation periods. However, the analysis that we performed at two structure levels rather clearly diagnoses the moment when AE energy release grows simultaneously, on the principle of resonance synchronization. This moment corresponds to the appearance of the point of bifurcation in the core matter. This is also valid to a certain degree for the dry core (see Figure 7.21).

In the process of trigger treatment, active fracturing and restructuring of the elastic medium matrix results in feeding the elastic waves with potential energy released by the Earth. When elastic oscillations propagate through the reservoir, the density of their total energy in terms of spectrum should diminish much less as compared to their propagation through the "passive" medium. In certain cases, it should even increase. Variation of the energy release with time in the process of Earthquakes confirms potential occurrence of such events. Thus, observation of a number of the most violent earthquakes allowed to establish that the P-waves consist, as a rule, of multiple outbursts of gradually increasing intensity that are originated not in the earthquake focus but by the subsources located on the wave traces.

To estimate the depth of the efficient wave treatment and space-energy distribution of the elastic oscillation field in the oil- and water-saturated reservoirs, including the trigger effects, the book authors simulated the elastic wave propagation in specific environment of their occurrence [14, 30]. The Monte Carlo simulation allows using real saturated-reservoir parameters such as porous-medium permeabilities, porosities, fluid dynamic viscosity, matrix and fluid densities, matrix bulk modulus, shear modulus for the dry porous medium, compressibility factors of the phases as well as using the real geometries of productive reservoirs, assign emitter parameters, elastic oscillation excitation power, and frequency. Probabilistic events of the additional acoustic-energy dissipation and radiation were derived from the power probability functions such as $p(E) \sim E^{-1-\gamma}$, where the fractal indicators and amplitude factors were based on the analysis of experimental data of the AE spectra acquired when running treatments in the wells.

Figure 7.22 shows the isolines of the oscillation acceleration and displacement threshold values based on the maps of space-energy distribution of the elastic oscillations in the reservoir. They were obtained at various times of the AE evolution calculated on the models using experimental data for the real media.

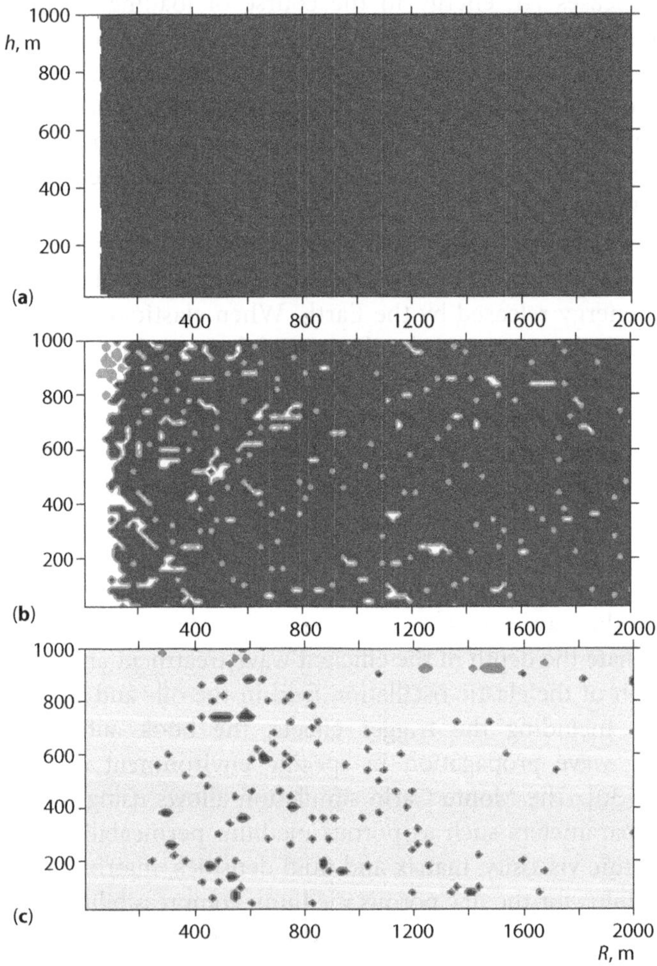

Figure 7.22 Isolines of threshold values of the oscillatory acceleration and displacement based on the maps of space-energy distribution of elastic oscillations in the reservoir given reservoir thickness $h = 20$ m, oscillation excitation frequency $f = 50$ Hz: (a) regardless of the reservoir seismic activity; (b) initial stage of the elastic oscillation excitation including the reservoir seismic activity; (c) full reservoir coverage with trigger effects at the advanced "pumping" stage; □—areas where the $\ddot{\xi}$ and ξ values exceed the threshold values of $\ddot{\xi}^* = 0.1$ g and $\xi^* = 2.0 \cdot 10^{-5}$ m; ■—areas where the $\ddot{\xi}^*$ and ξ^* do not reach threshold values.

The resulting estimates of the treatment depth including the observed specifics of the elastic oscillation propagation in the saturated-reservoir media demonstrate the specifics of the feeding with elastic oscillations as they go through the microseismically active reservoir medium.

The calculation results reveal that the required threshold parameters are achievable within the reservoir confines (to ensure efficiency) when performing wave treatments using oscillators at the well bottomholes, given the optimum selected treatment amplitudes and frequencies.

Annular subvertical fractured zones can be detected in the geological media that under productive reservoirs down to the crystalline basement depths and deeper. Those zones rise from the depths of the crystalline basement and penetrate into the oil and gas reservoirs [18, 19].

When elastic oscillations or pulses amplitudes of which exceed threshold values are used to treat local discontinuously strained areas that consist of alternating stress and destress areas, they initiate and intensify dynamic restructuring of the Earth matrix that accompany fracturing processes. These processes are associated with a strong phenomenon of high-pressure fluid degassing along the newly generated channels and fractures. It is based on the resonance synchronization principle and explains the high energetics and avalanche-like response, feeds energy to the propagating pulses from the potential energy due to the medium destress. Capillary imbibition starts in the saturation discontinuities of the compacted areas of subvertical zones, isolated fluid clusters are mobilized, and flow processes start and pick up speed. Saturation of the subvertical zones becomes continuous again; they are additionally saturated with deep-seated hydrocarbons that rise by the expanding and newly formed channels into the reservoir and are produced by the nearby wells.

The developed calculation procedure was tested in the oilfields of the Urals-Volga region. It is a constituent part of the software package to assess rational wave-treatment regimes for productive reservoirs.

Based on the extensive investigations aimed to optimize the wave treatment of the oil-reservoirs, the authors recommend the following sequence of operations:

1. *Run a geophysical study including reservoir tomography with seismic location of emission sources (SLOE) and seismic side scanner (SLOB) to detect fracturing fields and stress fields in the accumulation, to detect anomalously stressed zones, and to determine the main fracturing vector for the entire accumulation.*
2. *Assess conditions to run vibroseismic treatment; to determine the treatment exposure areas, when needed; to run operations*

to enlarge the areas of anomalous stress including cyclic water flooding; hydrodynamic treatment of the reservoirs within the accumulation.

3. *Determine threshold energy values of the flow and other effects of the wave treatment with elastic oscillations.*
4. *Justify the frequency mode and multiple frequencies of the wave treatment, to assess the depth of efficient treatment, and to determine the reservoir coverage including achievement of threshold values at which the effects manifest in the reservoir medium.*
5. *Run vibroseismic treatment, get a simultaneous response for the reservoir tomography and set the treatment parameters (source locations, frequencies, pulse delay times, etc.), and determine the optimum modes, including treatment feasibility focusing on maximum response initiation with maximum variation in fracturing and merge of abnormally stressed zones.*
6. *Run vibroseis treatments at optimum regimes and achieve release of the proper reservoir energy and emergence of new open-fracturing fields in the required reservoir zones.*
7. *Run once again the geophysical studies over the entire accumulation, including medium tomography (SLOE and SLOB), to assess the acquired results that are used to determine the new state of the fracturing fields and stress fields, and to adjust the main fracturing vector in the accumulation.*
8. *The main vector is used to detect the high-stress zones and to assess areas exposed to the treatment and, when needed, a hydrodynamic reservoir treatment is run once again. A vibroseismic treatment is run simultaneously with reservoir tomography (SLOB and SLOE). Optimum treatment modes are updated.*
9. *Run a vibroseismic treatment at the updated optimum modes to initiate the new fracturing zones in the vector's direction.*
10. *Run operations in items 6 to 9 until maximum treatment result is achieved and maximum possible open-fracturing fields are initiated in the productive reservoir of the accumulation.*

The acquired results serve as a basis to run a pilot project on vibroseismic reservoir treatment.

The authors used the investigation results discussed to develop:

- new equipment and technologies to improve well productivities through reservoir treatment with elastic oscillations;

- procedures to monitor wave processes while maintaining a feedback with the treated medium;
- pilot technology to enhance oil recovery using an integrated wave treatment.

Moreover, the acquired results were used in the INNT RAEN to justify and optimize the SLOB and SLOE technologies.

References and Bibliography

1. Arutyunov, S.L., Davydov, V.F., Kuznetsov, O.L. *et al.*, The phenomenon of infrasonic-wave generation by a hydrocarbon reservoir, in: *Otkrytiye (Discovery)*, vol. 109, RANS, 1999.
2. Ashchepkov, M.Yu., Ashchepkov, Yu.S., Berezin, G.V., *New resource-saving technology for enhanced oil recovery and stimulation of oil production*, p. 52, M, VNIIOENG, 2001.
3. Barabanov, V.L. and Nikolaev, A.N., Oil-recovery enhancement by low-frequency seismic reservoir treatment, in: *TechnoEcoGeophysics*, pp. 31–37, Ukhta, USTU, 2002, new technologies to produce mineral resources in the XXI century. Ed. by O. L. Kuznetsova, member of the RANS.
4. Baryshnikov, A.Ya., Creep of concrete and reinforced concrete under forced cyclic deformations. *Beton i zhelezobeton (Concrete reinforced concrete)*, 12, 58–61, 1967.
5. Belyakov, A.S., Lavrov, V.S., Nikolaev, A.N., Acoustic resonance of oil deposits. *Rep. Acad. Sci.*, 397, 1, 101–102, 2004.
6. Belyakov, A.S., Lavrov, V.S., Nikolaev, A.V., Khudzinsky, L.L., Changes in the phase of diurnal underground sound variations on the day of the vernal equinox. *Rep. Acad. Sci.*, 357, 4, 531–534, 2000.
7. Belonenko, V.N., Pavlov, N.V., Mashurov, A.D. *et al.*, Results of application of vibroseismic technology in the Sutorminskoye field. *Neftepromyslovoye Delo.*, 8, 18–22, 2000.
8. Beresnev, I.A. and Nikolaev, A.V., Study of the passage of long-term vibration signals through the geophysical environment. *Izvestiya AN SSSR. Phys. Earth.*, 1, 86–89, 1990.
9. Builo, S.I. and Kozinkina, A.I., Revisiting the issue of assessing the damaged accumulation and the moment of transition from scattered to localized defect formation based on restored values of the flow of the acoustic-emission events. *Fizika Tverdogo Tela (Solid State Phys.).*, 38, 11, 3381–3384, 1996.
10. Westermark, W., Brett, J.F., Meloni, D.R., Inflow stimulation by vibration treatment of the bottomhole area to enhance oil production. *Neft mira (World Oil). (Neftegazovye technologii) Oil Gas Technol.*, No, May–June, 2002.
11. Gadiev, S.M., *The use of vibration in oil production*, p. 154, Nedra, M, 1977.

12. Dryagin, V.V. and Igolkina, G.V., Application of acoustic well treatments to restore permeability and assess reservoir saturation. *Materials of the 1st All-Russian Geophysical Conference – Trade Fair*, Ukhta, pp. 57–61, 2002.
13. Dunin, S.Z. and Nagornov, O.V., Waves of finite amplitude in soft ground. *Izvestiya AN SSSR. J. Fluid Mech.*, 6, 142–145, 1984.
14. Dyblenko, V.P., Kamalov, R.N., R.Ya., Sharifullin, Tufanov, I.A., *Vibration-wave treatment to enhance well productivity and reactivate well production*, p. 382, Nedra, M, 2000.
15. Dyblenko, V.P., Marchukov, E.Yu., Tufanov, I.A., Shariffullin, R.Ya., Evchenko, V.S., *Wave Technologies and their Implementation to Produce Tight Reserves*, p. 338, RANS Publishing House, M, 2012.
16. Mandelstam, L.I., Complete collection of works. IV, in: *Lectures on Oscillations*, pp. 1930–1932, p. 512, Publishing house of the Academy of Sciences of the USSR, 1955.
17. Entov, V.M., Zak, S.A., Chensin, E.O., On two-phase flow in the porous medium exhibiting microheterogeneous wettability. *DAN SSSR*, vol. 274, pp. 1334–1337, 1984.
18. Muslimov, R.Kh., The pivotal role of permanent replenishment of the hydrocarbon deposit reserves, in: *Geologya i Geologorazvedka (Geology and Exploration)*, vol. 3, pp. 24–29, 2007.
19. Muslimov, R.Kh., The potential of the basement of oil and gas basins is a reserve for replenishing hydrocarbon resources in the XXI century. *Sci. Tech. J. "Georesursy"*, 4, 2–5, 2003.
20. Nikolaev, A.V., Problems of Induced Seismicity: Induced Seismicity, in: *OIFZ RAN (United Institute of the Earth Physics of the Russian Academy of Sciences)*, pp. 5–15, Nauka, M, 1994.
21. Nikolaevsky, V.N., Viscoelasticity with internal oscillators as a possible model of a seismically active medium. *Rep. USSR Acad. Sci.*, 283, 6, 1321–1324, 1985.
22. Nikolaevsky, V.N., The mechanism of vibration impact on oil recovery and dominant frequencies. *Rep. USSR Acad. Sci.*, 307, 3, 570–575, 1989.
23. Nikolaevsky, V.N., *Theory of nonlinear waves and characteristics of seismic signals: Nonlinear wave processes*, pp. 273–296, Mir, M, 1987.
24. Nikolaevsky, V.N., Vibration of rock massifs and ultimate oil recovery. *Izvestiya RAN. Mechanics liquid gas.*, 5, 110–119, 1992.
25. Carpinteri, A., Lachidonia, J., Pugno, N., Multiscale temporal effects: monitoring of fractured structures using the acoustic-emission method. *Fizicheskaya Mezomekhanika (Physical Mesomechanics).*, 5, 85–89, 2005.
26. Krutin, V.N., Energy ratios when elastic waves are emitted from a wellbore, in: *New geoacoustical methods of research, prospecting and exploration of mineral deposits: Collection of articles*, pp. 76–88, M, VNIIYAGG, 1982.
27. Kuznetsov, O.L., Dyblenko, V.P., Sharifullin, R.Ya., Tufanov, I.A., Energy criteria for the mechanisms of flow effects in productive formations exposed

to elastic vibrations: Oil recovery stimulation. *Materials of the Report at the 12th European Symposium*, Kazan, pp. 101–106, 2003.
28. Kuznetsov, O.L., Dyblenko, V.P., Chirkin, I.A., Sharifullin, R.Ya. *et al.*, Features of the mechanical-stress energy accumulation and anomalous seismoacoustic radiation in oil-saturated rocks. *Geofizika.*, 6, 8–15, 2007.
29. Kuznetsov, O.L., Chirkin, I.A., Dyblenko, V.P., R.Ya., Sharifullin, Ryazanov, E.G., Features of spectral characteristics of acoustic emission of oil-saturated rocks, in: *Scientific notes of the Physics*, vol. 6, Faculty of the Moscow State University, 2014.
30. Kuznetsov, O.L., Chirkin, I.A., Kuryanov, Yu.A., Dyblenko, V.P., Rogotskiy, G.V., *Seismoacoustics of the porous and fractured geological media*, vol. 2, p. 360, VNIIGEOSystem, M, 2004.
31. Kuznetsov, O.L., Chirkin, I.A., Yu.A., Kuryanov, Dyblenko, V.P. *et al.*, *Seismoacoustics of the porous and fractured geological media*, vol. 3, p. 434, M, VNIIGEOSystem, 2007.
32. Kurlenya, M.V. and Serdyukov, S.V., Nonlinear effects during the radiation and propagation of vibroseismic signals in the rock mass. *Phys. Tech. problems Min.*, 2, 3–10, 1999.
33. Oparin, V.N., Simonov, B.F., Yushkin, V.F., *Geomechanical and technical fundamentals of enhanced oil recovery in vibration wave technologies*, p. 404, Novosibirsk, Nauka, 2010.
34. Ryashentsev, N.P., Yu.S., Ashchepkov, Simonov, B.F. *et al.*, *Controlled seismic reservoir treatment*, p. 60, Novosibirsk, IGD SO RAN USSR, 1989.
35. Sadovsky, M.A. and Pisarenko, V.F., *Seismic process in a block environment*, p. 96, M, Nauka, 1991.
36. Sadovsky, M.A., Abasov, M.T., Nikolaev, A.V., Prospects of the oil-reservoir vibration treatment to enhance oil recovery. *Bull. Acad. Sci. USSR.*, 9, 95–99, 1986.
37. Haken, G., *Synergetics*, p. 300, M, Mir, 1980.
38. Fayzullin, I.S. and Shapiro, S.A., Scattering of seismic waves and fractal analysis of lithosphere inhomogeneities. *Izvestiya AN SSSR. Phys. Earth.*, 10, 43–49, 1989.
39. Beresnev, I.A. *et al.*, Elastic wave stimulation of oil production: A review of methods and results. *Geophisics.*, 59, 6, 1000–1017, June, 1994.
40. Volkov, A.V., Kuznetsov, O.L., Chirkin, I.A., Dyblenko, V.P. *et al.*, Determining the Fluid Saturation and Depth of Reservoirs by Passiv Seismic SLEC-Monitoring Technology. *EAGE 69th Conference @ Exhibition*, pp. 11–14, June, 5 2007.
41. Dullien, F.A.L., *Porous media, fluid transport and pore structure*, p. 567, Academic Press, New-York, 1979.
42. Kirkpatrick, S., Percolation and Conduction. *Mod. Phys. Rev.*, 45, 4, 574–588, 1973.
43. Takayuki, H., Fraktal structure of spatial distribution of microfracturing in rock. *Geophys. J. R. Astron. Soc*, 90, 369–374, 1987.

44. Takens, F., *Dynamical Systems and Turbulence*, pp. 366–381, Lecture Notes in Mathematics Berlin, Spring-Verlag, 1981.
45. Turpening, R. and Pennington, W., *Final Report. Calibration and Testing of Sonic Stimulation Technologies*, p. 83, Michigan Technological University, 2005, DOE Contract No. DE-FC26-01BC15165.

8
Types of Existing Treatments

The existing methods of using elastic oscillations includes the following [2] main types of treatments:

- *vibroseis treatment;*
- *vibrowave treatment;*
- *shock pulse treatment;*
- *dilatation wave treatment;*
- *seismo-acoustic treatment;*
- *acoustic treatment.*

A brief description of these treatments is presented in Table 8.1.

Results of classification of oscillatory near-wellbore zone and reservoir treatments are presented in Tables 8.2 and 8.3. They contain also the methods developed by the authors of this book or with their participation.

Table 8.2 shows the integrated vibrowave and pressure-drawdown/chemical treatment of the near-wellbore zone using low-frequency hydrodynamic generators, thermal gas pressure-drawdown and pulse method, and HydroVibroFrac.

Table 8.3 shows seismo-acoustic method using integrated oil-well units that ensure reservoir treatment at multiple frequencies, integrated approach using wave, hydrodynamic, physical and chemical treatments, and heat-transfer fluids.

The authors believe that, in the case of tight reservoirs, it is most reasonable to use such methods from among the above mentioned ones as vibrowave, vibroseis, seismo-acoustic, and dilatation wave methods.

In addition, in some cases, it is advisable acoustic and shock pulse methods to treat the near-wellbore zones that involve chemical and electrophysical sources. The priority targets in these fields are heterogeneous low-permeability shaled-out reservoirs, small accumulations, and depleted watered-out reservoirs and accumulations containing high-viscosity and bituminous oils featuring initial shear gradient. It is advisable to integrate these wave treatment methods combining them with physicochemical,

George V. Chilingar, Kazem Majid Sadeghi, and Oleg Leonidovich Kuznetsov. Acoustic and Vibrational Enhanced Oil Recovery, (261–310) © 2022 Scrivener Publishing LLC

Table 8.1 Wave treatments.

Vibroseis
• *frequencies of the oscillations generated by sources up to 40 Hz;* • *lengths of the generated waves are commensurate with or exceed the size of the reservoir block macrostructure;* • *covers productive reservoir areas.*
Vibrowave treatment
• *frequencies of the oscillations generated by sources up to 40 to 800 Hz;* • *implementation of the treatment is subject to both effects manifestation and depth of the near-wellbore area coverage in terms of frequency range;* • *coverage of the near-wellbore zone and far reservoir zones.*
Shock pulse treatment
• *primarily pulse repetition frequency 0.01 to 0.1 Hz;* • *mostly shock wave action is generated;* • *near-wellbore zone coverage.*
Dilatation wave treatment
• *creation of rock dilatation (loosening) around the active wells as well as variable elastic deformations with excitation frequency from fractions to several Hz;* • *cover productive reservoir areas.* • *one should bear in mind that different treatments generate in productive reservoirs waves of wider spectrum as compared to the frequencies of oscillations or pulses generated by the sources.*
Seismo-acoustic treatment
• *productive reservoir areas of great extent;* • *acoustic low-amplitude waves of seismic frequency range up to 100 Hz;* • *use of the earth energy potential and manifestation of initiated anomalous (trigger) effects in abnormally stressed reservoir areas;* • *integrated treatment using various sources, including onshore ones.*
Acoustic treatment
• *prevailing frequency range from 19 to 25 kHz;* • *elastic oscillations that feature well-formed wave pattern right in the near-wellbore zone, manifestation of non-linear effects within the local reservoir volumes near the sources at elastic-wave intensity more than 1 kW/m^2;* • *coverage of small near-wellbore zone areas.*

Table 8.2 Wave methods to treat the near-wellbore zone (aimed to enhance well productivity and oil recovery).

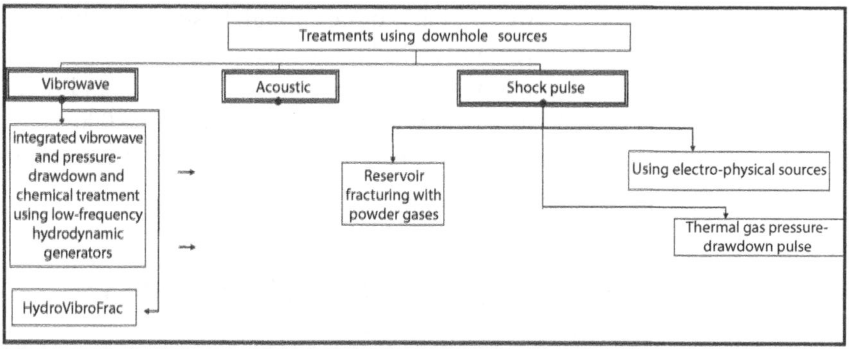

Table 8.3 Methods of wave reservoir treatment (aimed to enhance oil recovery).

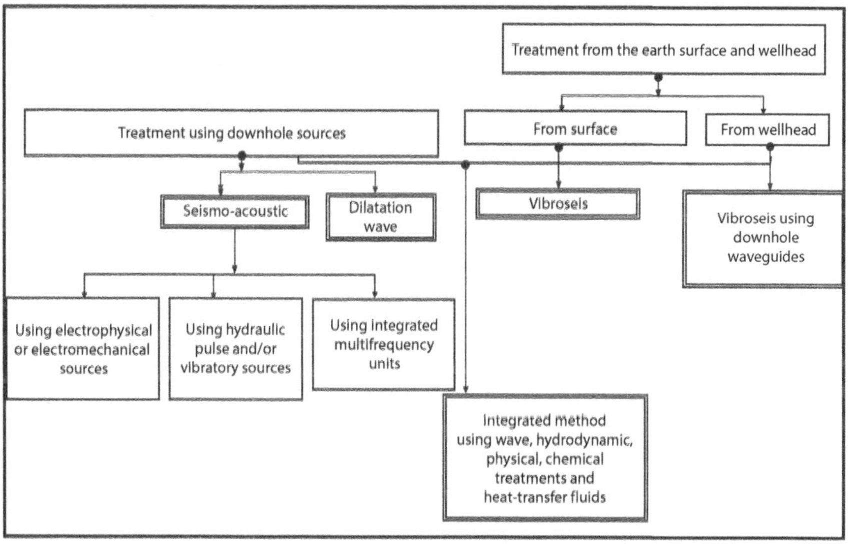

hydrodynamic, and thermal methods and applying them in the abnormally stressed areas.

Wave methods of near-wellbore zone treatment are the most mature ones, and their implementation is under way. Such methods were used to successfully treat tens of thousands of wells.

Methods to treat the near-wellbore zones with powder gases, electrical and hydraulic shocks, short-duration pulse pressure drawdowns, as well as vibrational and acoustic treatments find wide application. These methods are based on transfer of oscillatory energy from the downhole sources to the reservoir. Significant losses of the energy transferred and propagated in the saturated reservoir medium limit the treatment depth to several meters from the wellbore walls. However, that is quite enough to successfully clean the wellbore walls and the well bottomhole from bridging materials.

Elastic oscillations were first used to treat petroleum reservoirs in Russia in the early 1970s [1, 2]. A vibrator was used to treat more than 6,000 wells. This tool did not go any further as it consumed too much energy, had poor efficiency, and underperformed in difficult wells.

8.1 Integrated Technologies of the Near-Wellbore Zone Vibrowave Treatment

This section addresses new technologies developed by the authors of this book aimed at enhancing well productivity using integrated treatment by physical and physicochemical methods [3–7, 10, 11, 13].

The physical factor in these technologies is a remote low-frequency wave or pulse treatment of the productive medium with elastic oscillations combined at the same time with hydrodynamic and in some cases with thermodynamic treatments.

The physicochemical factor is implemented in two ways:

- chemicals are added to the circulating fluid;
- estimated volumes of the chemicals with special functionality, such as solvents, surfactants, alkali and acid solutions, heat-generating, and other active chemicals or their combinations, are successively injected into the reservoir through the down-hole equipment. This being the case, the most reasonable heat-generating chemicals are those that generate heat through the exothermal reactions between their components or between the components and the wellbore, or with reservoir fluids.

A new authors' procedure based on spectral, wavelet, and fractal real-time analysis of the disturbance noises is applied under these technologies. Those noises come from the treated reservoir. The new procedure allows diagnosing the treatment results and make adjustments to the operating practices.

To implement the developed set of treatment processes in every geological and physical environment, a normal series of high-efficiency hydrodynamic and gas-dynamic oscillation generators was created, their basic operation has no analogues all over the world. Absence of moving mechanical parts ensure high operation reliability and long service life. Prior to the reservoir treatment, the vibrowave equipment is run into the wellbore on the tubing string, coiled tubing or logging cable.

The technologies and the equipment used are patented in Russia and patent pending abroad.

The developed technologies are implemented in the following main options:

- *integrated vibrowave, pressure-drawdown, and chemical near-wellbore zone treatment (VDHV);*
- *vibrowave and foam treatment (VPV);*
- *deep chemical-wave reservoir treatment (GRVP);*
- *elimination of the water and gas zone isolation issues (OGVP);*
- *coiled-tubing wave technologies (KVT);*
- *hydraulic fracturing along with vibrowave treatment (HydroVibroFrac).*
- These technologies have the following features:
- they can be used in the fields where wells exhibit abnormal operating conditions, and traditional well production procedures are inefficient and not economical;
- integrated approach;
- multiple frequencies of near-wellbore zone treatment with elastic oscillations under pressure drawdown;
- integrated treatment through combination of pressure drawdown and overbalance;
- combination with injection of chemicals and heat-generating agents to the reservoir;
- application of reasonable treatment modes, including integrated treatment based on geological and physical environment of each specific target;
- improved both vertical and horizontal productive-reservoir treatment efficiency;
- treatment monitoring using both the standard and special software measurement systems that allow diagnosing the treatment processes by the reservoir-medium response in real time.

Treatment with elastic oscillations within the applied amplitude and frequency range is environment friendly and causes no damage to the well health.

8.1.1 Downhole Equipment

Certified downhole systems, STRENTER, were created and are used to run the near-wellbore zone treatments. They include hydrodynamic generators and jet pumps that operate together [8, 9]. In addition, downhole pulse systems UKVS [7] that operate together with a sucker rod pump and an electric submersible pump as well as systems to prepare emulsions in the oscillatory field, and other equipment have been developed and their field tests are under way.

A specialist software measurement system monitors the equipment performance. It includes standard instruments and a specialist signal-processing computer unit.

Distinctive features of the downhole equipment that allow its application to perform a wide range of operations and functioning in various geological environments including:

- vortex excitation of elastic oscillations;
- treatment modes are adapted to each specific well;
- high efficiency;
- compact design;
- no moving mechanical parts, reliable operation, and long motor operating life;
- high maintainability:
  - mutual compatibility of components of various equipment;
  - well cleanouts, injection of chemical solutions, and heat-generating agents without pulling the treatment equipment out of hole;
  - compatibility with the oil field equipment of local and foreign manufacture;
  - sour service (acids, solvents, etc.).

Technical specifications of the STRENTER equipment is shown in Tables 8.4 and 8.5.

Figure 8.1 is a typical schematic and outer appearance of the downhole hydrodynamic oscillation generators of the STRENTER system.

Swirled liquid flow forms in the vortex unit in the same way as in the known systems, such as the unit manufactured by the Institute of Machine Science of the Russian Academy of Sciences (IMASH RAN). However, the vortex plays a quite different role in the formation of the auto-oscillating system. In the known IMASH RAN facilities, oscillations are generated through amplification of the self-oscillations of the swirled liquid layers (in the kilohertz and higher frequency range) and creation of conditions for the resonances.

These resonances have sharp frequency tuning and slump when the wellbore environment changes and when the system geometry changes

Table 8.4 Types of generators, their technical specifications, and purposes.

Generator type	Hydraulic fluid consumption (dm^3/s)	Operating pressure differential (MPa)	Pressure amplitude (MPa)	Minimum tubing size (mm)	Main purpose
GD2V-3	2.5–3	7–12	3–4	60	Enhance productivity (injectivity) of horizontal wells, sidetracks, or other types of wells, having casing size restrictions, and going through terrigenous reservoirs that consist of poorly consolidated sandstones. Cleanout (screen reactivation) of water producers.
GD2V-4	3.5–4	7–8.5	4–5	60	
GD2V-5	5–5.5	9–12	6–7	73	Treat wells that go through carbonate reservoirs and shaly sandstones. Treat horizontal wells by injecting foam.
GD2V-6	6–7	6.5–9	6–8	73	
GD2V-10	10–12	8.5–12	8–10	89	Deep reservoir wave-and-

(*Continued*)

Table 8.4 Types of generators, their technical specifications, and purposes. (*Continued*)

Generator type	Hydraulic fluid consumption (dm^3/s)	Operating pressure differential (MPa)	Pressure amplitude (MPa)	Minimum tubing size (mm)	Main purpose
GD2V-20	20–25	5–7	7–9	102	chemical treatment, HydroVibroFrac, enhance well production.
GD2V-30	30–35	5–7	8–10	102	
GD2V-2K	2–2.5	10–12	2–3	48	Treat near-wellbore zones, clean out tubing and casing strings using 38.1-mm, 44.5-mm, and 50.8-mm coiled tubing.
GD2V-3K	2.5–3.5	10–12	3–4	48	
GD2V-1VSh	1.0–1.5	4.0–5.0	1.5–2.0	73	Water injection through a generator on constant hanger in water injectors; run in ESP pumping wells.
GD2V-2VSh	2.0–2.5	4.0–5.0	2.0–2.5	73	
GD2V-3VSh	3.0–3.5	4.0–5.0	2.5–3.0	79	
GD2V-4VSh	4.0–4.5	4.0–5.0	3.0–3.5	89	
GD2V-6VSh	6.0–7.0	4.0–5.0	6.0–8.0	89	

Tubing installed size varies from 42 to 102 mm; generator length from 1.2 to 2 m; weight from 6 to 20 kg.

Table 8.5 Downhole IS-type jet pumps.

Code	Maximum well total depth (m)	Maximum operating pressure (MPa)	Maximum diameter (mm)	Length (m)	Weight (kg)
IS-S	5,000	50	89	0.6	15.0
IS-R	5,000	50	73	0.6	8.0
ISV-K	3,000	25	54	0.3	3.0

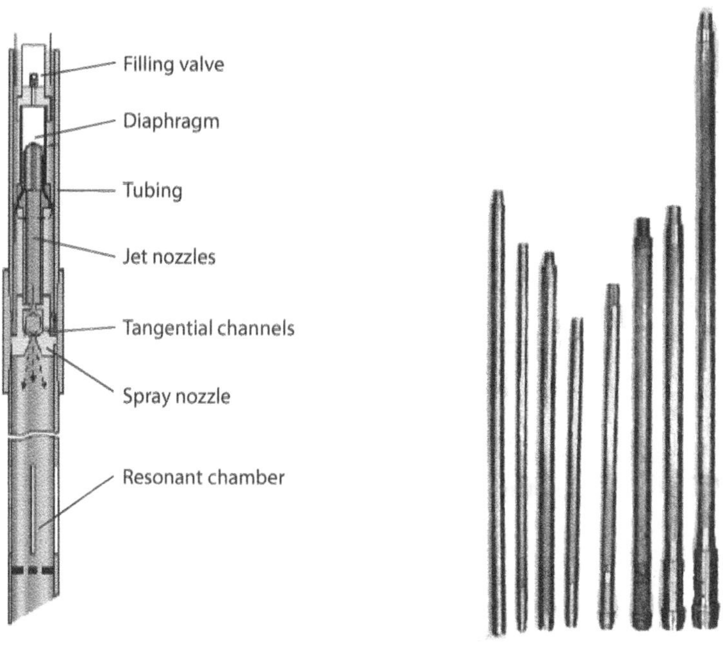

Figure 8.1 Typical schematic and outer appearance of hydrodynamic generators.

due to the cavitation wear. Energy release of such downhole systems is not stable and vanishes when off resonance.

In our case, the liquid swirl is an important controlling element of the liquid-flow low-frequency oscillations amplifier (flow rate and pressure). This allows reaching the oscillation amplitudes and frequencies required for the treatment, the energy release is proportional to the output of the pump-discharge line.

Figure 8.2 shows the configuration of the equipment that is tubing conveyed downhole for two technical solutions, with and without a jet pump.

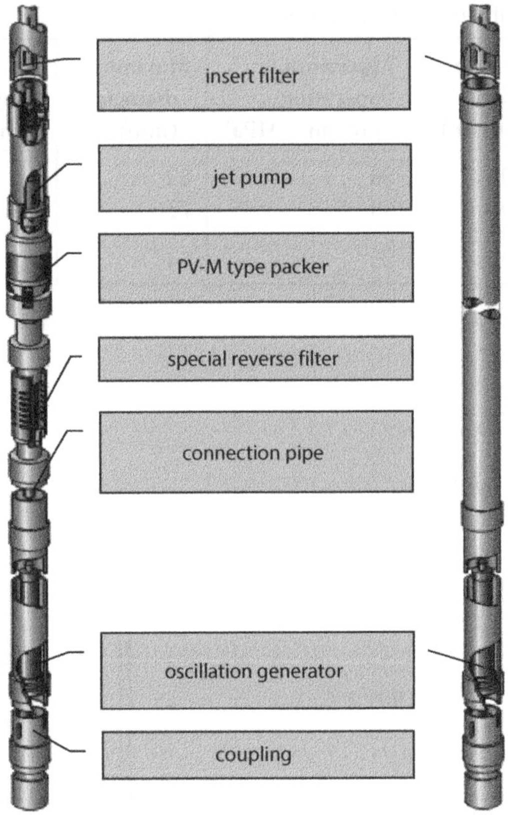

Figure 8.2 STRENTER system configuration.

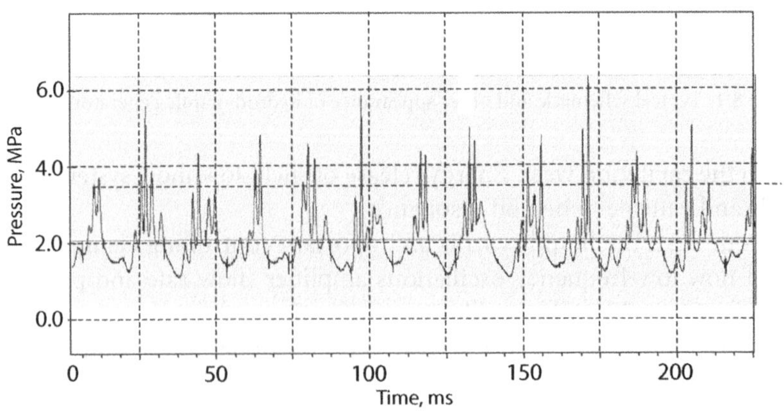

Figure 8.3 Bench performance of the GD2V-4,5 oscillation generator.

Table 8.6 Results of bench hydrodynamic generator tests.

Generator type	Liquid flow rate through generator (m^3/day)	Hydraulic fluid injection pressure (MPa)	Generator pressure drop (MPa)	RMS Value of oscillation amplitude (MPa)	Pressure oscillation amplitude (MPa)	Dominant oscillation frequencies (Hz)
1	2	3	4	5	6	7
GD2V-3	240	8.0	7.5	0.9	4.8 (6.4*)	20, 160
GD2V-6	500	8.0	7.0	2.1	8.0 (9.6*)	33, 250

* for individual wavelets.

Figure 8.3 shows bench operational dynamics of the oscillations generator GD2V-4,5, while Table 8.6 shows the results of the hydrodynamic generator tests.

- Frequency $f = 57$ Hz;
- Pressure drop in the generator $\Delta P = 10.2$ MPa;
- Working fluid consumption $Q = 4.5$ dm^3/s.

8.1.2 Integrated Vibrowave, Overbalance/Pressure-Drawdown, and Chemical Treatment (VDHV)

Scope of application includes vertical and inclined wells of all types exhibiting a skin-effect due to the formation damage caused by the various materials and due to blocking effect of the trapped phases, i.e., water, oil and/or gas, stable emulsions, and mud filtrate, including wells where productivity dropped over time after the frac job.

The essence of the well *reactivation and productivity enhancement (VDHV)* consists in integrated near-wellbore zone treatment with intense elastic oscillations at reasonable modes under controlled pressure drawdown alternating with the overbalance cycles along with physical and chemical factor [11]. The STRENTER system excites the high-amplitude pressure oscillations at the bottomhole, whereas local pressure drawdowns are created at the same time in the productive reservoir. Equipment layout scheme is shown in Figure 8.4. It includes generators GD2V-3, GD2V-4, and GD2V-5 and jet pumps IS-S or IS-R.

Special software determines the reasonable regimes of the vibrowave equipment operation. They are set subject to the geological and production parameters of each specific well. Reservoir treatment with elastic oscillations under pressure drawdown reduces the blocking effect of the

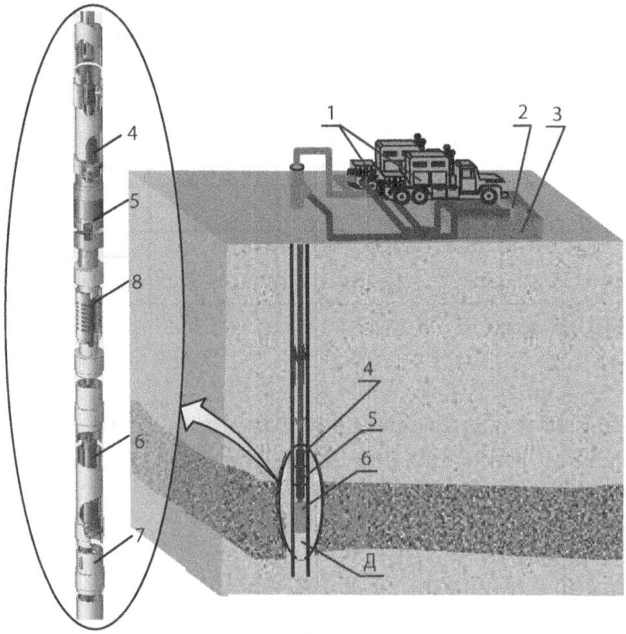

Figure 8.4 STRENTER system layout sketch to run near-wellbore zone treatment using VDHV technology: 1, pumping units AH-700; 2, separator; 3, holding tank; 4, injector; 5, packer; 6, oscillation generator; 7, resonant chamber; and 8, special filter.

trapped phases—water, oil, or gas; accelerates liquid flow; moves the bridging materials to the wellbore; improves penetration of the chemicals into the tight reservoir zones; enhances chemical reactions; and accelerates removal of their products. These phenomena stimulate oil inflow to the well, remove bridging materials from the reservoir and clean-out the entire near-wellbore zone. They normally bring about a multifold increase of the well productivities (both vertical and horizontal reservoir coverage).

Reservoir treatments based on this technology increase well productivities by 1.5 to 10 times; waterflooding sweep efficiency grows 50% to 90%; watercut of the produced fluid decreased by 5% to 30%; treatment efficiency is up to 95%; and effect duration is 12 to 24 months and more.

Note that ¼ of the treated production wells and ½ of the treated injection wells were previously shut-in wells, inactive or abandonment candidates, i.e., VDHV technology was actually used to reactivate them.

Figure 8.5 shows example of fluid-movement profiles for the Prirazlomnoye well 808[1] prior to and after the treatment as well as (see Figure 8.6) injectivity profiles for the Novo-Elkhovskoye well 3468 that show the improved sweep efficiency during well treatments.

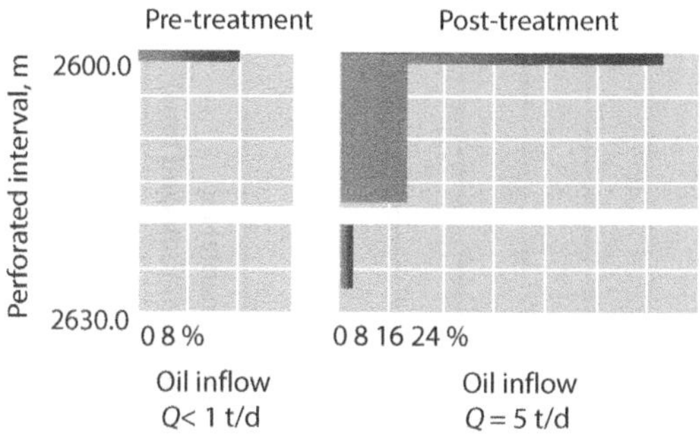

Figure 8.5 Inflow profiles for Prirazlomnoye well 808 before and after the VDHV treatment.

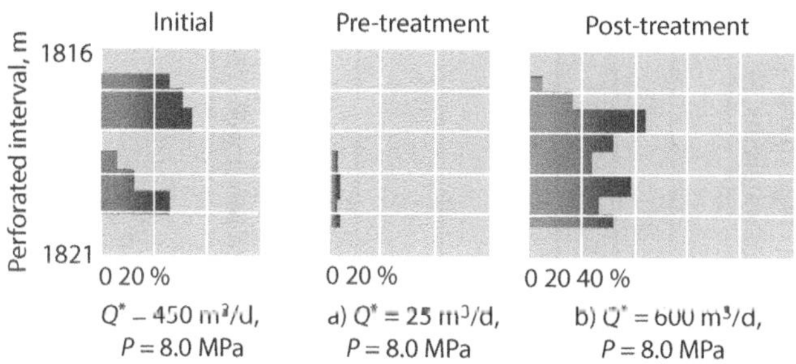

Figure 8.6 Injectivity profiles for the Novo-Elkhovskoye well 3468 prior and after VDHV treatment.

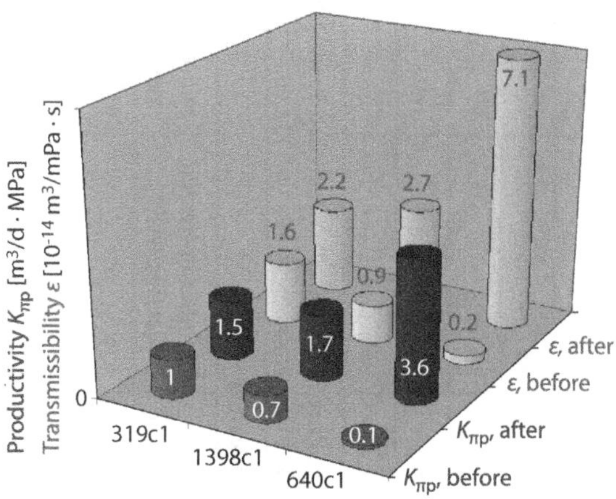

Figure 8.7 Near-wellbore zone flow properties and productivities of the Tuymazinskoye wells before and after VDHV treatment.

Table 8.7 The results of using VDHV technologies in various reservoirs of the Novo-Elkhovskoye field.

(A) Production wells

Type of reservoir	Number of wells	Success rate, %	Incremental oil production, t/d	Average changes in watercuts, %	Total incremental oil production (thousand tons)
Terr.	164	93	0.5–14	-5.6	169
Terr.	58	97	0.3–6.8	-2.8	108
Carb.	160	99	0.3–8.5	-3.4	214
Carb.	1	100	8.1	0.0	14
Total	353				505

Average duration of the effect exceeded 30 months.
Average incremental oil production per one well-treatment is 1,430 tons.

(B) Injection wells

Type of reservoir	Number of treated injecting wells	Success rate (%)	Incremental injectivity (m^3/d)	Average changes in watercuts of offset wells (%)	Total incremental oil production (thousand tons)
Terr.	129	95	30–300	95	194
Terr.	40	97	30–180	102	48
Carb.	13	100	50–450	137	80
Total	183				250

Average duration of the effect exceeded 19 months.
Total incremental water injection exceeded 2.1 mln m^3.
Incremental production per 1 well-treatment averaged 1250 tons.

Figure 8.7 shows well-test results for three Tuymazinskoye wells before and after the VDHV treatments. They indicate significant and stable improvement of the flow properties in the near-wellbore zones and well productivities.

Table 8.7 shows results of using VDHV technologies in various reservoirs of the Novo-Elkhovskoye field.

8.1.3 Vibrowave and Foam Treatment (VPV)

Important feature and main advantage of this technology is potential integration of the vibrowave, pressure-drawdown and physicochemical treatment in horizontal (HW), inclined horizontal wells and sidetracks (ST), i.e., in targets where the reliable bottomhole section sealing using standard packers is a technical challenge, and, therefore, it is not easy to use jet, pumps to create pressure drawdown.

The essence of the technology consists in an all new approach aimed to treat the entire reservoir by exciting elastic oscillations under the pressure drawdown, using continuous interval-by-interval reservoir treatment, taking advantage of the waveguide reservoir properties, combined with the injection of chemicals with a specified functional purpose, namely, alternating chemical the injection with injection of water-in-oil emulsion and foams when much fluid is lost to the formation or in the presence of bottom water zone.

The equipment layout scheme is shown in Figure 8.8.

Hydrodynamic STRENTER generators are used to treat a reservoir with elastic oscillations. They are run into the well on tubing or coiled tubing. They operate when gas-liquid mixtures (oil or surfactant-water solutions with nitrogen) are circulated through them.

Intense dynamic and vortex processes that accompany elastic oscillations, when small amounts of surfactants are added, producing very fine and active foam at the hydrodynamic generator outlets, that are used to fill the annulus. Hydrostatic-pressure drop in the well creates a rather long pressure drawdown, which, coupled with vibrowave treatment, provides favorable conditions needed to stimulate oil production and clean out the near-wellbore zone. Moreover, the foam produced at the generator outlets has high floatability properties and pronounced relaxing properties. Combined with the elastic-oscillation treatment, these foam properties allow the most efficient cleanout of the wellbore walls covered with mud cake while drilling was on, and eliminate the formation damage under the pressure drawdown.

Implementation of such operations based on this technology is well suited for the wells with the uncemented screen liner in the horizontal

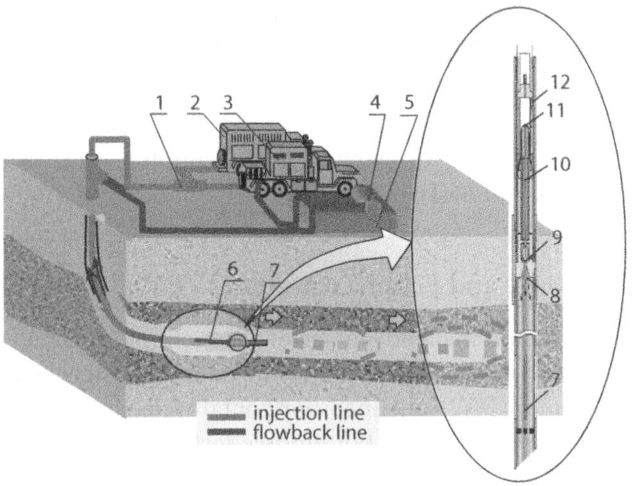

Figure 8.8 Equipment layout scheme and configuration of the downhole oscillation generator to treat near-wellbore zone using VPV technology. 1, aerator; 2, nitrogen compressor; 3, pumping unit; 4, separator; 5, tank; 6, oscillation generator; 7, resonant chamber; 8, generator spray nozzle; 9, tangential channels; 10, jet nozzles; 11, diaphragm; and 12, tubing string.

section of the well as well as for modern well designs with lateral sidetracks, especially radial one where traditional methods are not very efficient to remove the remains of the drilling mud.

Specific features of using downhole generators to excite elastic oscillations in the horizontal wells are used to determine optimum modes of the vibrowave equipment operation to reach waveguide propagation of the elastic oscillation energy and maximum treatment depth. Due account is given to the development of real modes in specific geological and physical environment of the reservoir.

VPV treatments increase well productivities 2- to 8-fold. Watercut of the produced fluids decreases by 10% to 50%. Treatment success rate up to 90%, whereas the effect duration is 12 to 24 months and more.

8.1.4 Deep Chemical-Wave Reservoir Treatment (GRVP)

The purpose the deep chemical-wave reservoir treatment is as follows:

- enhance or restore the connectivity of the wells with the reservoir in the deeply damaged near-wellbore zone and for the wells in the low-productivity carbonate reservoirs where fracture jobs are not economical;

- enhance productivities of the wells drilled in the carbonate reservoirs near-wellbore zones which were treated many times.

Essence:

- create (open) fractures and cavities in the carbonate reservoirs; improve permeability of the tight sandstone layers;
- inject successive repeated (more than 3 times) batches of chemicals into the reservoir through the generator;
- flow back the reaction products from the reservoir after injection of each batch of chemicals;
- when needed, inject in advance solvents or surfactant solutions through the generators;
- alternate injections of chemicals and water-in-oil emulsions (without round trips) in case of large fluid losses or in the presence of bottom water.

Distinctive features:

- deep treatment of near-wellbore zones of production and injection wells;
- enhanced injection of chemicals into tight layers to open additional channels and fractures;
- use of emulsions to shut off the most permeable flow channels;
- creation of fine hydrocarbon-acid emulsions, and deep treatment of carbonate reservoirs.

The above VDHV and VPV operations can be used to implement the GRVP technologies. Their selection is subject to the well type, well design and depth, reservoir pressure, etc. The GD2V generators of different capacity are used to implement GRVP technology. Generators are selected subject to the reservoir type, its shale, consolidation, reservoir thickness, etc. These are mostly generators GD2V-4, GD2V-6, and GD2V-8 controls degree of pump IS-R making part of the STRENTER system.

Figure 8.9 shows schematically the main stages of the treatment cycles.

The depth and success rate of reaction-product removal at each treatment cycle is subject to combination of pressure-drawdown and wave treatment of the near-wellbore zone using oscillators together with the jet pumps or filling the annulus with foam.

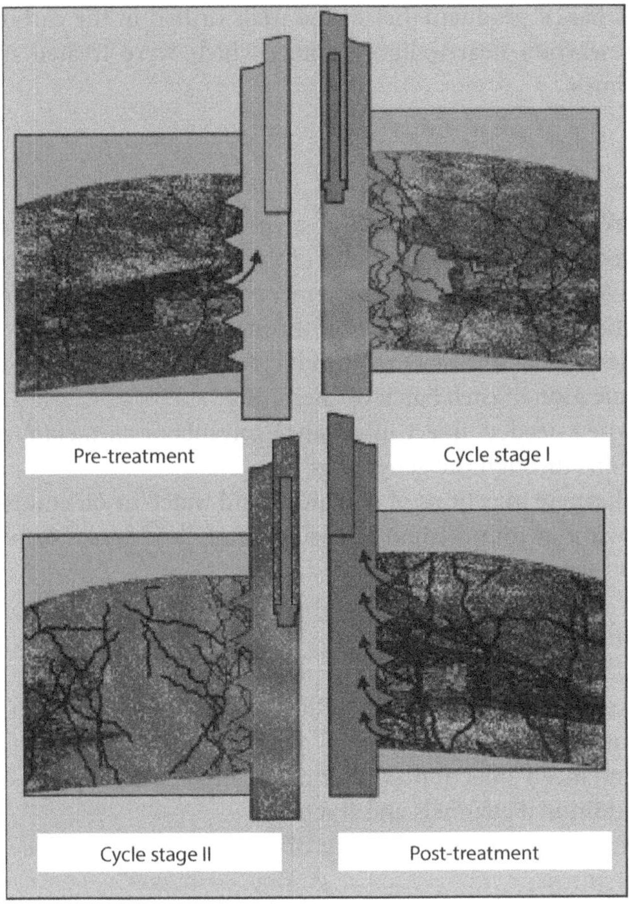

Figure 8.9 Main stages of the GRVP technology stages.

Elastic oscillations do not only initiate recovery of the bridging materials and reaction products from the productive reservoirs. They become also sort of regulators, because enhancement of flow processes increases also the depth of the response zone and the reservoir-treatment coverage in terms of thickness. Efficient recovery of reaction products that terminates each cycle of repeated treatment provides favorable conditions for the full and deep opening of the new channels and fractures at the following cycle of chemical-batch injection.

The simultaneous wave treatment initiates fracturing processes and rock restructuring; thus, reducing threshold fracture-opening pressure when working fluids are injected into the reservoir. This is especially favorable for the subsequent injection of high-viscosity emulsions. Treatment efficiency is enhanced due to the creation of new fractures and stimulation of existing ones, formation of cavities, uniform acid interaction with the rocks under reduced injection pressure, which contributes to subsequent rational use of acid and oil-acid emulsion rims. As a result of the cyclic GRVP operations, new fractures open and expand more uniformly over the entire near-wellbore zone, including the areas far-away from the wellbore. This reduces the hazard of water encroachment from the water-saturated interval. Solvent injection improves wax removal from the near-wellbore zone and contributes to bringing the well to stable production.

As an example of implementing the GRVP technology was well 7 of the Beloglinskoye Field that was drilled as an exploratory well. Upon completion, the well produced 14 m^3/d of liquid. Then, the well was temporarily abandoned. Eleven years later, it was brought to production; first, it flowed naturally and subsequently was converted to deep-pumping production. Before the GRVP treatment, the oil production rate was 0.8 t/d at 2% watercut and 23.9 MPa reservoir pressure.

The operations were carried out in five cycles. The downhole wave equipment included oscillation generator GD2V-5 and injector IS-R. The well treatment resulted in the oil production-rate growth to 40 to 50 t/d at 2% to 3% watercut.

Figure 8.10 shows variation of the additional incremental oil production and the number of the well-operations by years due to implementation of the GRVP treatments in the Novo-Elkhovskoye field.

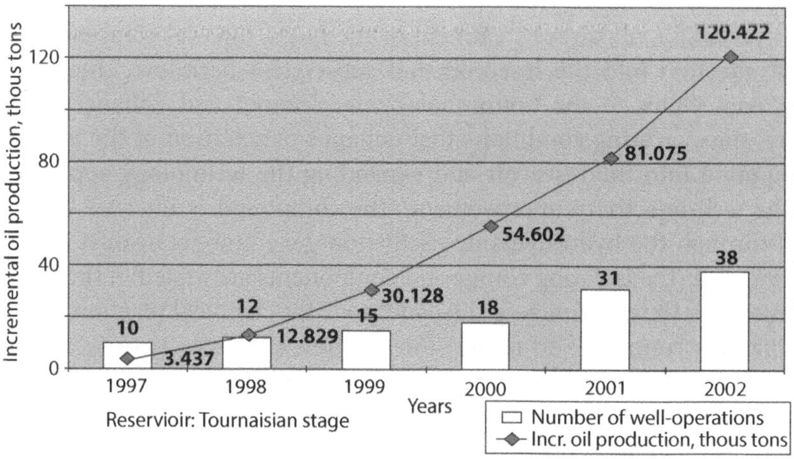

Figure 8.10 Cumulative incremental oil production and number of well-operations by years resulting from implementation of the GRVP technology in the Novo-Elkhovskoye field.

In total, more than 200 wells were successfully treated in the fields of the Urals-Volga region using GRVP technology. Treatment success rate was 95%. Well productivity increased 3- to 8-fold, for 15 to 30 months and more.

8.1.5 Remediation of Troubles When Shutting Off Water and Gas Entries

This technology is used in the wells drilled in layered heterogeneous reservoirs where water breaks through under waterflooding, or gas breaks through from the gas cap; wells drilled in the monolithic reservoirs with bottom water or close water-oil contact.

The essence of the technology consists in preliminary mixing the isolating compounds and injecting their mixture through the generator into the reservoir. However, prior to injection, the near-wellbore zone is treated with elastic oscillations; when needed, the bridging materials are removed from the near-wellbore zone by using vibrowave treatment and pressure-drawdown/overbalance treatment in order to prepare the pore and fracture surfaces for maximum adhesion with the isolating material [12].

Traditional isolation squeeze technologies often do not allow preliminary mixing of the chemicals as their viscosity increases sharply at the bottomhole when injected into the reservoir. Without mixing, the sequential penetration of the isolating-compound component rims into the near-wellbore zone causes the chemical reaction mostly in the areas where the mutual-displacement fronts dissipate and the resulting barriers are not sufficiently strong and long lasting.

This new technology to shut off water and gas entries into the wells involves first a vibrowave treatment of the target interval combined with fluid injection into the reservoir and subsequent backflow. This operation contributes to the bottomhole-zone cleanout and enhances injectivity; thus, creating conditions that enhance penetration of the isolating compound into the reservoir and expanding the technology application to the well-injectivity improvement. The compound is injected into the well through the hydrodynamic oscillation generator set against the target interval. The isolating compound components are mixed in the course of injection, thus, optimizing the conditions for chemical reaction. Elastic oscillations contribute to reduce the solution effective viscosity, disintegrate the particle aggregates in the pore channels, and initiate flow of the suspended particles. In addition, elastic oscillations contribute to penetration of the sediment forming mixtures into the pore channels, slow down the particle-integration processes and formation of impermeable cakes on the surface of the perforation tunnels in the wellbore vicinity.

Elastic oscillations reduce the amount of bound water in the near-wellbore zone, enhance chemical reactions, and significantly strengthen the "barriers" formed during the treatments. Injection of mixed isolation compound components combined with vibrowave treatment improves their penetration into the reservoirs where they fill the pore space deeper and more evenly. Moreover, injection pressure decreases, and injection rate increases. There is an opportunity to combine the compounds, add dispersed fillers, and save the main isolating materials. In general, the OVGP technologies have the following features and advantages:

- use of various liquid, solid, and gaseous agents such as emulsions, foams, gels, and compounds that polymerize and solidify in the reservoirs;
- production of high-dispersion and stable isolating compounds, water-oil emulsions, multiphase foams at the well bottomholes;
- easy penetration of isolating compounds; more even and deeper filling of the pore space and reservoir structure;
- high selectivity of the chemical injection into the reservoir;
- possibility to alternate rims of different isolating materials;
- use of 0.24% to 0.34% butyl-rubber polymer (BK-1675) solution in high-resin oil (or a mixture of stock-tank oil with bitumen) to create an isolating barrier in carbonate reservoirs. The advantages are as follows:
    - low polymer consumption (3 kg/m of the target-reservoir thickness);
    - creation in the reservoir of a molecular lattice (structure) that retains only water and gas, and lets the oil flow through;
    - low cost of the isolating compound;
- use of polymerizable compound (PC) based on mineral materials (5% to 12% silicates SiO_2+Al_2O_3+Na_2O+) and 6% to 12% hydrochloric acid (HCl) to create an isolating barrier in the terrigenous reservoirs. The advantages are as follows:
    - feasibility, high resistance to washout;
    - checkable properties before injection;
    - adjustable gelation time;
    - resistance to mechanical impacts (vibration, etc.);
    - potential barrier breakdown.

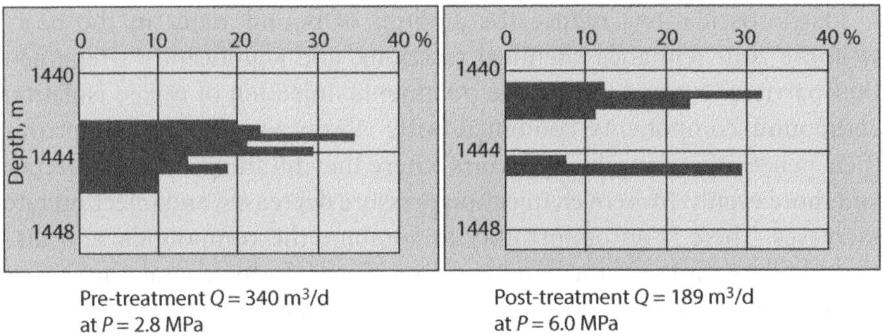

Figure 8.11 Injectivity profiles for well 757 of the Igrovskoye Field before and after injection of polymerizable compound to shut off water entries.

OVGP treatments reduce associated water production 1.5- to 3-fold; treatment success rate reaches 80%; effect duration is 10 to 15 months and more. Figure 8.11 shows injectivity profiles for well 757 of the Igrovskoye Field before and after the squeeze jobs were carried out. They show a barrier created, in the main intake layer, and an injectivity profiles expanding over the reservoir as a reservoir that was not previously completed was brought to production.

8.1.6 Coiled Tubing Wave Technologies (KVT)

Coiled tubing wave technologies KVT are applied in the wells having low (zero) productivity that reduced in the course of production, killing or intervention operations due to the deposits in the tubing, bottomhole, and (formation) damage.

Coiled tubing technologies aimed to enhance well productivity based on traditional bottomhole-treatment methods using chemicals, solvents, surfactants, and their combinations as well as using special packers and jet nozzles at the coiled tubing end to treat the openhole wellbores, etc. They are actually not only more economical but also often more efficient as they allow greater possibilities to combine various techniques to continuously treat the reservoir interval. However, bottomhole treatments of difficult wells are often not efficient enough; therefore, development and implementation of new bottomhole-treatment methods designed for and seamlessly compatible with coiled-tubing technology is a topical challenge today.

New coiled tubing wave technologies KVT are based on vibrowave treatment of the near-wellbore zone combined with other physico-chemical methods. They are the modifications of the VDHV, VPV, GRPV, and IVGP technologies adapted for the coiled tubing application.

Solid experience and exploratory studies enabled creation of hydrodynamic oscillation generators with compact nozzle generators that can be used with coiled tubing in the cased wells.

As to their design, they are made in the form of a nozzle attached to the coiled tubing with an adapter rolled at the end of the coiled tubing (see Figure 8.12). Subject to the tubing size or presence of a packer in the wellbore, the GD2V-2K or GD2V-3K generators can be run into the well. Their outer diameters are 42 mm or 51 mm, respectively. These generators make part of the STRENTER vibrowave system. This system includes also ISZ-K injectors; their design allows large pressure differentials in the high vortexes that excite elastic oscillations and simultaneously create a pressure drawdown in the reservoir-wellbore system.

Figure 8.12 The exterior of the GD2V-2K generator attached to the coiled tubing.

Coiled tubing KVT technologies allow continuous reservoir-by-reservoir bottomhole-zone treatments and efficiently clean out tubing strings and bottomholes while cutting down the required time and efforts.

8.1.7 Tubing and Bottomhole Cleanout Technology

This technology is applied to the wells having loose or firm deposits, such as wax, sand, silt, sulphides, and barites or their mixtures with oxygenated oil.

The principle is to clean out the tubing string through the generator at the flow rate that ensures the required level of cleanliness subject to both the tubing size, fluid flow rate, and type of deposits.

Specific features and advantages are as follows:

- improve the flow rate and cleanout efficiency due to pulsing flow and fluid flow twisting as well as axial and radial vibration of the generator frame operating downhole;
- generator-support control by the weight indication at the bottomhole flushing;
- use of special generator nozzles to improve the cleanout efficiency;
- use of oil-based fluids in production wells and water-based fluids in injection wells with such chemical additives as wax solvents, surfactant solutions, acid or alkali solutions, as well as specially selected compounds subject to the type of deposits.

8.1.8 HydroVibroSwabbing Technology

This technology is applied to the injection wells or naturally flowing production wells injectivity (productivity) of which was nonzero and reduced due to the formation damage caused in the course of well exploitation, killing operations, or interventions.

The essence is as follows:

- cyclic bottomhole-pressure increase above the reservoir pressure to create overbalance for a time long enough to accumulate a potential reserve of elastic energy generated by the fluid-and-rock compression in the most damaged near-wellbore zone. This is followed by subsequent creation of pressure drawdown along with bottomhole-zone treatment with elastic oscillations;

- combination of pressure-drawdown, overbalance, and wave treatments with chemical injection.

Specific features and advantages:

- continuous generator movement for detailed treatment of the perforated intervals;
- potential cleanout of both tubing string and bottomhole in one round trip.

Horizontal-well treatment technologies and fluid-loss control and conformance-control technologies are in essence similar to VPV and IVGP technologies, respectively.

The coiled-tubing wave technologies were successfully implemented in the fields of the Urals and Volga regions.

Success rate of the treatments was 90%. The specific injectivity index increased two- and threefold. The treatment efficiencies are 1.5- to 3-fold higher as compared to the traditional coiled-tubing technologies. One third of the wells were treated without acidizing. Injectivity was restored in a number of wells where previous standard bottomhole-zone treatments were fruitless.

8.1.9 Hydraulic Fracturing Technology Combined with Vibrowave Treatment (HydroVibroFrac)

This technology is applied to the wells that go through the low-permeability heterogeneous carbonate and terrigenous reservoirs.

As the authors created generators whose flow rates and pressures allow performing frac jobs, elastic oscillations can be used to improve frac job efficiency, reduce the maximum breakdown pressure, creating deeper and extensive network of fractures in the near-wellbore zone.

The essence of the new technology, named HydroVibroFrac, consists of carrying out a series a interrelated and mutually dependent processes as follows: a preliminary integrated vibrowave treatment of the selected reservoir interval; elimination of formation damage; opening of the existing and creation of new microfractures; hydraulic fracturing by injecting working fluids into the field of elastic oscillations and, thus, creating a network of extensive and deep fractures [11].

The known frac job technologies create fractures by default from the weakest reservoir interval (most probably from its top or bottom), and these fractures can extend to the adjacent nonproductive rocks. By contrast, the HydroVibroFrac method initiates fractures from a selected zone prepared in advance, and this ensures fracture propagation within the confines of the entire reservoir along its width and depth. Oscillatory treatment during chemical injection contributes to efficient filling of the largest and narrowest fractures, and ensures both vertical and horizontal treatment efficiency. The reservoir treatment parameters are aimed to not only increase the number of fractures and the depth of their propagation but also to control their formation and distribution within the reservoir.

The preliminary vibrowave treatment allows cleaning out the flow channels in the porous medium, initiates fracturing fluid penetration into the rock micropores and fractures, ensures the onset of the wedging effect, and weakens the surface rock strength and formation of the fracturing and microfracture the nuclei. Use of the acid solutions as working fluids enhances those processes in both lateral and vertical sense. When needed to improve the cleanout efficiency and weaken the rocks, treatment with elastic oscillations is combined with cyclic pressure-drawdown and overbalance treatment. This leads to the reduction of the breakdown pressure and formation of a network of deep and extensive fractures in the reservoir.

In addition, the above processes are accompanied by rock restructuring, rock stress redistribution, and stress relief in the fractured rocks. When combined with a chemical treatment, the created system of fractures and pore channels becomes irreversible. They do not close when the pressure drops afterward. In many cases, there is no need to pump special proppant into the reservoir fractures, and the frac jobs become easier to carry out.

Specific features and advantages:

- initiation and enhancement of microfracture and macrofracture network formation; residual stress redistribution and relief;
- efficient penetration of the working fluid (acid) into the formed flow channels;
- reduced breakdown pressure; creation of a deep and extensive network of fractures; their reduced tendency not to close even without proppant;
- use of advanced computer-aided measurement systems to monitor and control the hydraulic fracturing operation.

Two technology options are available. One is acid fracturing in sandstone reservoirs involving mud-acid injection. The other is acid fracturing in carbonate reservoirs involving hydrochloric-acid injection. Hydrodynamic generators GD2V-10, GD2V-20, and GD2V-30 are used as downhole equipment.

8.1.10 Hydraulic Fracturing Operations

Well treatment includes two main stages.

The first stage involves bottomhole-zone treatment with elastic oscillations combined with wellbore cleanout. Then, the well is periodically exposed to pressure-drawdown and overbalance treatments, when the bottomhole pressure is increased above the reservoir pressure. Afterward, it is decreased until the reservoir fluids flow to the wellbore. As the hydrodynamic generator is located in the selected reservoir interval, the maximum oscillatory impact is focused in the center of this interval. The bridging materials in this zone soften intensively, and the reservoir-fluid flow entrains them from the existing fractures and pore channels toward the wells. Such cleanout and recovery of the natural permeability along with overbalance and pressure-drawdown stages in the field of elastic oscillations weakens the rock matrix and structure while initiating a great number of fracture initiation nuclei in the central part of the reservoir. When the reservoir is exposed to overbalance and pressure drawdown along with elastic oscillations, they cause additional quick-changing elastic compaction-relaxation strains. This enhances formation of microfractures on the surface of the perforation channels that are perpendicular to the channel axis. In addition, long fractures propagate into the reservoir at large pressure gradients. Drop of pressure in the near-wellbore zone redistributes and reduces residual elastic stresses and reduces the fracture tendency to close; thus, increasing the surface of the pores open for the fluid flow.

After the target reservoir interval is isolated with a packer, the operator proceeds to cyclic pressure changes to expand and deepen the formed fractures. The subsequent injection of working fluid and chemicals along with elastic oscillations, reservoir pressure build-up, and increasing injection rate create the main and extensive network of fractures. The working fluid (acid) efficiently penetrates into the formed flow channels. The breakdown pressure decreases, and the fractures exhibit lesser tendency to close when the pressure drops.

HydroVibroFrac technology option with mud acid was tested in well 8030 of Romashkinskoye field. The well was to be converted to water injection into a shaly siltstone reservoir. A GD2V-20 generator was run into the center of the 6-m-thick perforated reservoir interval. Figure 8.13 shows the wellhead pressure diagram for this injection well during the above test.

If before the test the reservoir started to take water at pressure over 25 to 26 MPa, then after the first stage of the vibrowave treatment, the reservoir started to take water at 18 to 19 MPa. At subsequent stages, the reservoir was fractured with mud acid, the job was carried out stepwise and

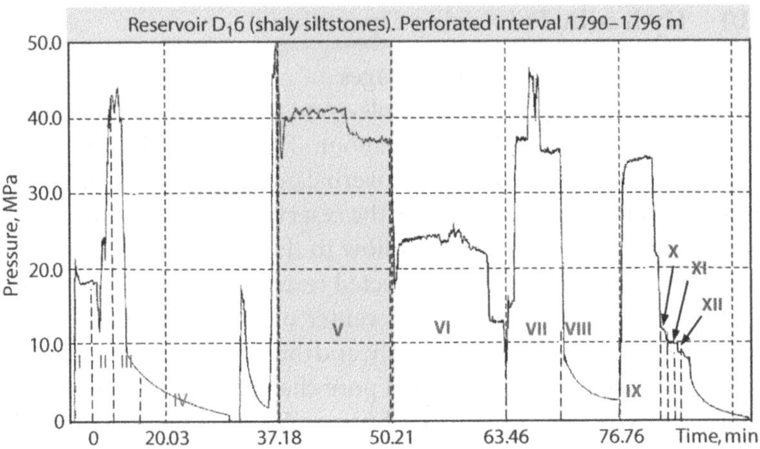

Figure 8.13 Wellhead pressure profile for the injection well 8030 Romashkinskoye Field during HydroVibroFrac treatment. Pre-treatment injectivity, $Q = 0$ m^3/d at $P = 20.0$ MPa; Post-treatment injectivity, $Q = 187$ m^3/d at $P = 9.5$ MPa. **I.** Preliminary vibrowave and overbalance/pressure-drawdown reservoir treatment. Injectivity $Q = 170$ m^3/d at $P = 18$ MPa (not shown in the diagram). **II.** Injectivity assessment, $Q = 6.4$ l/s. **III.** Packer integrity test, $Q = 16$ l/s. **IV.** Pre-treatment bottomhole-zone test. **V.** Fracture job, $V = 30$ m^3, $Q =$ 18.6 l/s. **VI.** Mud acid injection, $V = 11$ m^3, $Q = 12.5$ l/s. **VII.** Overflush, $V = 30$ m^3, $Q =$ 18.6–20 l/s. **VIII.** Soak period. **IX.** Squeze mud acid into the reservoir, $Q = 18.6$ l/s. Various water injection rates: **X.** $Q = 6.2$ l/s; **XI.** $Q = 5.2$ l/s; **XII.** $Q = 4.2$ l/s.

smoothly (see stages V and VI in Figure 8.13). The resulting well Injectivity was 187 m^3/d at pressure of $P = 9.5$ MPa. The final data shows that the treatment results manifested at several stages: the elastic oscillations contributed to preliminary reservoir cleanout and softening of the reservoir rocks within the generator location interval.

The pilot projects were carried out in carbonate reservoirs in 16 producers.

Desirable operational effects were obtained in all the wells; oil production rates increased at least 1.6-fold, averaging 4.1-fold. The gain was very significant in more than 2/3 of the cases. Oil production dynamics in the wells treated with the HydroVibroFrac technology allow expecting that the operational effect will last at least 18 to 20 months. At the same time, the watercut did not increase which confirms indirectly the formation of subhorizontal fractures. All the high gains were obtained in the producer's current reservoir pressure which is 80% and more of the initial one. Water injection system also has a significant influence. The gains were rather low in the wells with low reservoir pressure in tight reservoirs where water was not injected and did not exert any significant influence on the producer performance.

Figure 8.14 shows the bottomhole pressure and temperature profiles that describe the processes that occurred in well 3269 of the Yugomashevskoye Field during its treatment. The plot clearly shows that the pressure and temperature drop at stage V of oil-acid emulsion injection at a pressure of about 23.5 MPa (close to the overburden pressure) indicating that fracture opened. The oil-acid emulsion was injected into the formed fracture at the maximum rate (stages VI and VII). At that, while within a rather short time, the pressure went up to 24.5 MPa (equal to the overburden), one can identify four sections of small temperature drop that, apparently, correspond to discrete fracture extension stages. Another significant fracture opening is identified by temperature and pressure drop at stage VII, when oil-acid emulsion was squeezed into the reservoir at 24.5 MPa pressure. For reasons of economy, the treatment was terminated, though more oil-acid emulsion could have been injected into the open fracture to extend and expand it. Temperature drop is more indicative of the fracture formation

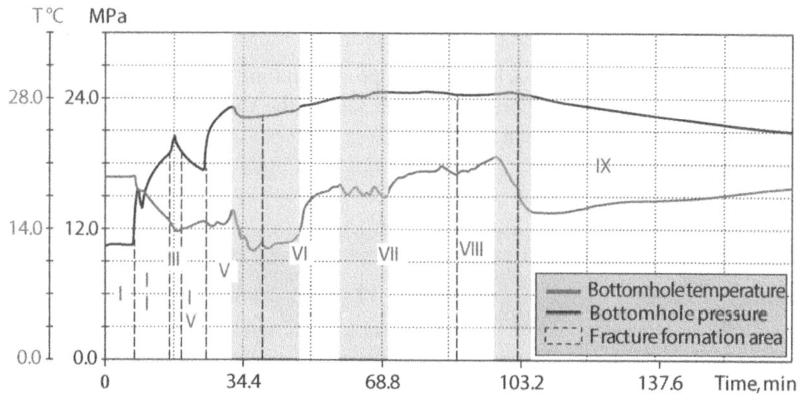

Reservoir S_2bsh (Bashkirian stage carbonates, perforated interval 1,110.0–1,117.6 m).

I. Preliminary vibrowave and overbalance/pressure-drawdown treatment of the target reservoir, HCl injection (P_{init} = 12 MPa, P_{fin} = 10 MPa, not shown on the profile).	
II. HCl injection, Q = 3 l/s. III. Injection of oil-acid emulsion, Q = 6 l/s. IV. Technical shutdown. V. Injection of oil-acid emulsion, Q = 6–7 l/s.	VI. Injection of oil-acid emulsion, Q = 11–13 l/s. VII. Overflush with oil, Q = 7–12 l/s. VIII. Overflush with oil, Q = 3–5 l/s. IX. Soak period

Figure 8.14 Bottomhole pressure and temperature profile for producing well 3269 of the Yugomahevskoye Field during HydroVibroFrac (HVF) treatment.

because it is sensitive to a greater volume of cold fracturing fluid injected into the reservoir from the surface. The fracturing process does not take much time, 1 or 2 hours, or one shift including the preparation and final operations.

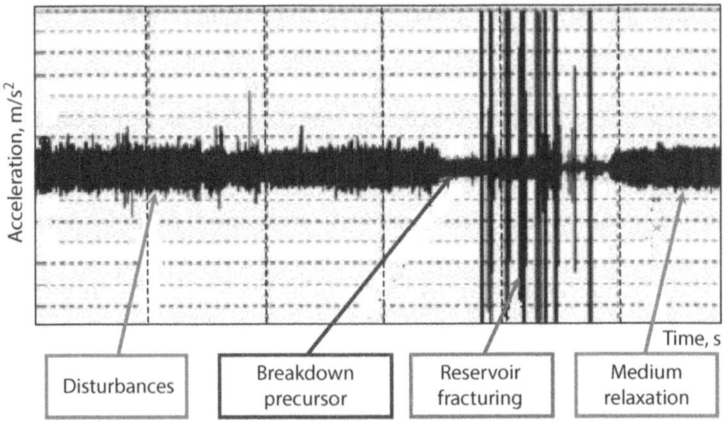

Figure 8.15 Progress of the oil-acid fracturing process in time based on analysis of the SAE signals from the near-wellbore zone in well 3260 of the Yugomashevskoye Field during HydroVibroFrac treatment.

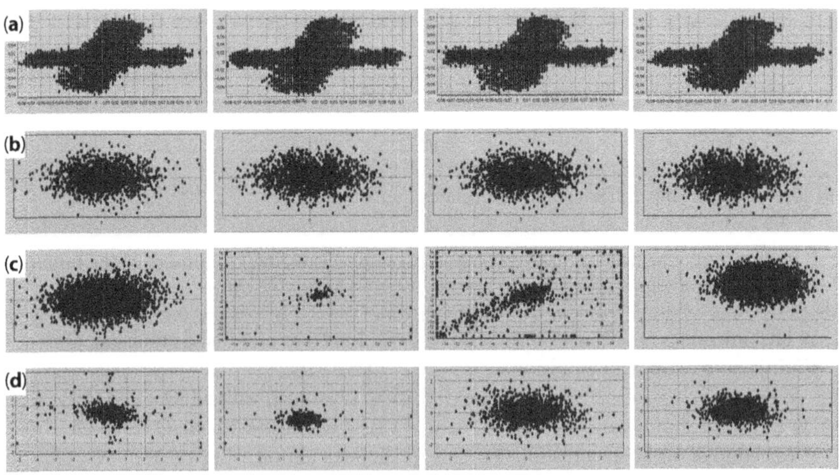

Figure 8.16 Stage diagrams of the hydraulic oil-acid fracturing progress in the near-wellbore zone of well 4269 of the Yugomashebskoye field. (a) Disturbances. (b) Breakdown precursor. (c) Fracturing. (d) Medium relaxation.

Table 8.8 HydroVibroFrac technology performance indicators.

Parameters	Pre-treatment (damaged zone)	Post-treatment (treatment zone)
(**A**) Near-wellbore zone parameters		
Damaged zone radius, m	min 0.5÷10	20÷50
Permeability, μm^2	0.01÷0.20	0.1÷1.0
Skin-factor	+5÷ (−1)	−2÷ (−5)
(**B**) Well production performance		
Oil production rate, t/d	0÷8	3÷30
Liquid production rate, m^3/d	0÷15	3÷50

A system is developed and tested in the wells to control and monitor the fracture-formation process by acoustic emission (AE) signals. Its aim was to identify a sufficient accumulation of AE events caused by fracture formation under vibrowave treatment in order to select the time the most favorable to start hydraulic fracturing. AE parameters are used to assess the future fracture characteristics (length, orientation, etc.). This method will be useful for the HydroVibroFrac treatments to create vertical fractures in deep wells while minimizing breakdown pressure and enhancing the operation efficiency.

As an example of monitoring, Figures 8.15 and 8.16 show the results of AE signal processing for the well 3269 of the Yugomashevskoye field. One can clearly monitor the key stages of the hydraulic oil-acid reservoir fracturing process.

Table 8.8 shows the aggregated performance indicators of the new HydroVibroFrac technology.

8.1.11 Integrated Treatment of Water Production Wells

Regularities of the formation damage and porous-medium cleanout processes have a common nature. Production and injection-well treatments based on wave impact aimed to clean out the well perforated intervals and reservoir porous medium from the impurities can also be applied to water production wells that underperform due to scale deposits in the perforated intervals, silting of the water-bearing reservoirs with fine mechanical and shaly bridging materials.

The authors have developed and tested a technology to treat water production wells using oscillation generator GD2V [3].

The essence of the technology is to excite elastic oscillations within the perforated intervals and in the bottomhole zone using generators by injecting water through them. Moreover, the water is aerated to reduce the bottomhole pressure. Massive reservoirs are treated interval by interval.

To determine the reasonable amplitudes and frequencies of the wave treatments for the water production wells, one should bear in mind the design features and higher flow properties of the water-bearing formations. The authors determined conditions for the optimum input of the oscillatory energy into the reservoir. Under such treatment regime, the productive reservoir interval absorbs maximum of the oscillatory energy. Such regime initiates pulsing flows in the porous medium that contribute to degradation and removal of the bridging particles from the perforated interval.

If one considers the contribution of such attenuation to the radiation that is useful in terms of cleanout effects, when the frequencies are lowered down below 100 Hz, the fraction of the energy that is radiated into the reservoir increases to a certain value that depends on the behavior of the attenuation factor curve. At frequencies higher than 100 Hz, this phenomenon does not actually contribute to the emissions from the wells, and no intense pulsing flows arise in the reservoir. The resulting data is used to set and select the operation mode for the hydrodynamic generators to treat the water production wells.

Application:

- Water production wells of all types up to 500-m-deep productivity of which decreased due to the silting and clogging of the perforations with bridging materials of various origins.

Piping schemes and technology implementation are similar to the schemes and operations used under VPV technologies. Very thick reservoirs are treated interval by interval. Once the lower interval treatment is over, the bottomhole of the well is washed with clean water pumped through the generator.

Sixteen water production wells in the Tulva Field in the Permian region were treated by this technology. The wells were 16-m deep; water-bearing formation was 6-m thick. Results of the treatments showed their high efficiency and success rate. Well production rates increased 3- to 5-fold.

Implementation of this technology increased the well productivities 2- to 10-fold.

8.2 Enhanced Oil Recovery Technologies Based on Vibroseismic Treatment (VST)

The oilfield VST operation results are covered in the literature in detail [2, 3].

Analysis of the pilot operations results shows that the VST operations degas the geological media, redistribute the stress-strain state of the rock massif, change physical properties of the reservoir fluids, increase well production rates, reduce production watercuts, increase the near-wellbore zone permeability, change water saturations and relative permeabilities, and bring the bypassed oil into production.

However, the known papers do not always give an unbiased assessment of the VST results. In some cases, they are ambiguous, and the technologies used require improvement and optimization

The essence of the vibroseismic treatment technology developed by the authors of the book consists in cyclic treatments of the individual areas of the field using mobile seismic vibrators that are held for a short time in singular points. Selection of the treatment points and mobility allow treating large areas [14–17]. The VST treatments take into account the field geology and the current state of the field production as well as stress-strain state of the fields, presence of faults. The VST operating methods contribute to the following:

- eliminate the formation damage;
- form the gas phase as a displacing agent;
- release and displace the trapped slow-moving and residual oil;
- activate the aquifer to replenish reserves of the reservoir elastic energy; enhance field production;
- initiate forced resuscitation and inflow of reservoir-fluids from deep formations.

The faults are detected as follows:

- construct depth structure maps of the tops of producing reservoirs based on drilling data;
- identify interblock structural boundaries and parcels inside the blocks;
- identify structural faulted zones in the reservoirs based on detailed seismic-profile shooting data;

- assess the fault inheritance degree in vertical sense based on the available data, including interpretation of the aerial and satellite survey data;
- analyze performance data on the wells located in the faulted areas.

Pilot VST treatments were carried out from time to time in the Yugomashevskoye, Chetyrmanskoye, and Tatyshlinskoye fields during 2 years. Terrigenous reservoirs of the Lower and Middle Carboniferous stage were treated. A block of the Earth's crust where the area of the oil field is located is described as an inhomogeneously stressed field where local areas of stressed and faulted state formed with alternating stress and relaxation zones.

Table 8.9 shows the geological and production data of the productive reservoirs and physical and chemical properties of the saturating fluids. The current watercut is 80% to 90%; oil production rate of the wells averages about 1.0 t/d.

Seismic vibrators MERTZ M/20 (see Figure 8.17 and Figure 8.18) were used to run the VST operations. Their specifications are shown in Table 8.10.

Operating regimes of seismic vibration treatment differed subject to the treatment point location and treatment parameters, i.e., cycle durations and times between the treatment cycles (see Table 8.11).

The working sweep-signal frequency range was 10 to 40 Hz. However, the optimum frequency in terms of the combined effect manifestation was about 20 Hz.

In the first year, treatments were mostly focused within the oil-bearing contour, while the following year, the VST treatments were aimed at the identified faulted elements in the reservoirs. Two cycles were carried out in each treatment point within the oil-bearing contour, including about 1 hour of the seismic vibrator continuous operation and a technology break of the same duration. The lapse of time between the cycles was 48 hours.

To enhance the treatment regime in one of the areas in the Yugomashevskoye field, two seismic vibrators were used simultaneously and synchronically. In the second year, the duration of the treatments in the faulted areas and the lapse of time between the cycles were increased to several days.

The annulus pressure, liquid level in the wellbore, liquid production rates, and watercuts were measured from time to time to monitor the well performance under treatment.

Table 8.9 The geological and production data of the productive reservoirs and physical and chemical properties of the saturating fluids.

Field	Main producing formations	Initial reservoir temperature (°C)	Physical and chemical oil properties			
			Permeability (μm²)	Pay thickness (m)	Viscosity (reservoir conditions) (mPa·s)	Bubble-point pressure (MPa)
Tatyshlinskoye	Middle Carbonif.	18–20	0.07	1.2–6.2	14.5	5.3
	Lower Carbonif.	22–24	0.40	1.5–9.8	7.4	2.7
Chetyrmanskoye	Lower Carbonif.	22–24	0.56	1.2–6.8	10.3	7.1
Yugomashevskoye	Middle Carbonif.	18–20	0.05	1.7–6.5	12.0	5.7
	Lower Carbonif.	22–24	0.53	2.1–11.2	21.5	7.0

Figure 8.17 General view of the seismic vibrator MERTZ M/20.

Figure 8.18 The exterior of the seismic vibrator MERTZ M/20 base plate.

Table 8.10 Seismic vibrator MERTZ M/20 technical specification.

Maximum force output, kN	270
Working frequency, Hz	7 to 250
Base plate area, m^2	2.62
Base plate weight, kg	4,082

Table 8.11 Operating regimes of seismic treatment.

Force output, kN		220	
Working frequency, Hz		10–40	
Motor hour	Continuous run time, min	40	
	Technology break, min	20	
Number of motor hours per day		10	
		Within oil bearing contour 1st year	Faulted areas 2nd year
Number of motor hours at one location		1–3	3–5
Interval between treatments at one location, hours		2–3	0–2

The VST treatments resulted in the following:

- periodic annulus pressure build-up;
- variation of the dynamic liquid level;
- liquid production surge (multifold increase of short duration);
- change in production rates and watercuts as compared to those before treatment.

One of the indirect but immediate indicator of the seismic treatment influence on the productive reservoir was to increase the annulus pressure observed in almost every observation well. In the observation wells of the Yugomashevskoye field, the annulus pressure went up to 3.8 MPa (well 4261) and exceeded 3.0 MPa in some other wells. The annulus pressure changes reached 3.1 MPa during the treatment. The dynamic levels in the producing wells did not vary much, even when the annulus pressure varied dramatically (in well 4129 from 0.3 to 1.5 MPa).

In many wells, noticeable variations of the dynamic level occur at the same time as the surges of the liquid production rate. Both single and multiple surges were observed in all the areas.

Sharp changes in oil production rates were reported in the course of the treatments. In half the cases, the wells exhibited stable higher production rates afterward, while the watercut decreased.

Subject to their variation in the wells that showed positive results, the following groups of wells can be identified:

- liquid production rate increase and watercut decrease (see Figures 8.19 and 8.20, wells 1337, 2400, 4282G, 4283);

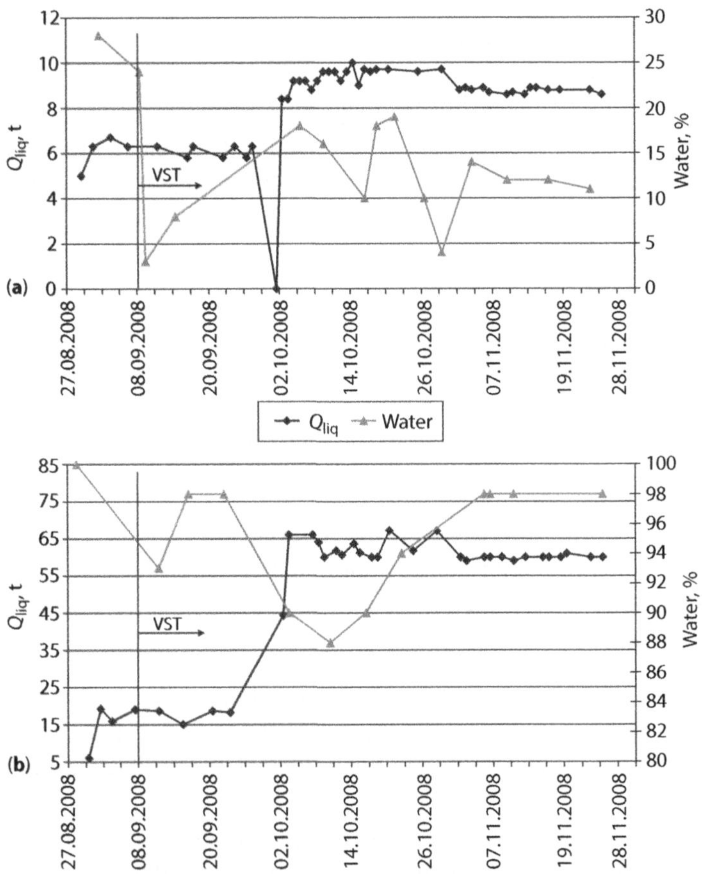

Figure 8.19 Performance behavior of the Yugomashevskoye wells, lower Carboniferous terrigenous formations. (a) Well 1337 (Area 2). (b) Well 2400 (Area 4).

- lower watercut/lower liquid production rate (see Figure 8.21, wells 3614 and 3192);
- drop of the liquid watercut at rather constant liquid production rate (see Figure 8.22, wells 1733 and 3126);
- higher liquid production rates and higher watercut (see Figure 8.23, well 2308).

The authors would like to highlight in particular the variation of performance indicators of well 1738. This well is located near the treatment area and near the identified en echelon fault. Well 1738 was drilled to the Lower Carboniferous terrigenous formations. Prior to the treatment, it produced

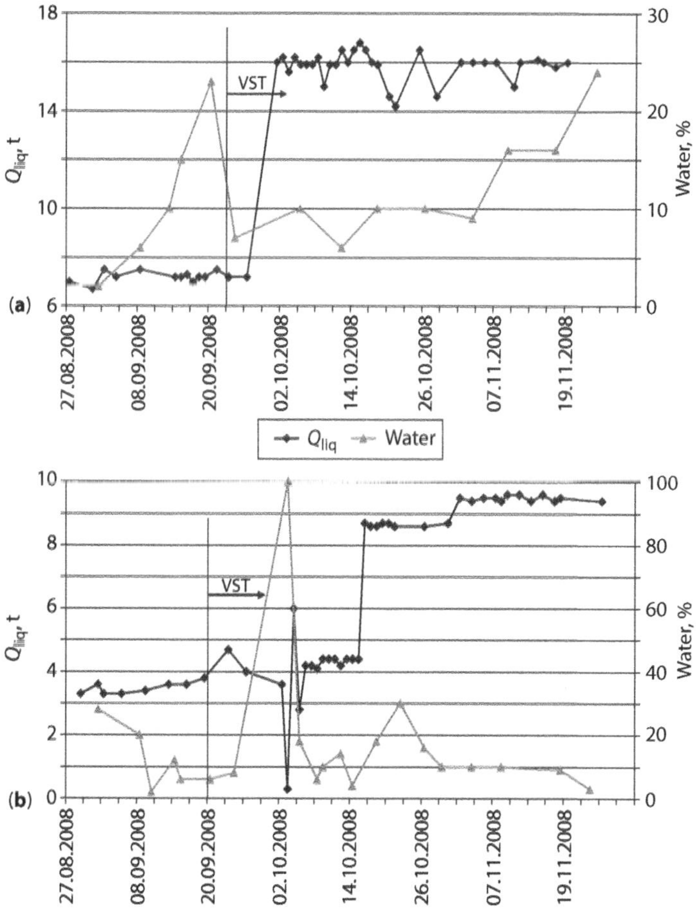

Figure 8.20 Yugomashevskoye wells performance, Area 2 (terrigenous lower Carboniferous formations). (a) Well 4282G. (b) Well 4283.

1.5 m3/d of liquid at 90% to 95% watercut. After the VST treatment, the well started to produce through the sucker rod pump at the initial rate of 130 m^3/d and 6.0 MPa tubing wellhead pressure. After bleeding off, the tubing wellhead pressure built up in 30 seconds. Subsequently, the liquid production rate slightly decreased (see Figure 8.24).

Unconventional response of the well 1739 production rate to the VST treatment suggests most probably that the deeper formations feed reservoir fluids through the faults. Low reservoir pressure in the vicinity of well 1738 (no water injection) contributed to the breakthrough of the accumulated hydrocarbon reserves after the VST treatment.

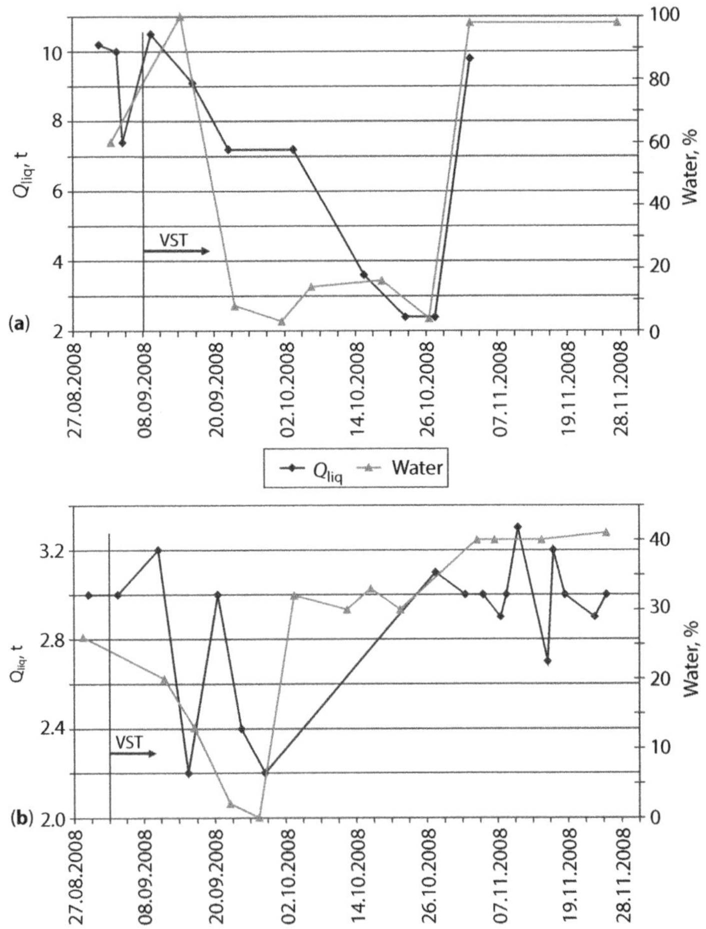

Figure 8.21 Yugomashevskoye wells performance (lower Carboniferous terrigenous formations). (a) Well 3614 (Area 1) middle Carboniferous. (b) Well 3192 (Area 4) (lower Carboniferous terrigenous formations).

Interesting results were obtained in the Tatyshlinskoye field. The VST treatments were carried out here both in the first and the second year in Area 4. The best positive effects were suddenly observed in a production well in the adjacent Area 5 where no VST treatment was carried out. Such results could be explained by the connectivity that established through the activation of the fault existing between for the two areas, and initiation of the ascending flows. Here, the TVDSS of the reservoir top in Area 5 are higher than in Area 4.

The final VST treatment results are summarized in Table 8.12. As shown in this table, the VST treatments carried out in the second year targeting the faulted elements of the reservoir structure resulted in a

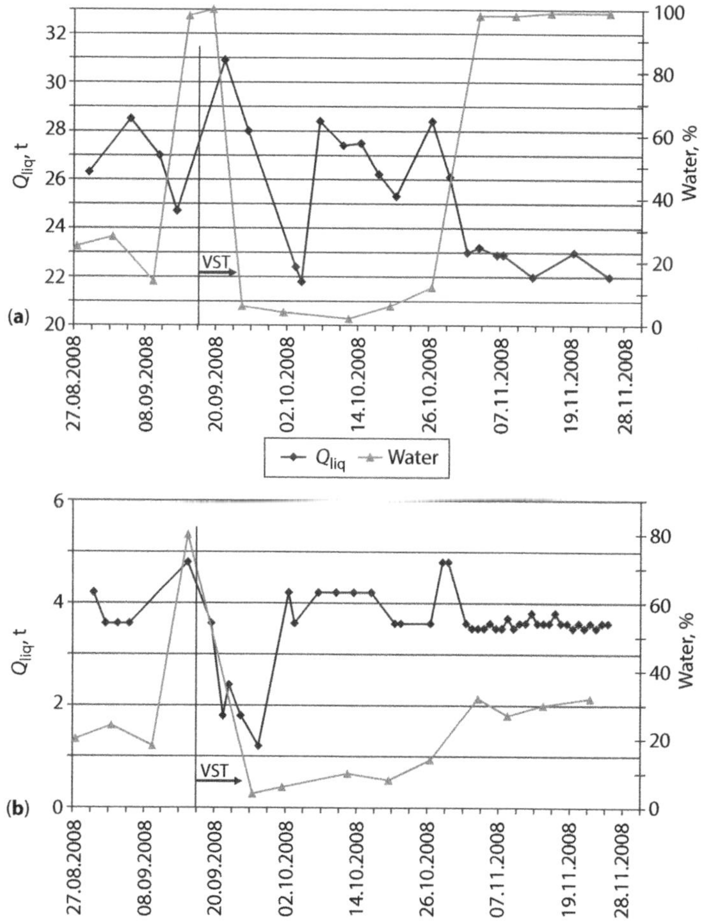

Figure 8.22 Yugomashevskoye wells performance, Area 1 (lower Carboniferous terrigenous formations). (a) Well 1733. (b) Well 3126.

higher incremental oil production in a shorter time as compared to the first year. The total amount of incremental oil produced from the targeted areas within the monitoring period exceeded 15,000 tons, including 71% from the Lower Carboniferous and 29% from the Middle Carboniferous. Thirty to forty percent of the treated production wells (subject to the reservoirs and fields) exhibited positive VST results. However, 15% to 20%

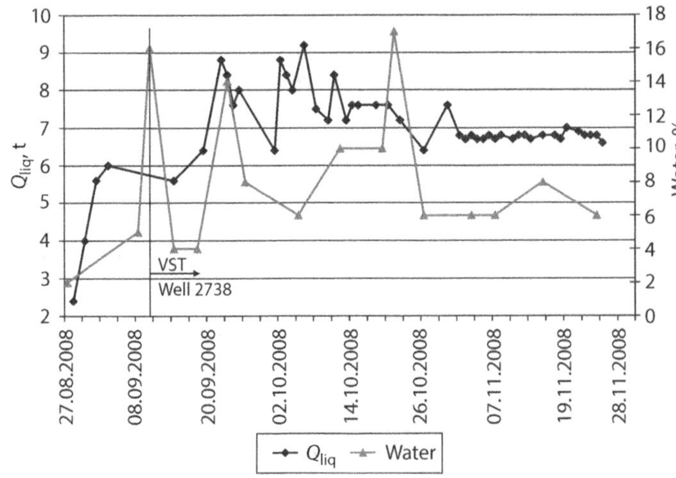

Figure 8.23 Yugomashevskoye well 2308 performance behavior, Area 1 (lower Carboniferous terrigenous formations).

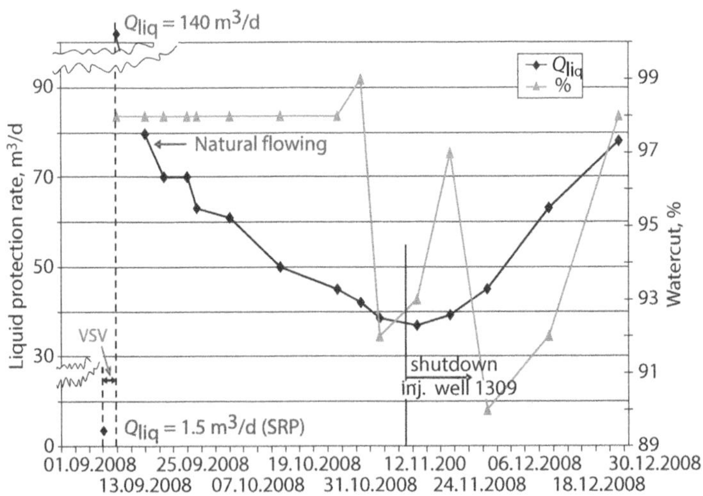

Figure 8.24 Performance history of well 1738, Yugomashevskoye field, Area 1 (lower Carboniferous terrigenous formation).

Table 8.12 Results of VST treatments in the fields of the Bashneft oil company.

Field	Number of responding wells		Treatment period (start – end)		Incremental oil production, tons	
	1st year	2nd year	1st year	2nd year	1st year	2nd year
Tatyshlinskoye	53	65	07.09–12.11	27.08–21.09	Over 7000*	Over 8000**
Chetyrmanskoye	20	344	16.10–24.11	02.10–26.10		
Yugomashevskoye	179	301	04.09–26.11	27.08–11.10		

* *In addition, a noticeable incremental oil about 1000 tons is produced as of 01.05.2008 in the area where 12 wells were drilled. The area is situated 2 km away from the VST site in Area 4 of the* Tatyshlinskoye *Field. One can presume there is a fault between those areas.*

** *Well 2738 of the Yugomashevskoye Field pumped with SRP at the rate of 1.5 m3/d of liquid (0.1 t/d of oil) started to flow naturally though the pump at the rate of about 130 m^3/d.*

of the above wells accounted for the most part of the enhanced recovery. Each of the wells produced at least 50 to 100 tons of incremental oil, which is equivalent to 50% or 100% increase in oil-production rate. As applied to all the responding wells, the average incremental oil production amounted to 10% to 12%.

Based on the results of the above oil field operations and the VST treatments in the Mancharovskoye Field operated by the Bashneft Oil Company, the authors have formulated the main requirements to meet when selecting the VST target reservoirs; have also developed the VST procedure that includes the following stages:

- *examine the reservoir geology features;*
- *identify faults, stress-strain reservoir zones;*
- *select treatment areas;*
- *select the required special equipment;*
- *estimate the expected technological effect;*
- *carry out the VST treatment;*
- *carry out the seismic monitoring of the treatment;*
- *control the well performance during the treatment;*
- *estimate the actual technological effect.*

Main requirements to select target reservoir for the VST treatment:

- *presence of faults within the confines or along the reservoir limits linked with the deep earth layers;*
- *target reservoirs or areas that are produced at low reservoir pressure (at depletion drive); low sweep efficiency; border zones;*
- *high energy potential of the aquifer zones and low connectivity with oil zones;*
- *reservoir porosity at least 10%; permeability about 0.01–0.15 mm^2, maximum oil viscosity 60 mPa·s;*
- *transport accessibility of the heavy-duty special vehicles to the target areas subject to the environmental legislation.*

Summary:

1. Analysis of the VST operations showed the efficiency of the developed procedure and viability of their further expansion.
2. By the end of following year after the VST treatment, the incremental oil production exceeded 15 thousand

tons, mainly (71%) coming from the terrigenous Lower Carboniferous formations. VST treatments improved well performance, namely: annulus pressure increased from time to time; dynamic levels changed; production water decreased; liquid production rates increased in 30% to 40% of the wells located within the VST treatment impact zone. The well production rates increased in average by 10% to 20%.

3. The greatest changes in production-well performance indicators in absolute values resulting from the VST treatments are achieved in the reservoir areas with low reservoir pressure and occurring at small true vertical depths subsea.
4. Well 1738 of the Yugomashevskoye Field exhibited an unexpectedly high response to the VST treatment. The well pumped with SRP and producing 1.5 m^3/d of liquid started to flow naturally at 130 m^3/d at the tubing wellhead pressure of 4.0 to 6.0 MPa. Most probably, the VST treatment enhanced the oil inflow from the deep formations through the faults.
5. It makes sense to use VST treatments to solve the following issues:

   - simultaneously eliminate formation damage for a group of wells without killing operations and round trips:
   - activate aquifer areas with the aim to replenish the reservoir reserves of elastic energy and to enhance production;
   - initiate, as well as enhance or reactivate inflow of reservoir fluids from the deep Earth formations through the faults and subvertical inversion-annular structures;
   - implementation of water-alternating-gas injection under favorable conditions (high gas content of formation water.

The technologies discussed in this chapter were successfully tested and are implemented in the fields with different geological and physical characteristics in the Urals-Volga and West-Siberia regions of Russia. Moreover, they proved their efficiency in the CIS countries (Kazakhstan and Ukraine) as well as abroad (Germany, China, Syria, Peru, and Ecuador).

Target reservoirs and conditions for implementation of the technologies are shown in Table 8.13.

The authors of the book engaged several service companies to test and implement the wave technologies.

Table 8.13 Target reservoirs and conditions for implementation of integrated technologies.

<table>
<tr><td>Wells</td><td>Well Categories
- prospecting;
- exploratory;
- development</td><td>Well's Purpose
- producing,
- injecting;
- special (including water production wells)</td><td>Wellbore Profile
- vertical;
- subvertical;
- inclined;
- horizontal;
- sidetracks</td><td colspan="2">Well Design
- depth from 50 to 5,000 m;
- size (nominal diameter):
(1) min 114.3 mm (4.5”) for the casing;
(2) min 101.6 mm (4”) for sidetracks;
- bottomhole: (1) cemented liner; productive interval is perforated with jet guns, drilling guns, sand-jet guns and mechanical guns with Mg or Zn plugs etc.; minimum 10 shots per foot;
(2) uncemented liner, screen liner;
(3) open hole in carbonate reservoirs.</td></tr>
<tr><td rowspan="2">Productive reservoirs</td><td>Lithology</td><td colspan="2">Extremely Admissible Reservoir Properties</td><td>Oil Viscosity and Oil Saturation</td><td>Reservoir Pressure Anomaly Ratio Ka, Fraction</td></tr>
<tr><td>- terrigenous: sandstones, siltstones, shaly sandstones;</td><td colspan="2">Porosity | Permeability | Volume of Shale
Minimum: | Minimum: | Maximum:
- for ordinary wells:
15% 0.010 μm² 10%
- for wells with propped hydraulic fractures:
8% 0.005 μm² 30%</td><td>Up to 200 mPa·s,
S_o min 60%</td><td>For reservoir tops up to:
Ka min:
down to 1,000 m - 0.25–0.35;
1,000–1,500 m - 0.40–0.60;
1,500–2,500 m - 0.70–0.75;
2,500–3,000 m - 0.75–0.80;</td></tr>
</table>

(*Continued*)

Table 8.13 Target reservoirs and conditions for implementation of integrated technologies. (*Continued*)

<table>
<tr><td rowspan="2"></td><td>- carbonates:
limestones,
dolomites.</td><td>- for ordinary wells:
7% 0.005 μm² 5%
- for wells with propped hydraulic fractures:
5% 0.002 μm² 15%</td><td>Up to 150 mPa*s,
So min 50%</td><td>3,000–3,500 m - 0.80–0.85;
3,500–4,000 m - 0.85–0.90;
> 4,000 m - 0.90–0.95
Static level Hst max 900 m.</td></tr>
<tr><td colspan="2">By type of voids
1/ pores; 2/ fractures; 3/ vugs; 4/ combination</td><td colspan="2">Reservoir temperature max 110°C (max 150°C when customized equipment is used)</td></tr>
<tr><td>Bridging elements</td><td>Inorganic
- clays;
- silts;
- scales;
- mud;
- corrosion products.</td><td>Organic
- ASPO;
- bacteria-induced;
- mud polymers.</td><td colspan="2">Other
- reaction products after chemical treatments;
- structured systems of technogenic and biogenic nature;
- emulsions</td></tr>
</table>

More than 5,000 wells were treated in 1985 to 2020. As a result, the vertical sweep efficiency increased by 50% to 90%; well productivities increased 1.5- to 8-fold; well utilization rate increased 1.5- to 3-fold. Well production watercut decreased by 3% to 30%. Success rate was 90%; the effect duration was 6 to 18 months and longer. Incremental oil production during those years exceeded 3 million tons. More than 30 million additional cubic meters of water were injected into the reservoirs.

In addition, a quarter of the producers and half the injectors were brought to production from among the previously inactive wells, temporarily abandoned wells, scheduled to be abandoned, i.e., they were actually reactivated.

More than 600 production wells and more than 500 injection wells were brought online using wave technologies, once the drilling was over. The most part of those well were horizontal wells and sidetracks in the terrigenous formations of the Devonian, Visean age and Vikulovskaya formations as well as carbonate reservoirs of the Famennian and Tournaisean ages, Middle Carboniferous and Lower Permian formations.

Inasmuch as some operations have injectivitiy problem with tight carbonate and sandstone resources, these operators can use technology proposed by Chilingar and Haroun, 2014 [18].

References and Bibliography

1. Gadiev S.M. The use of vibration in oil production. M.: Nedra, 154 p., 1977.
2. Dyblenko V.P. Wave treatment of oil reservoirs with unconventional reserves. Review and classification. M.: JSC VNIIOENG, 80 p., 2008.
3. Dyblenko V.P., Marchukov E.Yu., Tufanov I.A., Shariffullin R.Ya., Evchenko V.S. Wave technologies and their implementation to produce unconventional reserves. M.: RANS Publishing House, 338 p., 2012.
4. International Application PCT. RU No. 2010120080. Method of productive reservoir treatment and downhole equipment for its implementation. V.P. Dyblenko, O.L. Kuznetsov, Yu.V. Eremenko, R.Ya. Sharifullin *et al.* Publ. in B.I. 2011. No. 33.139. Patent No. 2258803 RF. Method of productive reservoir treatment. V.P. Dyblenko, I.A. Tufanov. Publ. in BI. No. 23., 2005.
5. Patent No. 2261990 RF. Method of the productive reservoir thermal gas dynamic treatment and solid-fuel charge for its implementation. V.P. Dyblenko, R.Ya. Sharifullin, A.P. Lysenkov *et al.* Publ. in BI. No. 28., 2005.
6. Patent No. 2228437 RF. Method to shut off water or gas entries or fluid-loss zones. V.P. Dyblenko, I.A. Tufanov, A.V. Ovsyukov *et al.* Publ. in BI. No. 13., 2004.
7. Patent No. 2366806 RF. Method of physical impact to produce hydrocarbon deposits and the downhole unit for its implementation. V.P. Dyblenko., A.P. Lysenkov, Yu.S. Ashchepkov *et al.* Publ. in BI. No. 25., 2009.

8. Patent No. 2087756 RF. Liquid-flow oscillations generation method and downhole unit for its implementation. E.Yu. Marchukov, V.G. Tarasenko, V.I. Zhdanov, V.G. Bazarov, V.P. Dyblenko *et al.* Publ. in BI. No. 23., 1997.
9. Patent No. 2144440 RF. Method to excite the fluid—flow oscillations and hydrodynamic oscillator. V.P. Dyblenko, E.Yu. Marchukov, V.I. Zhdanov, I.A. Tufanov. Publ. in BI. No. 2., 2000.
10. Patent No. 2366806 RF. Method of physical impact to produce hydrocarbon reservoir and the downhole unit for its implementation. V.P. Dyblenko., A.P. Lysenkov, Yu.S. Ashchepkov *et al.* Publ. in BI. No. 25., 2009.
11. Patent No. 2258803 RF. Productive-reservoir treatment procedure. V.P. Dyblenko, I.A. Tufanov. Publ. in BI. No. 23., 2005.
12. Patent No. 2228437 RF. Water or gas entry and fluid—loss zone shut off procedures. V.P. Dyblenko, I.A. Tufanov, A.V. Ovsyukov *et al.* Publ. in BI. No. 13., 2004.
13. Patent No. 2258803 RF. Productive-reservoir treatment procedure. V.P. Dyblenko, I.A. Tufanov. Publ. in BI. No. 23., 2005.
14. Patent No. 2349741 RF. Hydrocarbon-reservoir production procedure through geological-environment treatment. V.P. Dyblenko, V.S. Evchenko. Publ. in BI. No. 8., 2009.
15. Patent No. 2247828 RF. Oil-field production procedure. V.P. Dyblenko, O.L. Kuznetsov, R.S. Khisamov *et al.* Publ. in BI. No. 7., 2005.
16. Patent No. 2268996 RF. Hydrocarbon production procedure through the physical treatment of the geological environment. V.P. Dyblenko, O.L. Kuznetsov, R.S. Khisamov *et al.* Publ. in BI. No. 3., 2006.
17. Patent No. 2291954 RF. Hydrocarbon reservoir production procedure through the integrated physical reservoir treatment. V.P. Dyblenko, R.Ya. Sharifullin, I.A. Tufanov *et al.* Publ. in BI. No. 2., 2007.
18. Chilingar, G.V., and Haroun, M. Electrokinetics for Petroleum and Environmental Engineers. Scrivener Publishing (Wiley) New Jersey, 240 pp, 2014.

9
Laboratory Experiments

9.1 Laboratory Experiments

Laboratory experiments were done using (1) natural sandstone cores, (2) sandpacks prepared from quartz sand, and (3) sandpacks prepared from quartz and bentonite. The following aspects of production were studied:

1. Variation of residual oil saturation with time in natural core samples and sandpacks (a) in the presence of elastic waves generated by a vibrator and (b) when no vibrations were applied to the cores.
2. Variations in the speed of displacement of oil by water while subjecting, and not subjecting, cores to vibration.
3. Variations in the relative permeability to oil and to water due to the effect of elastic vibrations.
4. Change in water cut when imposing elastic waves.
5. Variations in the rate of degassing of fluids in cores and sandpacks due to action of elastic waves.

In all laboratory and field experiments, the frequency of the vibro-sources was in the range of 100 to 200 Hz, which is close to the range (10 to 100 Hz) utilized in seismic prospecting. As a result, seismic waves generated by these vibro-sources could travel considerable distances (in the range of 1 to 2 km) without significant attenuation.

9.1.1 Oil and Water Saturations of the Porous Medium Exposed to Elastic Waves

Core samples from the Elabuzhskiy oil field and sandpacks prepared from quartz sand were used. Kerosene ($\rho_k = 820$ kg/m^3) and water ($\rho_w = 1{,}000$ kg/m^3) were used as model fluids. The kerosene-water interfacial tension was 25 × 10 N/m, and the initial water saturation was 5%. Natural core

George V. Chilingar, Kazem Majid Sadeghi, and Oleg Leonidovich Kuznetsov. Acoustic and Vibrational Enhanced Oil Recovery, (311–320) © 2022 Scrivener Publishing LLC

samples were 0.05 to 0.26 m in length, with a diameter of 0.03 m. The permeability of core samples to nitrogen was 1.0 μm^2. Sandpacks had a length of 0.25 m, diameter of 0.02 m, and permeability of 9.3 μm^2.

Capillary saturation was measured at natural conditions and on applying vibro-energy at an acceleration of 7.8 m/s^2. After completion of each experiment, the initial conditions in the sandpacks were restored.

Figure 9.1 shows the effect of vibro-energy on the oil recovery and the rate of capillary water saturation. Two modes of capillary saturation were tested in combination with the applied vibro-energy and in its absence: (a) direct-flow capillary saturation when the direction of water movement in the process of capillary saturation coincided with direction of the fluid flow caused by the pressure gradient; and (b) inverse-flow capillary saturation when the direction of the water movement in the process of capillary saturation was opposite to the direction of fluid flow caused by the pressure gradient.

As shown in Figure 9.1, the rate of oil recovery and the displacement of hydrocarbons by water considerably increase in the case of direct-flow capillary saturation when vibrations are applied to the cores and sandpacks. In the absence of vibrations, oil recovery from sandstone was at a level of about 32% after 300 hours of monitoring oil saturation. In the presence of elastic waves, the oil recovery increased to 60% in 51 hours.

In the case of inverse-flow capillary saturation, vibro-energy caused not only a considerable increase in the rate of oil displacement but also resulted

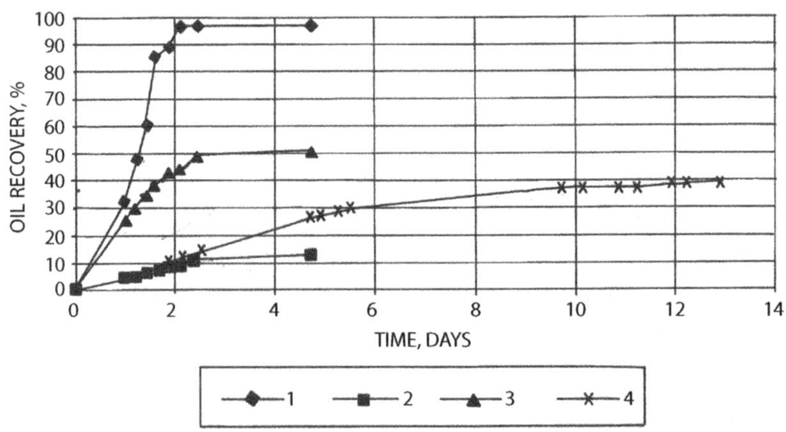

Figure 9.1 Rate of oil recovery from natural core samples in the presence and absence of elastic waves (vibration-stimulation). *1* = capillary saturation in the direction of fluid flow, no vibro-energy; *2* = capillary saturation in the direction of fluid flow, vibroenergy applied; *3* = capillary saturation against direction of fluid flow, no vibro-energy; *4* = capillary saturation against direction of fluid flow, vibro-energy applied.

in the recovery of practically all oil from the sandpacks. In the absence of vibrations, 56% of the initial oil was recovered from the sandpacks over 325 hours. In the presence of elastic waves, the oil recovery increased to 96% in 92 hours.

These experiments showed that, in both cases, direct-flow and inverse-flow capillary saturation, utilization of vibration energy not only increases the total oil recovery but also increases the rate of oil production. In the presence of vibrations, the speed of oil flow was about three times higher than that in its absence.

9.1.2 Rate of Displacement of Oil by Water and Effect of Elastic Waves on Relative Permeability to Oil

Some experiments were conducted using vertical sandpacks having a length of one meter and a diameter of 35 cm. The sandpacks, which were prepared from quartz sand, had a porosity of 0.33 and permeability of 5.4 μm^2. At first, the model was saturated with water (ρ_w = 1,000 kg/m^3), which was then displaced by kerosene (ρ_o = 820 kg/m^3). After that, oil was displaced by water to produce a natural level of residual oil saturation. After completing displacement of oil by water, the oil saturation stabilized at a level of 50%.

The model was installed vertically and the gravitational redistribution of oil in the two series of experiments was monitored when: (1) vibro-energy was not applied to the model (gravitational displacement of oil by water alone); and (2) elastic waves with an acceleration of 37.4 m/s^2 (combination of gravitational displacement and the effect of elastic waves) were applied.

In the absence of vibro-energy, the water/oil ratio was 0.6 at the top of the model and 1.37 at the bottom of the model after 20 days (see Kouznetsov *et al.* [11]).

When gravitational redistribution of oil and water phases was combined with the applied vibro-energy, a higher degree of oil-water gravitational separation was achieved in a time period about 500 times shorter than that in the case of natural gravitational redistribution alone. This illustrates a radical increase in the relative permeability of the porous medium to oil in the presence of elastic waves.

9.1.3 Degassing of Fluids by the Applied Vibro-Energy

Degassing of formation fluids under the influence of elastic waves is important because the volume of the free gas is increased. Figure 9.2 shows the dynamics of degassing of water containing CO_2 on application of elastic waves.

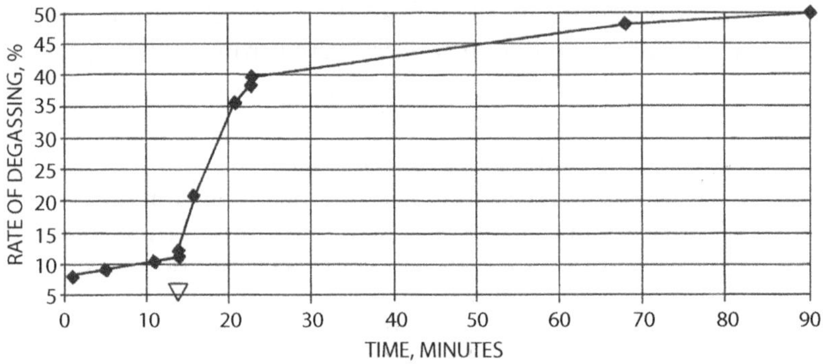

Figure 9.2 Degassing of CO_2-saturated water in the presence of elastic waves.

Before beginning the vibration, the volume of gas separated from the water grew at a rate of about 0.2% per minute. Immediately after initiation of vibrations, this rate increased to 3% to 4% per minute. After the relative volume of free gas reached 40%, the rate of degassing stabilized once again at the rate of 0.2% per minute.

Degassing is related to local instantaneous ruptures in the formation fluid due to the effect of elastic waves. Nuclei of bubbles are formed in these ruptures, and gas diffusion from fluid into these bubbles takes place. The bubbles periodically contract and expand due to the periodic variation of pressure caused by elastic waves. During contraction, gas migrates from the bubbles into the surrounding fluid, whereas during expansion, gas migrates from the fluid to the bubbles. Because during the expansion gas migration takes place through a larger surface area of the bubble compared to that during contraction, more gas moves into the bubbles from the fluid than out of the bubbles into the fluid. This causes a systematic increase in the size of the bubbles with a corresponding increase in the degassing rate.

Experimentally, degassing of oil in the field of elastic waves was evidenced by an increase (up to 20% to 30%) in the gas-saturation pressure. Gas bubbles released from the fluid due to seismic energy may plug some of the pore throats and channels and change the relative permeabilities to water and to oil. Whereas the relative permeability to water always decreases due to the release of free gas into the formation, the relative permeability to oil may either increase or decrease, depending on the water saturation.

Figure 9.3 shows dependence of relative permeability to oil and to water on water saturation at conditions when gas saturation is equal to (a) 0% and (b) 10%. For the same water saturation, the relative permeability to water is always lower when the gas saturation is higher. When water saturation is between 0% and 40%, the relative permeability to oil is lower

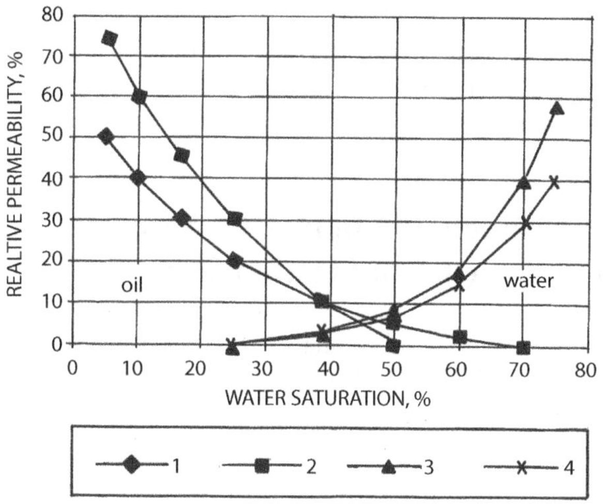

Figure 9.3 Relative permeabilities to oil and to water (three-phase fluid flow). *1* = gas saturation = 10% (permeability to oil); *2* = gas saturation = 0% (permeability to oil); *3* = gas saturation = 10% (permeability to water); *4* = gas saturation = 0% (permeability to water).

when the gas saturation is higher. When water saturation exceeds 40%, an increase in gas saturation leads to an increase in the relative permeability to oil. It is important to note that in the case of zero gas saturation, the relative permeability to oil decreases to zero at a water saturation of 50%, whereas for 10% gas saturation, the relative permeability to oil is zero at a water saturation of 70%. This means that, at high water saturation, degassing effects due to the application of elastic waves will increase the relative permeability to oil, which will lead to an increase in the oil production rate and a decrease in water cut. Figure 9.3 shows the relative permeabilities of oil and water in the three-phase fluid flow.

9.2 Displacement of Oil by Gas-Free Water in the Presence of Elastic Waves

The effect of elastic waves on the rate of oil displacement by water was studied using transparent sandpack cylinders 0.35 × 0.95 × 0.01 m in size. The models were filled with quartz sand resulting in a porosity of 45% and permeability of 18 to 20 μm^2. Initially, the models were saturated with gas-free formation water, which was then displaced by oil. Experiments were conducted in two steps. First, vibration energy was not applied to the model. Oil displacement continued until the oil output stopped. After that, the same model was subjected to the vibration energy which resulted in the additional displacement of oil by water additional recovery. An additional 10% of oil was recovered from the model, which is equal to about 20% of the residual oil remaining in the formation after waterflooding.

9.3 Displacement of Oil by CO_2-Saturated Water in the Presence of Elastic Waves

Experiments were conducted to analyze the effect of vibro-energy on the rate of oil recovery when oil is displaced by CO_2-saturated water. As a first step in these experiments, oil was displaced by CO_2-saturated water, and the displacement process continued until oil content in the produced fluid decreased to zero. After that, the process of oil displacement resumed with vibro-energy applied to the model. Elastic waves caused extensive release of carbon dioxide, leading to the complete degassing of water. Figure 9.4 shows dependence of oil recovery from the quartz sandpack on the cumulative fluid produced. Percentage of the recovered oil is presented as a function of the total cumulative volume of fluid for two cases: a. when oil is displaced by gas-free water; and b. when oil is displaced by CO_2-saturated water. In both cases, the displacement process was started in the absence of vibro-energy and then continued with vibro-energy applied to the sandpack. In the case of CO_2-saturated water, magnitude of the oil recovery was systematically higher than that in the case of gas-free water. Subjecting the

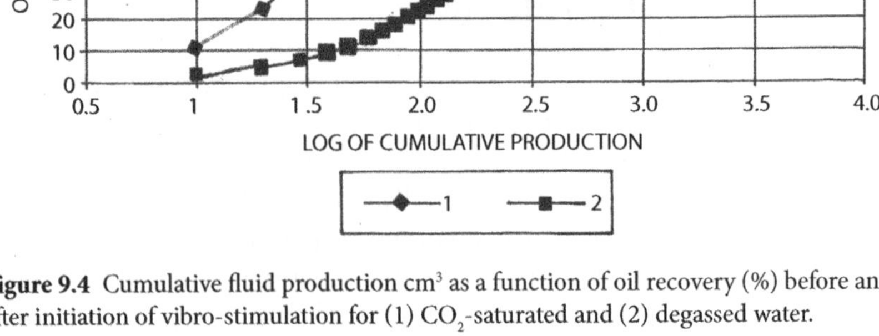

Figure 9.4 Cumulative fluid production cm³ as a function of oil recovery (%) before and after initiation of vibro-stimulation for (1) CO_2-saturated and (2) degassed water.

sandpack to elastic waves resulted in a further increase in oil recovery than in the cases of gas-free and CO_2-saturated water.

9.4 Modeling of Oil Displacement by Water in Clayey Sandstones

Analysis of the effect of elastic waves on displacement of oil by water in clayey sandstones was done using sandpacks of 40 cm long and 2.8 cm in diameter. The sandpacks were prepared with quartz sand having grain sizes of 0.05 to 0.2 mm and a bentonite admixture. The permeability of the sandpack was 0.2 to 0.4 μm^2, whereas porosity was 25%. Distilled water with dissolved NaCl (30 to 200 g/l) was used for displacement of oil, which was derivative by a mixture of kerosene with transformer oil having viscosity of 5.0 MPa s.

Prior to the experiments, residual water saturation was attained in the sandpacks. Saline water (200 g/l) was displaced by the oil until water was no longer produced from the sandpack. The experiments showed that increasing the salinity of water to 150 g/l led to lower swelling of clay than that at a salinity of 50 to 100 g/l. Lower swelling results in an increase of the permeability to oil and to water. Thus, subjecting clayey sandstone to elastic waves leads to more effective oil displacements in the case of 150 g/l water salinity. Experiments showed that the oil recovery at this salinity level reached 86.8%.

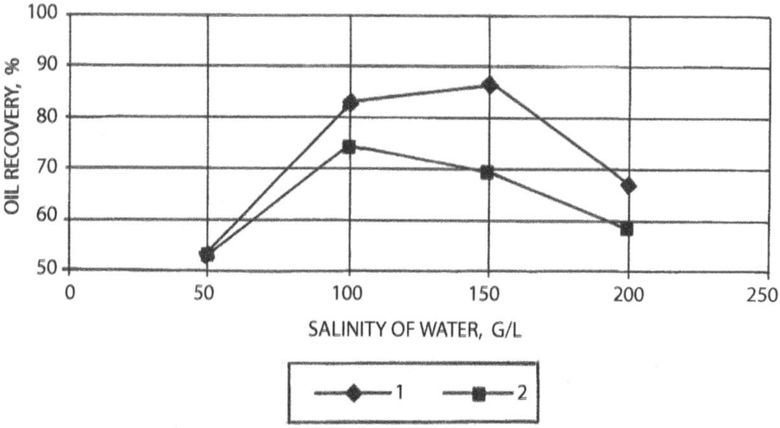

Figure 9.5 Dependence of the oil recovery on water salinity (1) in the presence of elastic waves and (2) in the absence of elastic waves.

Dependence of the rate of oil recovery on the salinity of water (50, 100, 150, and 200 g/l) that displaces the oil was also studied. The intensity and frequency of the elastic waves were the same for all salinity values. Experimental results showed that the influence of elastic waves on the oil production changed with changing salinity of the injection water, with maximum oil recovery attained at a water salinity of 150 g/l. Effect of elastic waves on the oil recovery is illustrated in Figure 9.5, which shows dependence of oil recovery on the water salinity in the presence and absence of elastic waves. One can observe that the application of elastic waves increases the oil recovery at all salinity levels.

References and Bibliography

1. Kouznetsov, O.L. and Simkin, E.M., Transformation and Interaction of Geophysical Fields in the Lithosphere, p. 271, *Oxford and IBH Publ*, 1994.
2. Simkin, E.M., Hydrodynamic description of oil displacement by water process within the elastic vibration field, in: *Proc. Int. Conf. Flow through Porous Media: Fundamentals and Reservoir Engineering Applications*, pp. 171–175, 1992.
3. Simkin, E.M. and Surguchev, M.L., Advanced vibroseismic technique for waterflooded reservoir stimulation, mechanism and test results. *Sixth European Symp. on Improved Oil Recovery*, 1991.
4. Surguchev, M.L., Kouznetsov, O.L., Simkin, E.M., Hydrodynamic, Acoustic, and Thermal Cyclical Processing of Oil Layers, p. 320, *Nedra Publ.*, Moscow, 1975.
5. Surguchev, M.L., Zheltov, U.V., Simkin, E.M., Physicochemical Microprocesses in Oil-Gas-Bearing Formations, p. 215, Nedra Publ., Moscow, 1984.
6. Vahitov, G.G. and Simkin, E.M., Utilization of Physical Fields for Enhancement of Oil Recovery, p. 231, *Nedra Publ.*, Moscow, 1985.
7. Fedin, L.M., Fedin, K.L., Fedin, A.K., *Basics of Increasing Heavy Oil Recovery*, p. 112, Dolyam, Simpheropol, 2013.
8. Beresnev, I. and Johnson, P., Elastic-wave stimulation of oil production: A review of methods and results. *Geophysics*, 59, 1000, 1994.
9. Beresnev, I., Theory of vibratory mobilization of nonwetting fluids entrapped in pore constrictions. *Geophysics*, 71, 2006.
10. Kouznetsov, O.L., Simkin, E.M., Chilingar, G.V., Katz, S., Improved oil recovery by application of vibro-energy to waterflooded sandstones. *J. Petrol. Sci. Eng.*, 19, 191–200, 1998.
11. Kouznetsov, O.L., Simkin, E.M., Chilingar, G., *Physical Bases of Acoustic and Vibrational Effects on Oil and Gas Deposits*, p. 260, Izd. Mir, Moskow, 2001.

12. Kouznetsov, O.L., Simkin, E.M., Chilingar, G.V., Gorfunkel, M., Robertson, J., Seismic techniques of enhanced oil recovery: Experimental and field results. *Energy Sources*, 22, 877–890, 2002.

10
Oil Field Tests

10.1 Abuzy Oil Field

The first field experiments with vibrators located at the surface were started in 1987 at the Abuzy oil field in the Krasnodarskiy Region of Russia. Two 50-kW vibrators were placed 50 m apart. Water saturation of the reservoir rock was 97%, whereas the depth of the oil-bearing formation was 1,500 m. These experiments were the first ones to demonstrate the improvement in oil production due to use of vibro-energy. On the average, the water cut decreased by 20% to 26%, whereas for some boreholes, the drop in water cut reached 50% and more. Degassing of formation oil was observed:

1. Volume of the hydrocarbon gases in the air samples at a depth of 45 to 100 m increased as much as 10 to 110 times.
2. Volume of the hydrocarbons in gas samples at a depth of 1,150 to 2,500 m in the oil-bearing formations increased 28 to 40 times.
3. The values of the light absorption coefficients for the oil samples decreased from the producing wells in productive intervals of 1,150 to 2,500 m.

10.2 Changirtash Oil Field

Another cycle of experiments was conducted at the Changirtash oil field, Kirgizstan, in 1986. The Changirtash oil field is one of the oldest fields in Kirgizstan, where oil production started in 1938. Its length from north to south is 4.5 km, whereas its width is 3.1 km. The oil-bearing formation is represented by the light-gray, fine-grained sandstones with thin layers of clay. Average depth is 410 to 570 m, porosity is 10.5% to 23%, permeability is 0.1 to 30 μm^2, and the initial oil saturation was 60%. Total thickness of the oil-bearing formation is 25 m, whereas the pay zone thickness is

George V. Chilingar, Kazem Majid Sadeghi, and Oleg Leonidovich Kuznetsov. Acoustic and Vibrational Enhanced Oil Recovery, (321–326) © 2022 Scrivener Publishing LLC

5.7 to 6.4 m. Initial production rates of the boreholes were 2 to 3 tons/day, whereas at the time of these experiments, they were 0.5 to 1 tons/day. There were 89 wells across the oilfield, and waterflooding was done through 25 boreholes. Total oil production from the oil field was 1,150,000 tons of oil, when water saturation reached 90%; the oil recovery was 31%. Experiments with vibrators were conducted in two areas:

1. The size of the first study area was 600 × 600 m. The oil-bearing formation is represented by homogeneous fine-grained sandstone with permeability of 0.2 μm^2. The depth of the oil-bearing formation is 240 to 448 m and its thickness is 4 to 10 m with a dip of 35°. The oil-bearing formation was penetrated by six boreholes with oil production rates ranging from 0.08 to 0.9 tons/day. Total yearly production in this area was 800 tons, with a water cut of 69%.

Two surface vibrators were installed at the center of the area, 20 m apart. Power of the vibrators was 50 kW each, and elastic waves were continuously generated during a one-month of October, 1988. During this time and later, samples of oil and gas were taken and oil production rates were recorded. Considerable increase in the content of light components was reported in the gas samples; for example, percentage of ethane increased from 45.3% to 53.3%. In the oil samples, the initial boiling point decreased and the amount of residuals increased due to the effect of elastic waves. This indicates enrichment of oil in light fractions. This phenomenon is similar to the one recorded at Abuzy oil field. The water cut decreased by 25% to 30%, and the additional oil produced was greater than 300 tons. It is also important to note that this effect continued after vibrations were discontinued.

2. In the second field experiment, the area was larger (1,500 × 1,500 m), the depth of the oil-bearing formation was greater (511 × 708 m), and the permeability (0.15 μm^2) of the producing formation was lower. The formation was penetrated by 16 boreholes having a total oil production of 2,540 tons. Water cut for the recovered oil was 89%. For the efficient transfer of vibration energy from the surface-located vibrators to the oil-bearing formation, a special wave-direction technique was used. Loss of energy using this technique ranged from 5 to 10% for 1 km of depth.

Figure 10.1 shows the relationship between the cumulative oil production and cumulative water production for two conditions: (a) absence of vibrations and (b) presence of vibrations. The start of vibrations is marked by ∇. Data are presented for both pilot tests and show the relationship between the cumulative oil production and cumulative water production. For both pilot tests, the decrease in water cut due to the effect of elastic waves was in the range of 20% to 25%. Total increase in the oil production

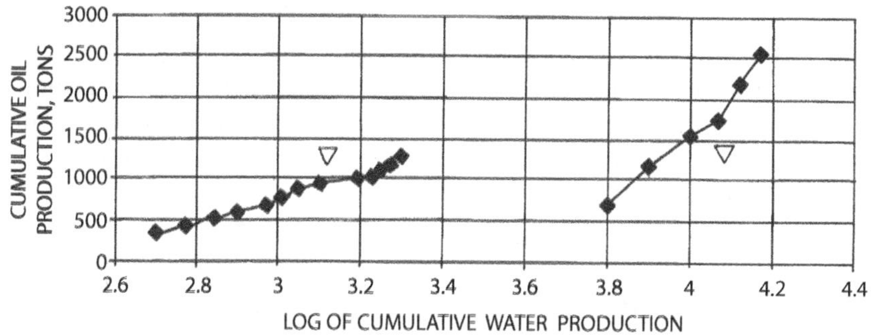

Figure 10.1 Cumulative oil production as a function of cumulative water production before and after stimulation by vibro-energy at Changirtash oil field at two pilot sites.

for the first pilot test for a period of 4 months from the beginning of vibration was greater than 300 tons, whereas for the second pilot test, the additional production exceeded 1,000 tons.

10.3 Jirnovskiy Oil Field, First Stage

First stage experiments at Jirnovskiy oilfield (Volgogradskiy region, Russia) were conducted between 1991 and 1993. According to the previously conducted hydrodynamic investigations, productive layer in the pilot experimental area, in spite of high water saturation in the oil-producing boreholes, contained an oil lens with total oil reserves of about 250,000 tons. During experiments, the vibrator was placed at a borehole located approximately at the center of the area. The vibration energy was transmitted from the surface to the oil-bearing layer (depth of about 1,000 m) via a specially designed wavetransmitting system which was lowered into the borehole. The monitoring radius of vibrations was 2.2 km. During the period of vibration, 35 wells located in the vicinity of the vibrator were monitored, and variations in the oil recovery were recorded for each borehole. Among these, 26 boreholes showed improved oil recovery. The remaining nine boreholes did not react.

During the first stage of experiments on the effect of vibro-energy on oil production, measurements of the intensity of transmitted seismic energy to the producing horizon were conducted. Recording of the seismic vibrations was done in a well located 850 m from the well through which the seismic energy was transmitted to the formation. The magnitude of seismic waves during vibration at the level of oil-producing formation was about 200 times greater than the background noise. In the neighboring formations, the magnitude of seismic waves was about 1.5 to 2 times lower than that in the oil-producing formation. This illustrates the efficiency of transmitting the seismic energy from the surface down to the geologic section and of focusing the energy on the oil-producing formation. After vibrations were stopped, the level of seismic energy in the oil-bearing formation slowly decreased.

10.4 Jirnovskiy Oil Field, Second Stage

The second-stage experiments at the Jirnovskiy oil field were conducted during the period of May 18 to July 13, 1992. The length of the period of

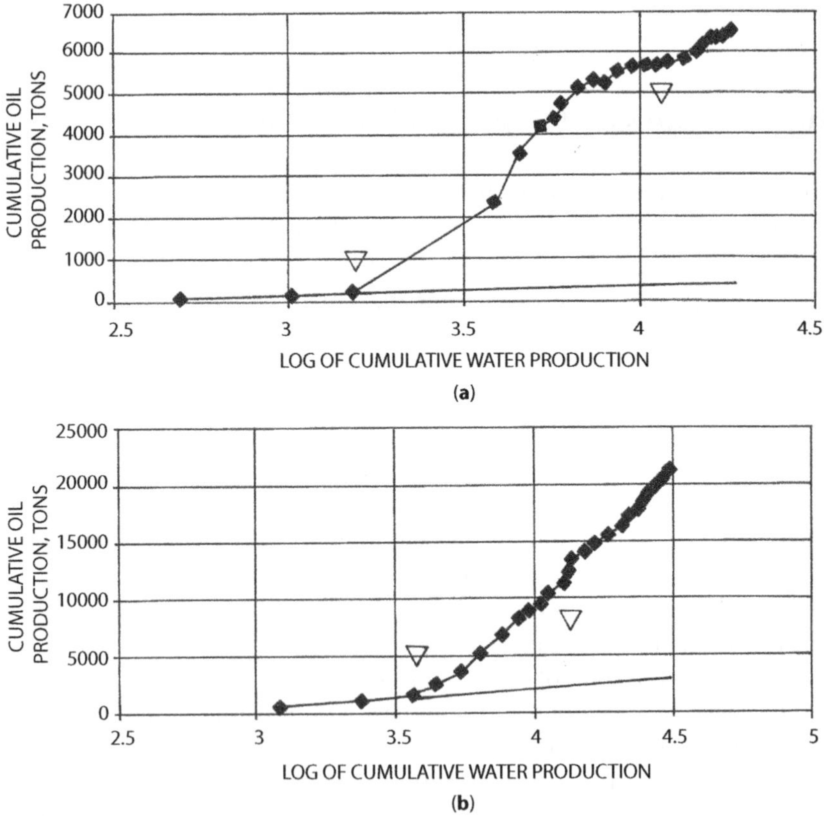

Figure 10.2 Cumulative oil production as a function of cumulative water production before and after start of vibro-stimulation for boreholes no. 231 (a) and no. 1056 (b).

generation of vibro-energy was the same as that used during the first-stage experiments at this field.

Analysis of results during the first and second stages of experiments showed that the oil production increased in about 40 wells on application of vibro-energy. In some boreholes, the decrease of water cut was 2- to 4-fold, whereas oil production increased 2- to 5-fold. Figure 10.2 shows two examples (wells 231 and 1056) of decrease of water cut due to the effect of elastic waves. Before initiation of vibration, the recovered oil/water ratio was very low. Soon after starting the vibration this ratio increased considerably.

Average data for 40 boreholes are given in Figure 10.3, which shows that during 2 years after starting vibration, the recovered oil/water ratio remained much higher than that prior to the vibro-stimulation. Total increase in oil production was about 250,000 tons. Figure 10.4 shows the average change in the water cut for the Jirnovskiy field (during approximately 1 year). One can see that as a result of application of vibro-energy, average water/oil ratio decreased from 9.7 to 8.3, although later, possibly due to exhaustion of the oil-producing lens, water/oil ratio increased.

It is important to note that the increase in oil production owing to the effect of vibration energy is due to change in the relative permeabilities (and decrease in the water cut) and due to reduction in the contact angle, interfacial tension between oil and water and greater dispersion of oil in water. For theoretical discussion, see Langnes *et al.* ([11], pp. 225–230).

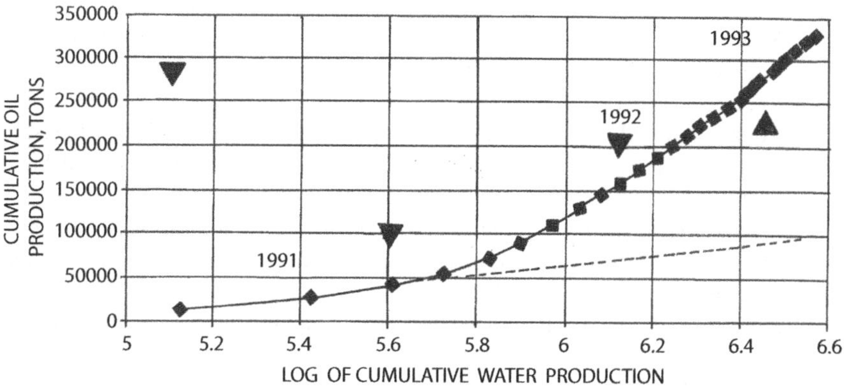

Figure 10.3 Cumulative oil production as a function of cumulative water production before and after initiation of vibro-stimulation at Jirnovskiy oil field (first pilot test). ▼ is the start of the vibrostimulation.

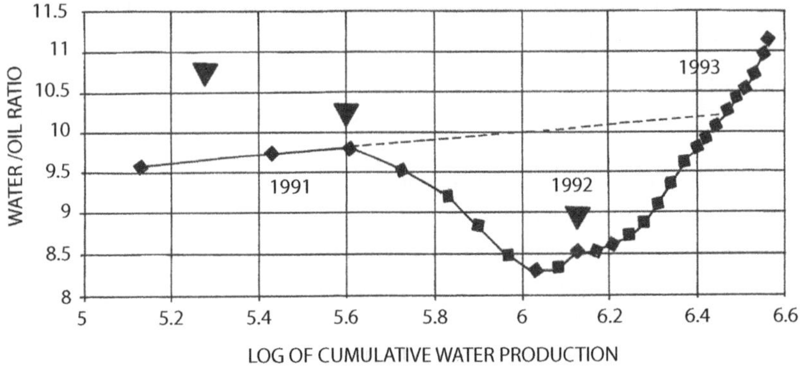

Figure 10.4 Water/oil ratio as a function of cumulative water production before and after start of vibro-stimulation at Jirnovskiy oil field (first pilot test). ▼ is the start of the vibro-stimulation.

References and Bibliography

1. Beresnev, I. and Johnson, P., Elastic-Wave Stimulation of Oil Production: A Review of Methods and Results. *Geophysics*, 59, 6, 1000:10017, 1994.
2. Beresnev, I., Theory of Vibratory Mobilization of Nonwetting Fluids Entrapped in Pore Constructions. *Geophysics*, 71, 6, 2006.
3. Fedin, L.M., Fedin, K.L., Fedin, A.K., *Basics of Increasing Heavy Oil Recovery*, p. 112, Dolyam, Simpheropol, 2013.
4. Granat, N.L., Settled Oscillation of Particles in a Two-Phase Mixture. *Iz. Akad. Naux SSR*, *5*, 70–75, 1964.
5. Kouznetsov, O.L., Simkin, E.M., Chilingar, G.V., *Physical Bases of Vibrational and Acoustic Effects on Oil and Gas Deposits*, p. 260, Izd. Mir, Moscow, 2001.
6. Kouznetsov, O.L., Simkin, E.M., Chilingar, G.V., Katz, S.A., Improved Oil Recovery by Application of Vibro-Energy to Waterflooded Sandstones. *J. Pet. Sci. Eng.*, 19, 191–200, 1998.
7. Kouznetsov, O.L., Simkin, E.M., Chilingar, G.V., Gorfunkel, M., Robertson, J., Seismic Techniques of Enhanced Oil Recovery: Experimental and Field Results. *Energy Sources*, 22, 877–890, 2002.
8. Lysenko, V.D., *Theory of Oil Field Development*, p. 310, Nedra., Moscow, 1993.
9. Pride, S., Flekkoy, E., Asutjo, O., Seismic Stimulation for Enhanced Oil Recovery. *Geophysics*, *73*, 23–35, 2008.
10. Simkin, E.M., Hydrodynamic Description of Oil Displacement by Water Process within the Elastic Vibration Field, in: *Proc. Int. Conf. Flow through Porous Media: Fundamentals and Reservoir Engineering Applications*, 171–175, 1992.
11. Langnes, G.L., Robertson, J.O., Chilingar, G.V., Secondary Recovery and Carbonate Reservoirs, p. 304, *Elsevier*, New York, NY, 1972.

11

Electrokinetic Enhanced Oil Recovery (EEOR)

Electrokinetic Enhanced Oil Recovery (EEOR) is an emerging technology that could significantly improve oil recovery, at cost below other secondary and tertiary oil recovery technologies, in environments where other technologies either do not work or are not attractive. EEOR has a non-existent water demand, a smaller carbon footprint than the traditional Enhanced Oil Recovery (EOR) technologies, such as steam injection, and does not involve injection of hazardous liquids, as in the case for caustic and/or co-solvents flood. EEOR requires minimal surface facilities. No steam generator, compressors, surface working fluid pumps, and/or hazardous material storage tanks are required. EEOR involves passing direct current (DC) electricity between cathodes (negative electrodes) in the producing reservoir and anodes (positive electrodes) either at the surface and/or at depth.

11.1 Introduction

Considerable hydrocarbon resources, of all grades, remain as residual oil in many older "depleted" fields, which are being abandoned. Depending upon the reservoir type:

- Light oil (i.e., greater than 20° API gravity) primary and secondary recoveries range from 25% to 75%. This means that at abandonment, as much as 75% of the *Stock Tank Original Oil In Place* (STOOIP) is left behind.
- Heavy oil (i.e., less than 20° API gravity) crude recovery is severely limited because of the difficulty in extracting these high viscosity liquids from oil reservoirs. Primary recoveries as low as 10% of the STOOIP reserves are common. Enormous heavy oil resources exist throughout the world (Attanasi and Meyer, 2010 [10]; Meyer *et al.*, 2007 [64]).

George V. Chilingar, Kazem Majid Sadeghi, and Oleg Leonidovich Kuznetsov. Acoustic and Vibrational Enhanced Oil Recovery, (327–380) © 2022 Scrivener Publishing LLC

Both of these types (i.e., light and heavy) of residual oil resources are known but are lacking a viable technology for their efficient and economic recovery.

The shallowest unrecoverable oil reservoirs, such as the Shallow Athabasca Tar Sands, of Alberta, and McKittrick Field, of California, may be mined, if surface hydrocarbon extraction from the mined rock is economic. Various forms of steam-assisted recovery may work well for deeper heavy oil reservoirs. Steam-assisted recovery has also been utilized as a finishing technique for lighter oil reservoirs (Blevins *et al.*, 1984 [16]).

Steam-assisted oil recovery, however, may not always be an option. The maximum depth for steam extraction has been increasing but the practical limit still appears to be between 2,000 and 2,500 ft. Deeper unrecovered oil resources require alternative technologies.

Steam-assisted oil recovery may also not be appropriate for very shallow resources (e.g., less than 500 ft), because of the potential danger of steam break-through to the surface (Staff, 2011 [83]).

Many reservoirs also contain reactive matrix minerals, which can cause irreversible reservoir damage from steam recovery technologies. Finally, steam-assisted oil recovery may face environmental hurdles, in areas where water is in short supply, wastewater disposal is a problem, and/or steam generators, compressors, and pumps require air permits.

Direct Current (DC) Electrokinetically Enhanced Oil Recovery (DCEOR) may be a viable alternative to steam-assisted oil recovery because of the following:

- It is operational outside of the above steam-assisted oil recovery depth windows.
- It has a negligible water demand.
- It generates no wastewater streams, beyond that of normal water cut.
- It appears to actually reduce water cut, below primary recovery levels.
- It has negligible to positive impact on reactive minerals.
- It may actually clear pore throats of colloidal and fine-grained particles.
- It has negligible air quality impacts.
- It may also reduce the production of noxious and toxic gases, such as H_2S.

11.2 Petroleum Reservoirs, Properties, Reserves, and Recoveries

Before discussing Electrokinetically Enhanced Oil Recovery, we need to briefly review petroleum reservoirs, reserves, and multi-phase saturation, as well as primary, secondary, and enhanced recovery. This background is needed to better understand the impact of direct electrical current on oil production.

11.2.1 Petroleum Reservoirs

Unfortunately, some jurisdictions refer to petroleum reservoirs as *Pools*, which conjures up images of vast underground lakes of petroleum. In reality, most petroleum reservoirs consist of porous rocks, which more closely resemble fine-grained sponges, rather than caverns filled with gas or lakes of liquid petroleum.

11.2.2 Porosity

Reservoir rocks may be considered to be made up of solid (matrix) and voids (pores). The percentage of rock volume occupied by pores is called porosity, Ø.

11.2.3 Reservoir Saturations

The pore spaces in reservoir rocks are filled with (usually) water, as well as oil and/or gas, as well as non-economic and toxic gases. Economic hydrocarbons are light (C_1–C_3) gases, liquid (C_4–C_{10}), and heavy ($>C_{10}$) hydrocarbon liquids. Common non-economic gases are air, N_2, and CO_2. Toxic gases include H_2S. Liquid petroleum is made up of the heavier (i.e., C_4 and above) hydrocarbon chain and ring molecules. Liquid contaminants are sulfur and heavy metal hydrocarbon compounds

Oil and water are immiscible, fluids, with neither particularly soluble in the other. Gas is soluble, to a point, in oil. Beyond that point, however, it becomes a separate phase. The same is true for gas and water.

Oil is (usually) less dense than water and gases are less dense than either oil or water. Because of these density differences, if all three are present in a given reservoir, they will segregate on the basis of density, with gas at the top and water at the base (i.e., much like Jell-O 1-2-3®). Pore throat surface effects at the contact between the fluids and rock matrix, however, will insure that these contacts are not always very sharp.

The types and amounts of the fluids, which will co-exist, at any given depth, will depend upon reservoir temperature and pressure, as well as pore space capillarity, surface tension between the immiscible fluids, wettability between the rock matrix and fluids, and the contact angle between the different immiscible fluids and the matrix.

Clastic (sand and shale) reservoir rocks tend to be preferentially *water wet*, because of the polar nature of water and the net negative charges on the pore throat surfaces, due to cation substitution in some silicate minerals. This usually results in some water, coating the pore throats, throughout the reservoir, making it preferentially water wet. Carbonate reservoirs tend to be preferentially *non-water wet*, as the cation substitution prevalent in clastic minerals is not present in carbonates. If, however, the carbonate reservoirs are lined with clay minerals they will also become partially to preferentially water wet. Finally, some clastic reservoirs also can be either one.

The percentages of the pore space filled with water, oil, and/or gas, at any depth, is called water, S_w, oil, S_o, and/or gas, S_g, saturation, respectively.

11.2.4 Initial Reserves

Initial reserves are the amounts of oil and/or free gas in the reservoir, before any production has taken place. For oil, the definition is

$$STOOIP = \frac{7758Ah\phi(1-S_w)}{B_{oi}}, \tag{11.1}$$

where *STOOIP* is stock tank original oil in place, in bbls.

> *A* is the reservoir closure area, in acres.
> *h* is the average reservoir (net) thickness, in feet.
> φ is the average reservoir decimal porosity.
> S_W is the average reservoir decimal water saturation.
> B_{oi} is the initial oil formation volume factor.
> 7758 is the acre-ft to bbls conversion factor (*different conversion constants will be required, for other unit systems*).

A similar relationship also exists for gas reserves.

11.2.5 Primary Oil Production and Water Cut

Wells are drilled into petroleum reservoirs to recover the oil and/or gas. Reservoir energy determines whether an oil well will flow (*without assistance*) or must be pumped. Flowing reservoir energy is provided by the dissolved gas, mechanical compaction, and/or hydraulic pressure within the reservoir. As a reservoir is depleted, more than one fluid type will be produced. For oil reservoirs, not only oil will be produced but also associated gas (*gas cut*) and/or water (*water cut*).

11.3 Relative Permeability and Residual Saturation

Figure 11.1 is a generic representation of an oil/water (two phase) relative permeability determination, for a preferentially water-wet reservoir rock. For multi-phase systems, the relative permeability is a measure of the amount of each fluid, which can be produced from a reservoir and under what conditions. Laboratory relative permeability measurements are utilized by reservoir engineers to aid in depletion planning.

The oil/water ($k_{o/w}$) relative permeability curves are complex functions of the following:

- The pressure gradient within the reservoir;
- The relative viscosities of the oil and water phases;
- The interfacial tension, between the two immiscible phases;
- The matrix wettability;

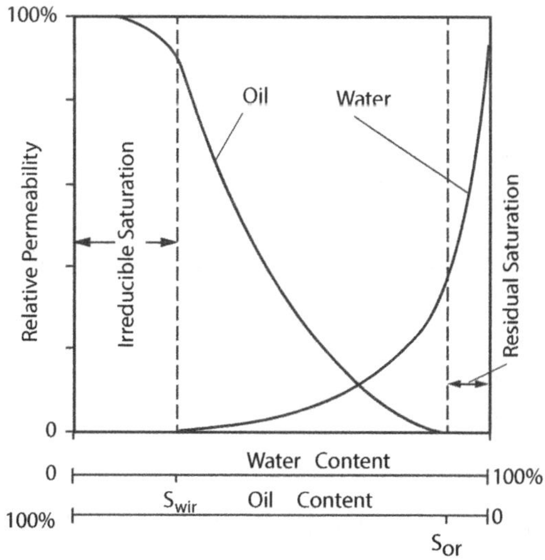

Figure 11.1 Generic relative permeability curves, for preferentially water wet reservoir (after Cohen and Mercer, 1992 [27]).

- The capillary forces at the pore throat restrictions, within the matrix;
- The amounts of oil and water (S_o and S_w) in the reservoir rock.

At irreducible water saturation (S_{wir}), only oil will be produced. At residual oil saturation (S_{or}), only water can be produced. Comparing the relative permeability values at intermediate S_w and S_o values provides estimates of the oil production and water cut.

For depletion management, the goal is to keep as high on the oil saturation (S_o) curve as (economically) possible, by managing what can be managed (usually field balanced production rates) for primary recovery. In this fashion, oil production is maximized and the water cut is minimized, increasing income and reducing disposal costs.

During primary production, the reservoir drive declines and secondary measures, such as gas lift, gas flood, miscible gas injection, and water flood are utilized to maintain production rates. However, even these measures will still leave considerable residual oil behind.

11.4 Enhanced Oil Recovery

The goal of *Enhanced Oil Recovery* (*EOR*) technologies is to reduce S_{or} and shift the $k_{o/w}$ curve to the right, as illustrated in Figure 11.2, by changing the physical conditions of the reservoir, resulting in increased oil recovery, reduced water cut and lower S_{or}. This is accomplished by modifying one, or more, of the following reservoir matrix and/or fluid properties:

- The viscosities of the oil and water phases;
- The interfacial tension, between the two immiscible phases;
- The matrix wettability;
- The capillary forces at the pore throat restrictions within the matrix.

Different EOR technologies modify different combinations of the above reservoir matrix/fluid properties.

11.5 Electrokinetically Enhanced Oil Recovery

Electrokinetic electrical enhanced oil recovery involves inducing electrical current, from antennas, or passing electrical current, between electrodes,

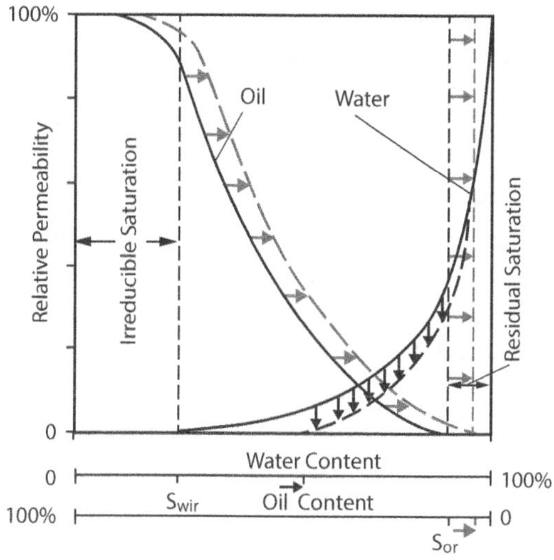

Figure 11.2 Enhanced oil recovery modification of oil/water ($k_{o/w}$) and water/oil ($k_{w/o}$) relative permeability, as well as residual oil (S_{or}) and irreducible water (S_{wir}) saturations (after Hill *et al.*, 2010 [50]).

in the producing reservoir and either at the surface and/or at depth. These applications have seen several different incarnations. When electrical current is introduced into the ground, several electrochemical processes can and usually do occur. Optimum application of electrical current for maximum effectiveness resides in the art of application.

11.5.1 Historical Background

Several groups and individuals have considered using electrical current for enhanced oil recovery (EOR). One of the earliest proposed electrical EOR technologies was by Workman (1930 [108]). His patent (which does not appear to have progressed beyond hypothesis) proposed conducting electrolysis in the water legs of oilfields to release gases (presumably hydrogen and oxygen):

> "...*which permeate, or are absorbed by the oil to render the same lighter and more mobile which facilitates its release from the sand and causes it to rise to the top of the water in the well*".

There is no documentation that this inventor ever attempted to evaluate or implement his ideas. The reference is included only as a benchmark that someone had considered using electricity for EOR as early as 1930.

Sarapuu (1957 [79]) proposed using high-density electrical currents to "*carbonize*" hydrocarbons to enhance extraction via well bores. Bell (1957 [15]) proposed using high-density DC current to drive oil from reservoirs into wells. Again, there is no evidence that either of these technologies progressed beyond the patent stage.

Gill (1970 [38], 1972 [39]) and Crowson and Gill (1971 [29]) proposed using alternating current (AC) electricity to heat oil reservoirs, which would decrease the oil viscosity and enhance its recovery. Low-frequency AC heating was evaluated in Canadian heavy oil fields (McGee, 2008 [62]; McGee *et al.*, 1999 [63]). Electromagnetic (EM) heating was evaluated by Cute *et al.* (1987 [26]), Chute and Vermeulen (1988 [25]), and Sahni *et al.* (2000 [77]). Radio frequency (RF) induction EOR was evaluated by Kasevich *et al.* (1994 [53]). All of these authors have advocated EM for near wellbore heating to reduce oil viscosity. AC current, however, poses problems, such as corrosion of all electrodes, inductive energy loss, and shallow penetration away from the electrodes or antennas. By contrast, DCEOR does not experience any of these problems

Downhole resistive heaters have also been proposed for heating the near wellbore (Sandberg *et al.*, 2013 [78]; Vinegar and Bass, 2007 [92]; Vinegar *et al.*, 2006a [93], 2006b [95], 2007 [94]).

Several major oil company research and development (R&D) subsidiaries and national laboratories have investigated various AC, EM, RF, microwave, and downhole heaters for EOR (e.g., Sahni *et al.*, 2000 [77]).

11.5.2 Geotechnical and Environmental Electrokinetic Applications

Electrokinetics has been applied for geotechnical and environmental restoration problems, as well as EOR. Knowledge gained in any of these applications can, and should be, applied to understanding the others.

Leo Casagrande pioneered the use of electroosmosis for construction soil dewatering and stabilization (Anon, 1966 [1]; Casagrande, 1952 [18], 1959 [19]; Casagrande *et al.*, 1961 [20]). Farris (1968 [35]) described the use of anodes and cathodes in a formation as a mechanism for dewatering it. Mitchell (1970 [65]) also described using electroosmosis to dewater soil and was actually retained as a consultant for stabilization attempts for the Campanile at the Cathedral of Pisa, Italy. Gray and Mitchell (1967 [40]) provided the current electrochemical basis of electroosmosis: i.e., that the electroosmotic membrane is not a physical membrane but a cation-selective membrane due to the Helmholtz double layer filling pore-throat constrictions. Mitchell (1993 [66]) also has an extensive discussion of the electrochemical basis for electroosmosis. Gray and Schlocker (1969 [41]) described electrokinetic alteration of clay minerals. Tchillingarian (1952 [87]) described the use of electrophoresis to move fine-grained clays through porous rocks.

Acar and Alashawabkeh (1993 [2]) and Acar and Gale (1992 [3]) discussed using electrokinetics to decontaminate heavy metal contaminated soils. Bruell *et al.* (1992 [17]), Hill (1994 [45], 1997 [46], 1998 [47]), and Hill *et al.* (1997 [49]) utilized electroosmosis to remove gasoline hydrocarbon and TCE contamination of soils. Chilingar *et al.* (1997 [24]) utilized electrokinetic remediation of hydrocarbon and metallic soil contamination. Döring (1993 [30], 1996 [31]), Döring *et al.* (2000 [32], 2001 [33], 2003 [34]), Iovenitti *et al.* (2003 [52]), and Lageman *et al.* (1989 [57]) also applied electrokinetics for remediation of both organic and inorganic soil contamination, with some rather dramatic case histories. Döring (1993 [30], 1996 [31]) and Döring *et al.* (2000 [32], 2001 [33]) demonstrated destruction of organic contamination by breaking down complex organic molecules into simpler daughter products. Wittle and Pamukcu (1993 [107]) and Pamukcu (1997 [72]) utilized electrokinetics for treatment of soils and sludges contaminated with hydrocarbon and metals. Another approach for *in-situ* remediation of contamination soils has been applied (Rahner and Röhrs, 2002 [74]; Röhrs *et al.*, 2002 [75]).

11.5.3 Direct Current Electrokinetically Enhanced Oil Recovery

DCEOR was originally conceived at a unit of General Electric (Bell and Titus, 1973 [11], 1974 [13]). The DCEOR technology is currently being evaluated and commercial applications developed by startup organizations in the U.S., Canada, and Argentina (Bell *et al.*, 1985 [14]; Hill *et al.*, 2010 [50]; Stainoh, 2011 [84]; Titus *et al.*, 1985 [90]; Wittle and Bell, 2005 [99], 2008 [100]; Wittle and Hill, 2006a [101], 2006b [102]; Wittle *et al.*, 2008a [103], 2008b [104], 2008c [105], 2011 [106]). These private organizations, collectively, have spent several years and millions of U$S in R&D to evaluate and overcome field operational difficulties encountered during development. Their combined efforts resulted in successful field DCEOR demonstrations at California, Alberta, and Argentina heavy oil fields.

George V. Chilingar and his students, at the University of Southern California (USC), conducted numerous laboratory tests involving electrokinetics, in 1950s to 1960s, which suggested that this low power drain DC electrical mechanism could be used for EOR (Ace, 1955 [4]; Amba *et al.*, 1964a [6], 1964b [7], 1965 [8]; Chilingar *et al.*, 1968 [22], 1970 [23], 1997 [24]). Tikhomolova (1993 [89]) described similar studies conducted at the University of St. Petersburg and also suggested electroosmosis as a potential EOR technology. Sibel Pamukcu and her students, at Lehigh University, have conducted numerous laboratory tests involving investigating potential DCEOR mechanisms and effects (Al Shalabi *et al.*, 2012 [5]; Ghazanfari *et al.*, 2012 [37], Ghazanfari *et al.*, 2013 [36]).

11.6 DCEOR (EEOR) and Energy Storage

Several authors (Chelidze, 1969 [21]; Hill, 1972 [44]; Howell and Licastro, 1961 [51]; Katsube, 1975 [54]; Keller and Licastro, 1959 [55]; Rokityansky, 1959 [76]; Scott *et al.*, 1967 [80]; Ward and Fraser, 1967 [97]) have used Lossy dielectrics as analogs for the electrical behavior of rocks. Rocks and soil can store electrical energy, much like a battery or supercapicator (Conway, 1999 [28]; Mayper, 1959a [60], 1959b [61]; Wait, 1959 [96]; Smythe, 1968 [82]). If a step potential is introduced between two electrodes, in the ground, then the current passing between them will not follow a step function but will rather spike and then decay exponentially, with time, to some steady state value. If the step potential is removed, then the current will change polarity, spike, and then decay to zero. This is shown schematically in Figure 11.3.

The potential difference between two potential electrodes (i.e., NOT the two current electrodes) will rise asymptotically to some limiting value. If the current is then cut off, then the potential will drop sharply, to some intermediate value and then decay exponentially to zero, as shown

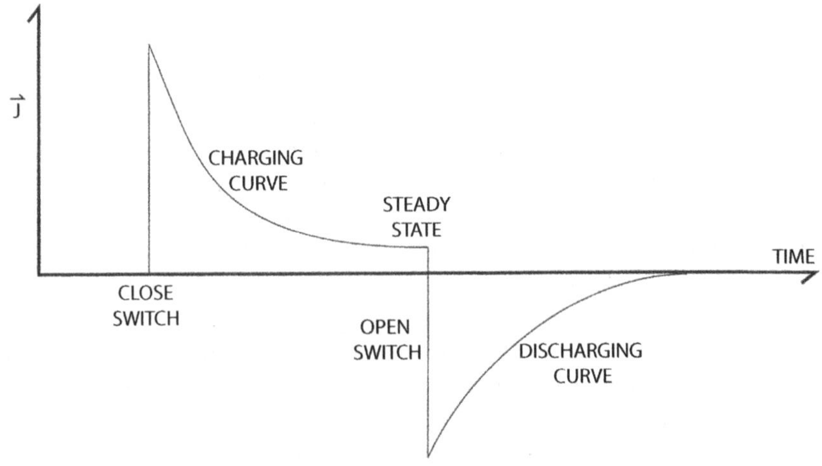

Figure 11.3 Generalized rock/soil charging and discharging current (after Hill, 1999 [48]).

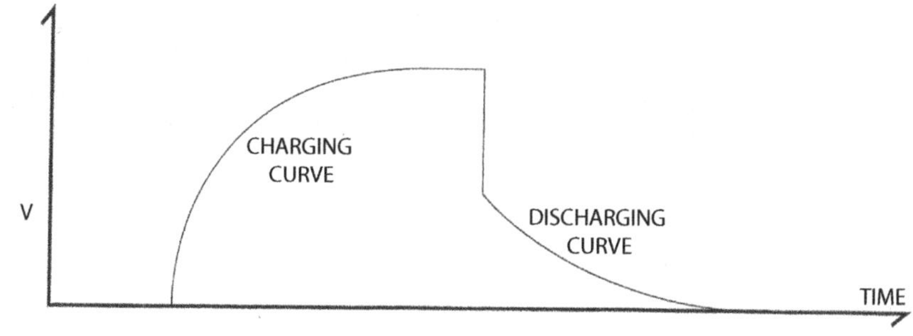

Figure 11.4 Generalized rock/soil charging/discharging potentials (after Hill, 1999 [48]).

in Figure 11.4. The earth responses, shown in Figures 11.3 and 11.4, are similar to that of a capacitor in an electrical circuit and are indicative of (electrical) energy storage (Hill, 1999 [48]).

Conrad Schlumberger first observed the phenomena shown in Figures 11.3 and 11.4 and disclosed his observations in a 1912 German Patent (Sumner, 1976 [86]). Schlumberger named the phenomenon *polarization provoquée* (*PP*), which has the excepted translation of "Induced Polarization" (IP) in the geophysical community (Schlumberger, 1920 [81]; Sumner, 1976 [86]). Schlumberger attributed the phenomenon to the presence of metallic mineralization, but claimed that background effects often masked metallic mineral PP (IP) anomalies, to the extent that they were difficult to identify. Later studies identified the source of these anomalies as due to *Membrane Polarization* in clay mineral sand mixtures (Hill, 1999 [48]; Komarov and Shmarova, 1961 [56]; Marshall and Madden, 1959 [59]; Mayper, 1959a [60], 1959b [61]; Sumner, 1976 [86] Vacquier *et al.*, 1957 [91]). Faced with the *PP* signal/noise problem, Conrad Schlumberger concentrated on using surface equal potential and resistivity mapping to locate ore bodies and later gravitated to making wireline measurements in oil wells, co-founding the oilfield services firm, Schlumberger, which still bears the family name.

11.6.1 Mesoscopic Polarization Model

Consider that the subsurface rocks can be thought to host a mesoscopic (i.e., pore scale and larger) distribution of polarization sites. In the absence of an electrical field, these polarization sites remain unpolarized (see Figure 11.5). In the presence of an electric field, however, these sites become polarized, over time, as shown in Figure 11.6, resulting in a volume polarization field, in opposition to the electrical field. If the electrical field is removed, then the polarization sites gradually decay, resulting in a current in opposition to the charging current. This discussions, as well as Figures 11.5 and 11.6, provide a qualitative explanation of the observations, as shown schematically, in Figures 11.3 and 11.4.

Note: The current discussion and the schematic models of Figures 11.5 and 11.6 are mechanism independent.

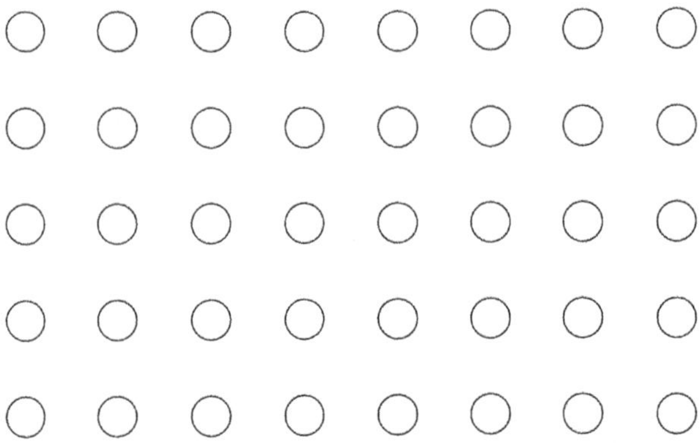

Figure 11.5 Schematic representation of polarizable rock/soil as a mesoscopic distribution of unpolarized polarization sites (after Hill, 1999 [48]).

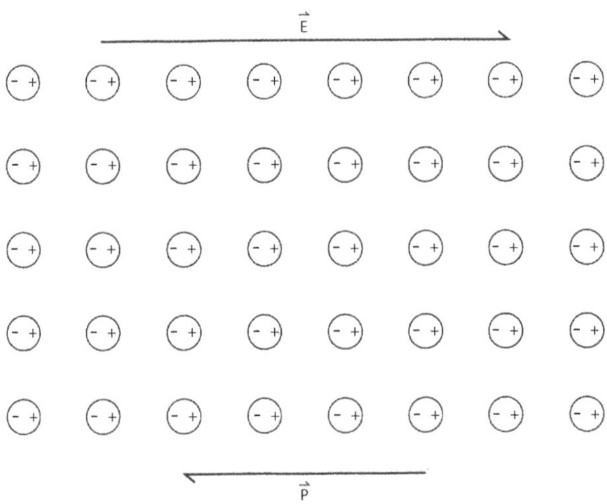

Figure 11.6 Schematic representation of Figure 11.5. Rock/soil polarized by an electric field, E (after Hill, 1999 [48]).

11.7 Electrochemical Basis for DCEOR

When electrical current is passed between electrodes embedded in the earth, several phenomena can and, usually do, occur. The best way to treat this situation is via the concept of coupled flows, introduced by Lars Onsager (1931a [68], 1931b [69]), for which he received the 1968 Nobel Prize in chemistry (Onsager, 1969 [70]).

11.7.1 Coupled Flows and Onsager's Principle

Five mechanisms appear to be operative, in DCEOR, based on available field and laboratory data:

- Joule heating
- Electromigration
- Electrophoresis
- Electroosmosis
- Electrochemically enhanced reactions

These mechanisms are often collectively called *Electrokinetics.* Figure 11.7 is a qualitative representation of the relative energy requirements for these mechanisms (with the middle three merged as "Electrokinetic Mass Transport"), as well as that required for electrically fusing (melting) the rock.

All of the above five mechanisms are coupled reactions. Nourbehecht and Madden (1963), as well as Mitchell (1993) provide simple linear algebra and tensor representations.

In matrix notation, Onsager's relationships can be represented as

$$\begin{bmatrix} J_1 \\ J_2 \\ J_3 \\ J_4 \\ J_5 \\ J_6 \end{bmatrix} = \begin{bmatrix} L_{11} & L_{12} & L_{13} & L_{14} & L_{15} & L_{16} \\ L_{21} & L_{22} & L_{23} & L_{24} & L_{25} & L_{26} \\ L_{31} & L_{32} & L_{33} & L_{34} & L_{35} & L_{36} \\ L_{41} & L_{42} & L_{43} & L_{44} & L_{45} & L_{46} \\ L_{51} & L_{52} & L_{53} & L_{54} & L_{55} & L_{56} \\ L_{61} & L_{62} & L_{63} & L_{64} & L_{65} & L_{66} \end{bmatrix} \begin{bmatrix} \nabla\phi_1 \\ \nabla\phi_2 \\ \nabla\phi_3 \\ \nabla\phi_4 \\ \nabla\phi_5 \\ \nabla\phi_6 \end{bmatrix} \tag{11.2a}$$

or, using the simpler tensor notation, they are

$$\mathbf{J}_i = \sum_{j=1}^{6} \mathbf{L}_{ij} \nabla \phi_j . \tag{11.2b}$$

where J_i are generalized flow, or flux vectors.

$\nabla Ø_j$ are generalized potential gradient, or force, vectors.
L_{ij} are generalized conductivity, or coupling coefficient (second rank) tensors.

The direct flow, or main diagonal terms, L_{ii}, of Equation (11.2a) relate non-coupled fluxes to their potential gradients, whereas the off diagonal terms, L_{ij}, relate coupled fluxes.

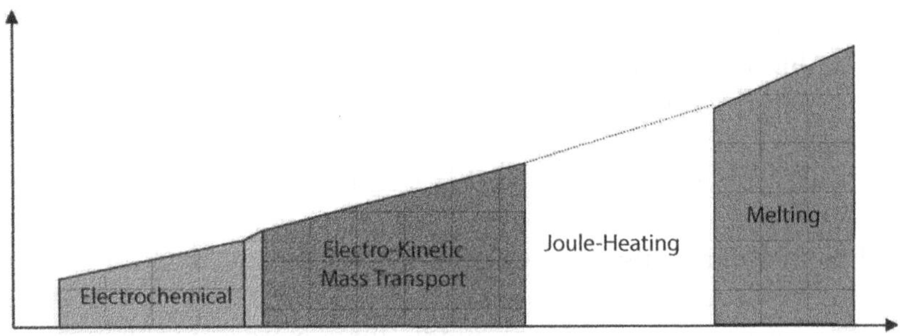

Figure 11.7 Schematic representation of energy requirements of various electrokinetic mechanisms and electrical melting (after Wittle *et al.*, 2008c [105]).

Reviewing Equations (11.2a) and (11.2b), one can observe the following:

- If J_1 is electrical current density and $Ø_1$ is electrical potential, then L_{11} is the electrical conductivity tensor, σ, whereas the coupled flow coefficient, $L_{1,3}$, includes junction potential and the coupled flow coefficient, $L_{1.5}$, includes Streaming Potential and Membrane Potential.
- If J_2 is heat flow and $Ø_2$ is temperature, then L_{22} is thermal conductivity tensor, K, while the coupled flow coefficient, $L_{2,1,}$ includes Joule Heating.
- If J_3 is ionic flux and $Ø_3$ is ionic concentration, $L_{3,3}$ is (ionic) diffusion coefficient, D_i, then the coupled flow coefficient, $L_{3,1}$, includes electromigration.
- If J_4 is charged particle flux and $Ø_4$ is the charged-particle suspension density, $L_{4,4}$ is the (charged particle) diffusion coefficient, D_p, and the coupled flow coefficient, $L_{4,1}$, includes electrophoresis.

- If J_5 is fluid flux and $Ø_5$ is pressure, then $L_{5,5}$ is k/μ, k is the absolute permeability tensor, μ is the fluid viscosity, and the coupled flow coefficient, L_{51}, includes electroosmosis.
- If J_6 is chemical reaction product flux and $Ø_6$ is the chemical reactant concentration, then $L_{6,6}$ is the chemical reaction rate constant, and the coupled flow coefficient, $L_{6,1}$, includes electrochemically enhanced reactions.

If the units of the left-hand side (LHS) fluxes J_i and the right-hand side gradients, $\nabla Ø_j$, of Equations (11.2a) and (11.2b) are such that their products, $J_i \bullet \nabla Ø_j$, represent free energy dissipation, then, by *Onsager's Principle*:

$$\mathbf{L}_{ij} = \mathbf{L}_{ji}. \tag{11.3}$$

Onsager's Principle [Equation (11.3)] is extremely useful because it is often easier to measure one of the pair of coupled coefficients than the other.

The five DCEOR (electrokinetic) mechanisms listed above, as well as Streaming and Membrane Potentials, are all [Equations (11.2a) and (11.2b)] off diagonal and, as such, are coupled mechanisms. Lorenz (1952 [58]) discusses the relationship between two coupled flows: electroosmosis and streaming potential. The primary DCEOR mechanisms are described below.

11.7.2 Joule Heating

The formation is heated by the passage of the electrical current through conducting fluids and solids within the reservoir. Electrical heating elements, such as stovetop burners and space heaters, are based on Joule heating. Joule heating of the reservoir matrix and fluids raises the reservoir temperature and lowers the viscosity of the crude, making it easier to produce. As shown in Figure 11.7, Joule heating requires the largest energy dissipations and, consequently, the greatest potential (voltage) gradients of the five DCEOR mechanisms.

11.7.3 Electromigration

Electro migration is the movemenof dissolved ionic components, such as (positively charged) cations and (negatively charged) anions under the influence of an applied electric field. Electromigration forms the basis of electrorefining and electroplating metallurgical processes, where metallic cations migrate from anode to cathode. Asphaltenes, which are negatively charged, will migrate toward the anode.

11.7.4 Electrophoresis

Charged colloidal particles, suspended in solution, will also migrate under the influence of an electrical field, similar to electromigration. In as much as clays are generally negatively charged, they will move away from the cathode and toward the anode. This process may clear blocked pore-throats and improve permeability, resulting in increased oil production (Tchillingarian, 1952) or be used to transport organisms and nutrients (Chilingar *et al.*, 1997).

11.7.5 Electroosmosis

Helmholtz double layers formed in the connate waters adjacent to clay minerals form cation selective membranes in narrow pore throats, allowing hydrated cations and water molecules to pass through the pore throats, but blocking hydrated anions. The membrane, in this case, is not a physical membrane as is in membrane filters, but an electrochemical one. This mechanism has been utilized to dewater expansive clays and increase apparent reservoir permeability, resulting in increased oil production. This is essentially the definition of electroosmosis used by Leo Casagrande (1952, 1959), as well as that used by Gray and Mitchell (1967), Mitchell (1993), and Tikhomolova (1993).

11.7.6 Electrochemically Enhanced Reactions

Reactions between the pore fluids and matrix materials enhanced by Eh/pH changes brought about by the passage of electrical current. Electrochemically enhanced reactions appear to enhance the breakdown of complex hydrocarbons, resulting in simpler (hydrocarbon) daughter products, which have lower viscosities than the original components, and increase reservoir energy. Both of these conditions will enhance oil production. Figure 11.7 shows electrochemically enhanced reactions. Field experience has demonstrated that this mechanism only occurs in very low potential (voltage) gradient situations (Döring, 1993 and 1997; Döring *et al.*, 2000, 2001, and 2003).

Of the five electrokinetic mechanisms, Joule heating requires the largest amount of energy and, consequently, the highest potential (voltage) gradient, while the electrochemically enhanced reactions require the lowest. The other three mechanisms: electromigration, electrophoresis, and electroosmosis, all have intermediate energy and voltage gradient requirements.

11.7.7 Role of the Helmholtz Double Layer

11.7.7.1 Dissociation of Ionic Salts

Water is a very unique fluid. The covalent hydrogen-oxygen bonds result in an H-O-H angle of 105°. Because both hydrogen ions are on the same side of the oxygen ion, the water molecule is strongly polar (dielectric constant of 78.54 at STP). It quickly dissociates electrolyte salts (e.g., NaCl) into positive ions, or cations (such as Na^+) and negative ions, or anions (such as Cl^-). The Na^+ ion is one electron short of electrical neutrality, whereas the Cl^- ion has one excess electron. The net result is the Cl^- ion is about twice the size of the Na^+ ion and one-half the size of water molecules. Each of the dissociated (positively charged) cations and (negatively charged) anions, in aqueous solutions, are hydrated, surrounded by envelopes of water molecules with poles of opposite charge (from that of the ion) facing the ions, attempting to neutralize the ionic charges (as shown, schematically in Figure 11.8).

In free solutions, the hydrolyzed (ions in water molecule clouds) are free to move and their relative numbers are comparable to what they were in the solid (i.e., crystalline) state. However, all of this changes near the matrix of clastic rocks.

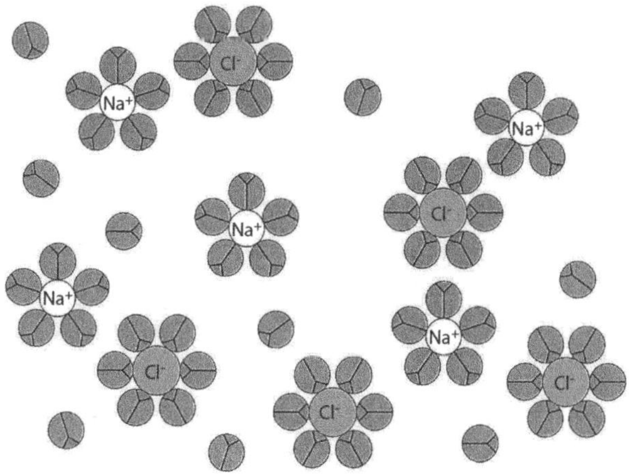

Figure 11.8 Schematic representation of NaCl dissolved in water (after Hill *et al.*, 1997). Water molecules (in blue), with positive poles adjacent to Cl^- anions and negative poles adjacent to Na^+ cations.

11.7.7.2 Silicates

The most common minerals in clastic rocks are silicates, cations surrounded by silicate radicals. The basic silicate building block is a tetrahedron, with a silicon ion at the center and oxygen ions on the four corners. These silica tetrahedra are the building blocks of silicate minerals, much like methane tetrahedra are building blocks of organic molecules (see Figure 11.9). Unlike hydrocarbons, most of which are either gases or liquids, at room temperature, however, silicates are all solid crystals.

In addition to the simple silicate tetrahedra, silicates also form single and double chains, as well as sheet structures and three-dimensional networks (see Figure 11.9). The common feature, of silicate crystal lattices is that they all consist of cations surrounded by anions and/or silicate radicals. This means that the negative portions of the (neutral) crystals are on their external surfaces.

Many rock-forming minerals do not have fixed chemical formulas but allow substitutions of cations, within their crystal lattices. In many

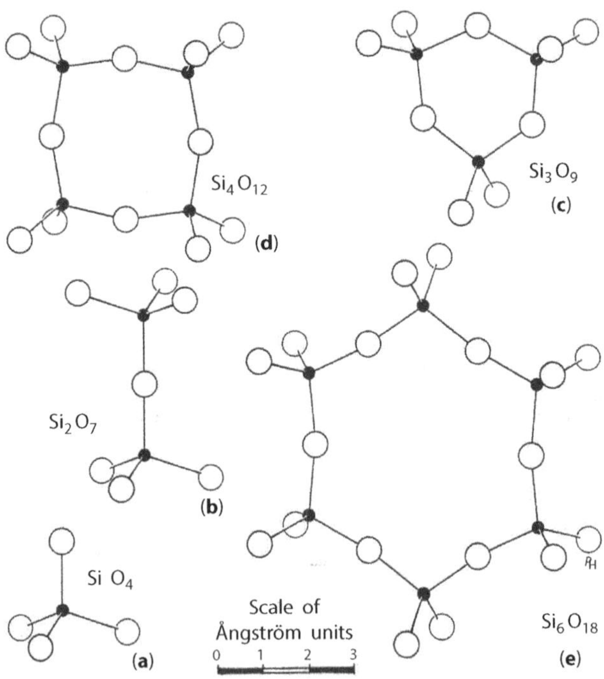

Figure 11.9 Silicate tetrahedra linkage (after Berry and Mason, 1959 [12]). (a) Independent tetrahedra; (b) Double tetrahedra; (c–e) Ring structures.

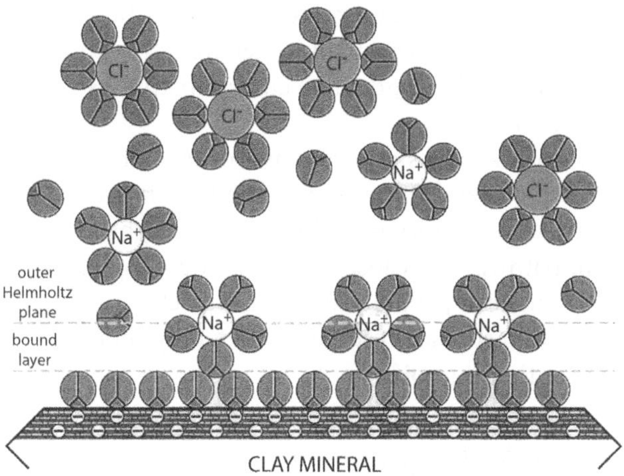

Figure 11.10 Schematic representation of the formation of the Helmholtz Double Layer adjacent to a clay mineral surface (after Hill *et al.*, 1997).

silicate minerals, this cation substitution involves substitution of divalent (e.g., Fe^{+2}) for trivalent (e.g., Al^{+3}) cations or tri-valent (e.g., Al^{+3}) for quadra-valent (e.g., Si^{+4}) cations, in the crystal lattice, resulting in a net negative surface charge on the crystal surface. This net negative charge density on the mineral crystal surfaces attracts the positive poles of water molecules and dissociated cations, while repelling the negative poles of water molecules and dissociated anions in solution (as shown in Figure 11.10).

11.7.7.3 Phillosilicates and Clay Minerals

Reservoir rocks often contain significant amounts of clays (clay-sized particles, and/or clay minerals) within their matrixes. Clay particle size grains are usually phillosilicates or sheet-silicates, such as micas and clay minerals, which have net negative surface charges.

Phillosilicate and clay mineral crystals are composed of composite sheets of AlO_6 octahedra and SiO_4 tetrahedra layers. Substitution of divalent for trivalent cations, in the octahedra layers or trivalent for quadra-valent cations, in the tetrahedra layers leaves strong net negative charge densities on the flat mineral phillosilicate crystal surfaces. Different clay minerals have differing amounts of excess negative surface charge. This variation in net negative surface charge affects the number of dissociated cations, which will attach themselves to the clay minerals.

This discussion covers silicate minerals and clastic (sand/shale) reservoirs, which tend to be preferentially water wet. This condition results in continuous electrolytically conductive current pathways along the walls of the reservoir pore throats.

Some of the world's most significant hydrocarbon reservoirs are carbonates, however, which tend to be preferentially *non*-water wet. Except for very high S_w, these reservoirs would not expect to have continuous electrolytic current pathways, because of the hydrocarbons covering the pore-throat walls. If, however, there are significant clay minerals coating the pore throats of the carbonate reservoirs, then they should begin to behave much like the clastic reservoirs.

11.7.7.4 Cation Exchange Capacity

In addition to the net surface charge, phillosilicates and clay mineral crystal structures are composites of layered silicates. The space between the composite layers can be filled with water and/or hydrated cations, as shown in Figure 11.11. A measure of the interlayer cation attraction is called cation exchange capacity (CEC) and is measured by determining the number of cations, which can be exchanged on the clay minerals (Mitchell, 1993). CEC is commonly measured in meq/gm. A more useful measure of CEC, for determining the electrochemical and flow behavior of shaly-sands silts,

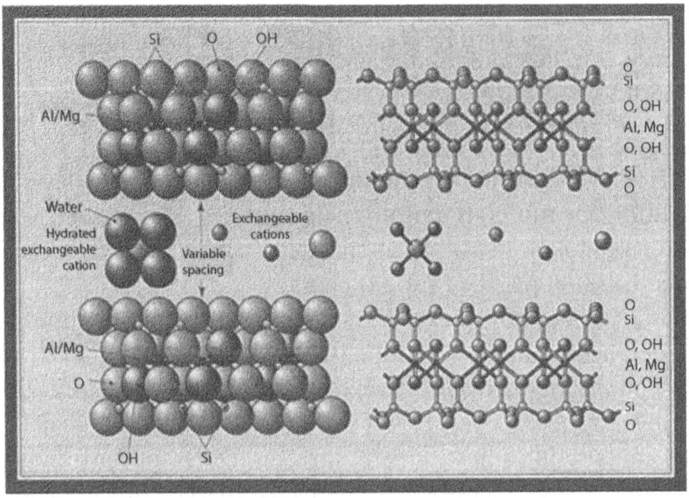

Figure 11.11 Montmorillonite crystal structure showing interlayer water and exchangeable cations (after UBC website).

and, perhaps, even shaly carbonates, is the CEC/unit pore volume, Q_v, of Waxman and Smitts (1968 [98]):

$$Q_v = \frac{CEC}{V_p}, \tag{11.4}$$

where Q_v is CEC per unit Pore volume, in meq/ml.
CEC is cation exchange capacity, in meq/gm.
V_p is the pore volume, in ml/gm.

The Q_v measure of CEC converts it from a mass measure to a specific volume measure. It also provides a direct link between CEC and

- The Membrane spontaneous potential;
- Resistivity behavior of shaly-sands;
- Streaming potential;
- Electroosmosis;
- Induced Polarization *Normal Effect* in shaly-sands.

Shaly sands with high Q_v tend to have lower permeability than those with low Q_v.

11.7.7.5 Electrochemistry of the Double Layer

Water is such a strong polar molecule that it preferentially (over non-polar and weakly polar petroleum molecules) wets the clay mineral surfaces, making them hydrophilic, or preferentially water-wet. The bound layer of water molecules is held tightly enough by the electrostatic attraction from the excess negative charges in the clay mineral surfaces that these water molecules do not move under normal sub-surface forces.

When water comes in contact with a clay, or other phillosilicate, mineral surface the water molecules, closest to the surface, align themselves with their positive (hydrogen ion) pole opposite the negative charges on the surface of the silicate minerals (Figures 11.10 and 11.12). This region of water molecules (which may be from a single-molecule to several-molecules thick) is tightly bound to the surface of the clay minerals and is variously called the "bound layer", "Stern layer", or "inner Helmholtz layer". The inner layer may or may not contain cations from ionic salts in solution but ***will not contain any anions.***

Figure 11.12 is a schematic diagram of a pore throat, illustrating the Helmholtz double layer lining. It is formed, in preferentially

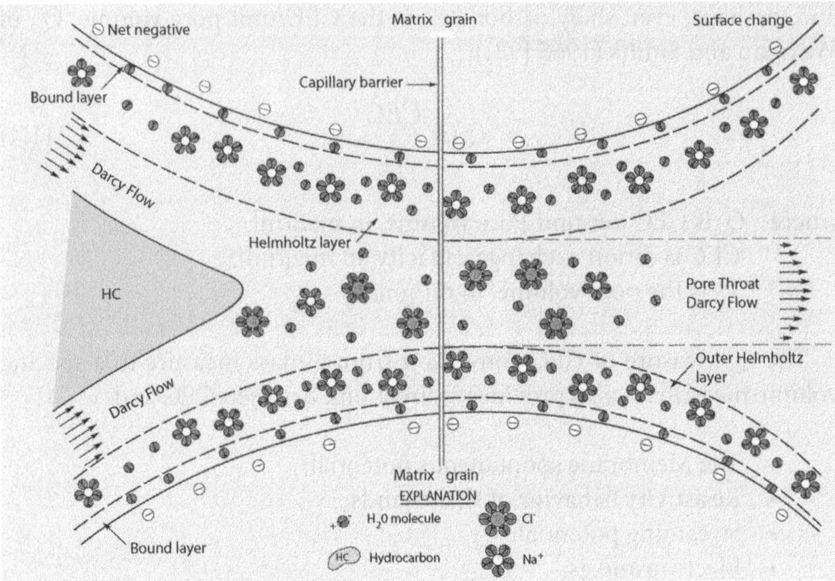

Figure 11.12 Pore throat schematic showing double layers forming adjacent to clay mineral surfaces and restricting darcy flow through the center of the pore throat (after Hill *et al.*, 1999).

water-saturated rocks, by the attractive forces between the electrically negative surfaces of minerals and the positive poles of water molecules and positive (cat) ions, in aqueous solution (connate waters). There are no negative anions within the Helmholtz double layer.

This electrical double layer is divided into two electrochemical regions:

- A region of fixed water molecules and cations (Bound, Stern, or Inner Helmholtz layer) held by strong electrostatic forces to negatively charged mineral surfaces.
- A more mobile region of loosely bound cations and water molecules (Outer Helmholtz, or Gouy layer), which are partially shielded from the negatively charged mineral surfaces by the water molecules and cations of the Stern layer. These loosely bound cations and water molecules can move in the presence of an electrical external field.

If the pore throats are large enough, then a third electrochemical region (free fluid) exists, containing water molecules, as well as both anions and cations. In preferentially water wet rocks, this is the portion of he pore throats, where hydrocarbons will reside. If the pore throats are small enough, however:

- The Helmholtz double layers completely fill the pore throat.
- The pore throat becomes an electrical cation selective membrane (see Figure 11.13).

Even if the Helmholtz double layers do not completely fill the pore throats, capillary forces will restrict the non-wetting fluid from entering the pore throats.

There is a shear, or slipping plane surface between the Stern and Gouy layers, which allows movement of the Gouy layer cations without disturbing the Stern layer. The potential difference between this plane and the free fluid is the Zeta Potential (Chilingar *et al.*, 1970).

Upon application of DC current, the mobile Gouy layer migrates, in the direction of the electrical field, toward the negative (cathode) electrode. This motion of the water molecules and cations within the Gouy layer, results in the following:

- The motion of the water molecules and cations within the Gouy layer effectively opens the diameter of the pore throat to the bound, Stern layer, as shown in Figure 11.14.
- Cations, anions, and other fluids in the free fluid will be dragged through the (now enlarged) pore throat.
- Both the wetting and non-wetting fluids will be produced.

Extensive experimental work, conducted at The University of Southern California (Ace, 1955; Amba *et al.*, 1964, 1965; Chilingar *et al.*, 1968, 1970,

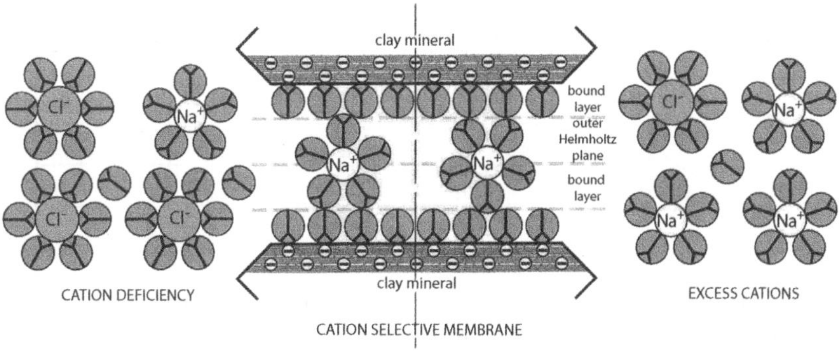

Figure 11.13 Schematic representation of the formation of a cation-selective membrane in a constricted pore throat (after Hill *et al.*, 1997).

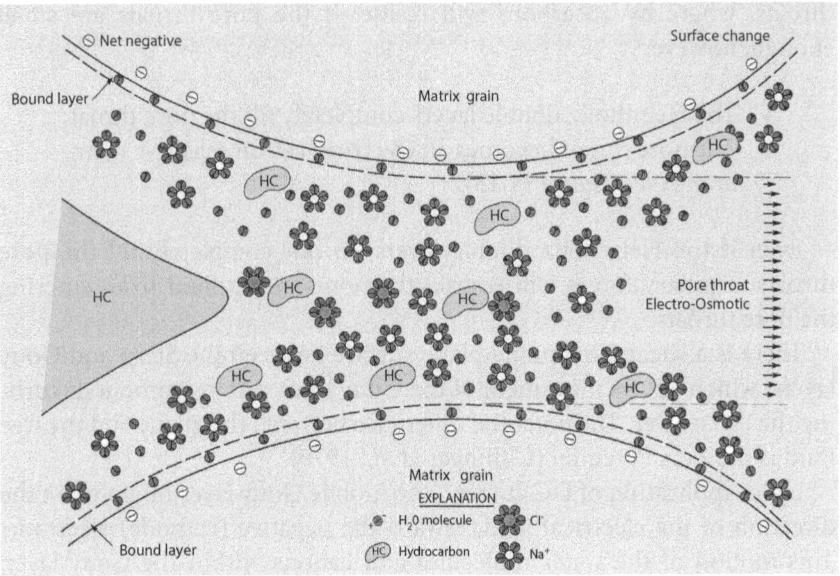

Figure 11.14 Pore throat schematic showing electrokinetic transport of the outer Helmholtz layer, modifying the effective pore throat size for fluid flow (after Hill, 1999).

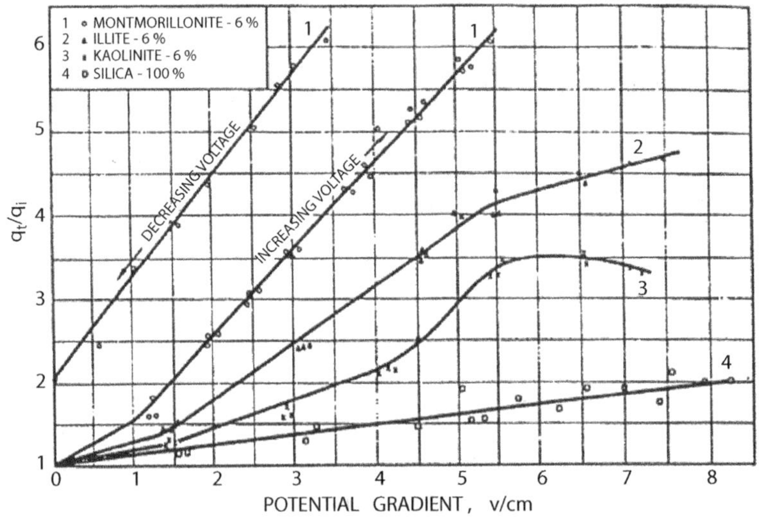

Figure 11.15 Relationship between potential gradient and normalized electrokinetically induced fluid flow, for different silica-clay mineral mixtures. Core No. 1: 94% silica sand and 6% montmorillonite clay. Core No. 2: 94% silica sand and 6% illite clay. Core No. 3: 94% silica sand and 6% kaolinite clay. Core No. 4: 100% 200-mesh silica sand (after Chilingar *et al.*, 1970).

1997, 1997; Tchillingarian, 1952) showed up to 40-fold volumetric fluid flow increases in cores containing clays, compared to only two- to three-fold increases in pure silica cores, as shown in Figure 11.15. Chilingar *et al.* (1970) speculated that this latter (silica core) fluid flow might be due to a thermal effect. Mitchell (1993) maintained that all silica minerals show increased flow, under DC fields, but that those with high CEC exhibit the greatest flow increases. The results in Figure 11.15 agree with Mitchell's (1993) argument.

For those pore throats, completely filled with the Helmholtz double layer, the application of current will develop salinity gradients across the cation selective membranes, which will store energy. Any fluctuation in the local electric fields will result in discharge of this stored energy, similar to discharging a super capacitor (Conway, 1999), potentially inducing break-down of complex hydrocarbon molecules, resulting in simpler daughter product molecules.

11.8 DCEOR Field Operations

Figure 11.16 illustrates one possible DCEOR field implementation, where anode(s) are placed near the ground surface and cathode(s) are placed in, or near, producing well(s). An alternative configuration utilizes both

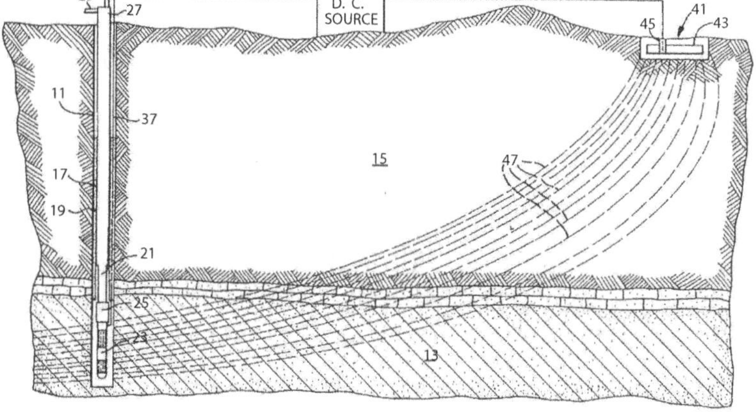

Figure 11.16 DCEOR Field Operation Schematic Rendering (after: Amba *et al.*, 1965; Titus *et al.*, 1985).

anode(s) and cathode(s) in the subsurface. Successful field pilots have utilized anode-cathode separations of as much as 400 m or 0.25 mi (Wittle and Hill, 2006a, 2006b; Wittle *et al.*, 2008a, 2008b, and 2008c; Wittle *et al.*, 1011).

11.8.1 Three-Dimensional Current Flow Ramifications

A common misconception is that three-dimensional electrical current flow in the subsurface can be represented by one-dimensional linear (Kirchhoff) circuit theory. For *one-dimensional linear circuit theory*, Ohm's law is

$$I = \frac{V}{R} \tag{11.5}$$

or

$$\Delta V = IR \tag{11.6}$$

Joule heat loss *across a given circuit element*, of resistance R, is given by

$$P = I\Delta V = I^2 R, \tag{11.7}$$

where P is the power loss, over the *individual* circuit element, of resistance, R.

ΔV is the voltage drop across the *individual* circuit element.
R is the *individual* circuit element resistance.
I is the current through the *entire* circuit, controlled by the ***total voltage drop across the entire circuit and the sum of all of the resistances*** in the circuit.

In three-dimensions, however, Ohm's law [Equation (11.5)] becomes

$$\mathbf{J} = \sigma \mathbf{E} = \sigma \nabla \phi = \rho^{-1} \nabla \phi, \tag{11.8}$$

or

$$\nabla \phi = \rho \mathbf{J}. \tag{11.9}$$

In three-dimensions, the ***local*** *Joule heating power loss* is given by

$$p_{x,y,z} = \mathbf{J} \bullet \mathbf{E} = \mathbf{J} \bullet \nabla \phi = |\mathbf{J}|^2 \rho, \tag{11.10}$$

where p_{xyz} is the *local* power loss, at the point of interest.

E is the *local* electric field vector, at the point of interest.
$\nabla\emptyset$ is the *local* electrical potential field gradient, at the point of interest.
σ is the *local* electrical conductivity tensor, at the point of interest.
ρ is the *local* electrical resistivity tensor (inverse of σ), at the point of interest.
J is the local current density vector, at the point of interest.

I is a constant, in Equations (11.5) to (11.7), dependent only on the total voltage drop across the entire circuit and the sum of the circuit element resistances in the circuit. For a heterogeneous Earth model, however, all of the parameters in Equations (11.8) to (11.10) are functions of position, with the vector and second rank tensor properties, directional, in nature.

Electrical current densities follow paths of least resistance, such that the total Joule power loss:

$$P_T = \oiiint p_{xyz}\, dx\, dy\, dz, \tag{11.11}$$

for the *entire (earth) system*, is minimized. This means that, in regions of high resistivity, |**J**| is much lower than in regions of low resistivity. Since p_{xyz} depends upon the local $|\mathbf{J}|^2$, Joule power loss is much less in regions of high resistivity than in regions of low resistivity. The exceptions, to this rule of thumb, are in the vicinity of the power electrodes, where the current density is controlled by the geometry of the power electrodes. Even in these situations, however, DCEOR electrode arrays are designed to maximize Joule heating where it is desired and minimize Joule energy loss, where this result is the desired outcome.

11.8.2 Electric Field Mapping

The primary objective of DCEOR is to pass electrical current through the oil reservoirs in an optimum manner to achieve the above mechanisms. This is achieved by electrode array design. Figure 11.17 shows a simple three-layer earth model, used to design electrode arrays. Figures 11.18 and 11.19 illustrate in cross section and plan view, respectively, the equipotential surfaces about parallel sets of three anodes and cathodes, in the model of Figure 11.17.

11.8.3 Joule Heating and Energy Loss

Electrode arrays are specifically designed, in DCEOR systems, to produce high current densities, where *high* Joule heating is desired, and low current

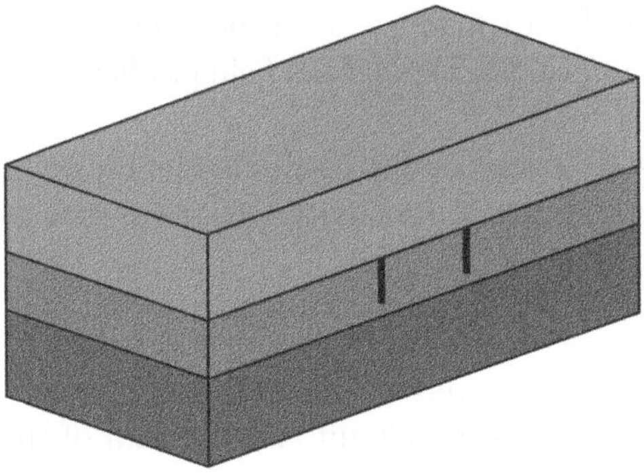

Figure 11.17 Three-layer earth DCEOR field mapping model (after Wittle and Hill, 2006b; Wittle *et al.*, 2008a, 2008c).

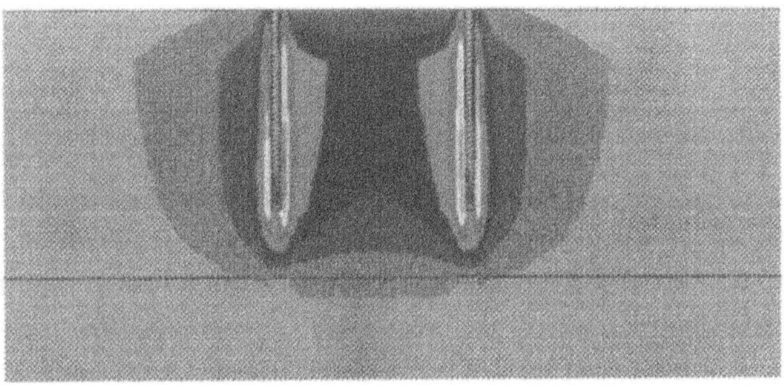

Figure 11.18 DCEOR Figure 11.18 model simulation equipotential field cross-section (after Wittle and Hill, 2006a; Wittle *et al.*, 2008a, 2008b, 2008c).

densities, where *low* Joule *only* power losses are desired (i.e., high Joule power losses are to be avoided). Figure 11.20 illustrates Joule heating (***NO conductive or convective heat transfer***) simulation results in the vicinity of a down-hole electrode within a heavy oil reservoir. Even after 7.5 months of DCEOR, the temperature beyond 14 ft of the electrode surface is essentially unchanged. Convective and radiative heat transfer, however, would increase the effective heating radius.

11.8.4 Comparison of DC vs. AC Electrical Transmission Power Loss

The power losses, as discussed above, depend only upon resistance [Equation (11.7)] or resistivities [Equations (11.10) and (11.11)], ***for DC electrical current***. For ***AC electrical current***, however, they ***also depend upon capacitive***

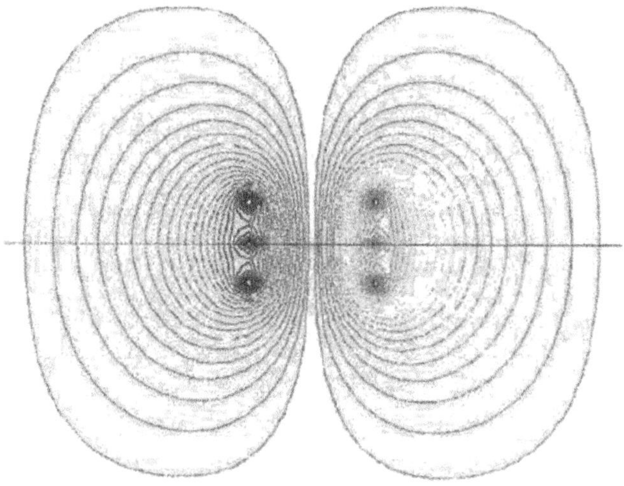

Figure 11.19 DCEOR Figure 11.17 model simulation equipotential field plan view (after Wittle and Hill, 2006; Wittle *et al.*, 2008a, 2008b, 2008c).

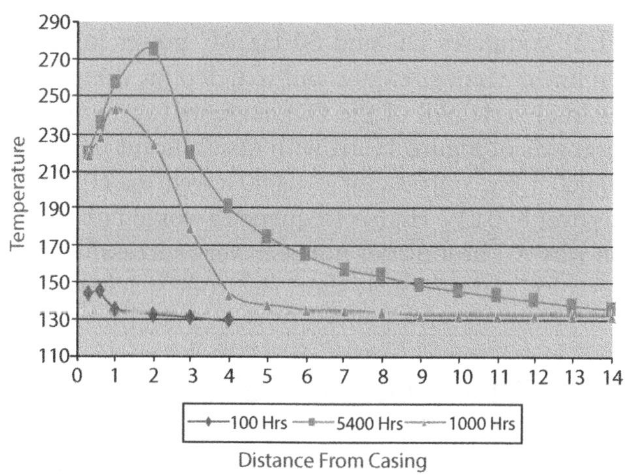

Figure 11.20 DCEOR simulation results: Reservoir temperatures after 100, 1,000, and 5,400 hours of stimulation. Temperatures (vertical axis) are in °F and distances (horizontal axis) are in radial feet, from the Casing (after Wittle and Hill, 2006; Wittle *et al.*, 2008a, 2008b, 2008c).

and inductive losses. Capacitive loss, in DCEOR, usually involves current leakage to the formation and is generally not a significant concern. Inductive losses for AC signals, however, can become considerable, because of inductive coupling to the earth, from the long cables and/or casing between the generators, at the surface, and the downhole and/or surface electrodes.

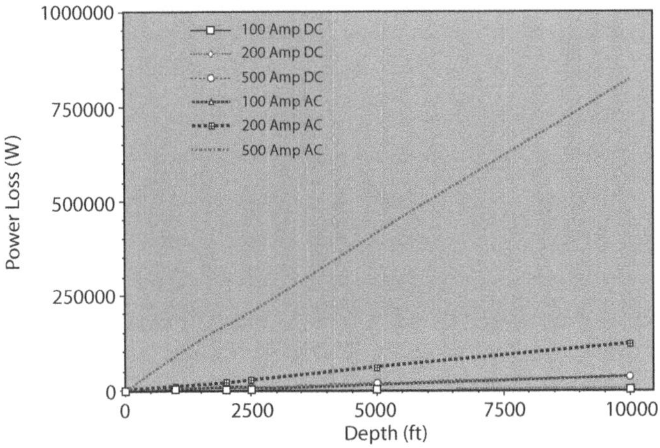

Figure 11.21 DCEOR simulation results: Comparison of AC and DC electrical power loss for various casing depths (after Wittle and Hill, 2006; Wittle *et al.*, 2008a, 2008b, 2008c).

Figure 11.21 compares DC and 60-Hz AC power losses for different transmission line current strengths and well depths, using a common vertical (power loss) scale. All of the DC currents (up to 500 A) fall along the horizontal axis of Figure 11.21, with insignificant power loss, even to depths of 10,000 ft. By contrast, for a 10,000 ft well, the 60 Hz 500 A system power loss is over 850 kW. Higher frequencies would fare even worse. This is the reason why AC heating has not been very successful, except for very shallow reservoirs, and extremely short electrode spacings.

11.9 DCEOR Field Demonstrations

Multiple DCEOR field demonstrations have been conducted at heavy oil fields in the Santa Maria Basin of California (USA), Lloydminster Heavy Oil Belt of Alberta and Saskatchewan (Canada), and Golfo San Jorge Basin of Santa Cruz (Argentina).

11.9.1 Santa Maria Basin (California, USA) DCEOR Field Demonstration

The Santa Maria Basin, California (USA) field DCEOR demonstrations, including breaks for data analyses and laboratory testing, lasted over 6 years. Figure 11.22 shows a typical short-term field production test. Within

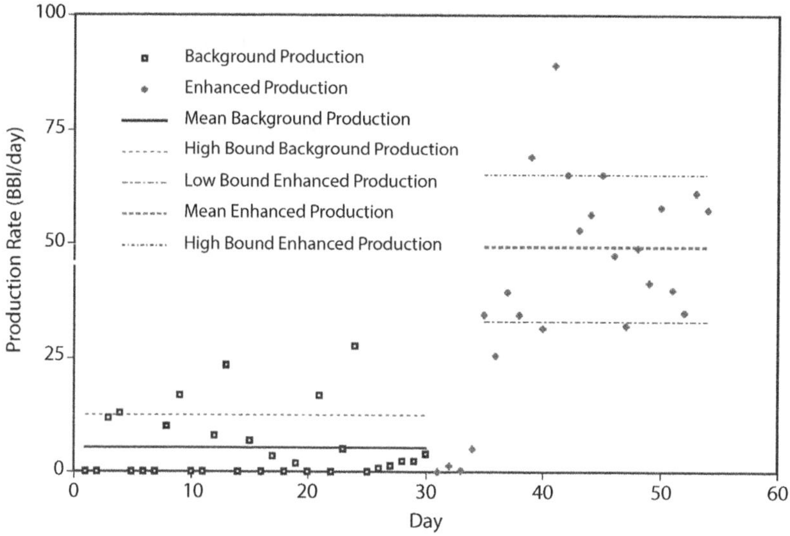

Figure 11.22 Santa Maria Basin (California, USA) short-term DCEOR demonstration, (after Wittle and Hill, 2006a, 2006b; Wittle *et al.*, 2008a, 2008b, 2008c, 2011).

5 days of initiating DCEOR, the daily production rate improved approximately 10-fold.

The Santa Maria Basin demonstration reservoir was a 100-ft-thick unconsolidated sand at a depth of approximately 2,800 ft below ground surface (BGS). Prior *unsuccessful EOR technologies* had included steamflood and cyclic (Huff and Puff) steam injection.

Table 11.1 summarizes the Santa Maria Basin field DCEOR performance, compared to the baseline production. DCEOR increased the oil production by an order of magnitude (10-fold). At the same time:

- The water cut dropped to about 25% of the baseline performance.
- API Gravity increased from 8.1° to 10.7°.
- Oil viscosity decreased significantly, with the DCEOR produced crude viscosity only about 41% of the baseline value.
- Gas production essentially doubled from the baseline value, with DCEOR.
- The produced gas energy content (PGEC) increased by a factor of 1.5.
- The H_2S production dropped from 2,290 ppm to 2 to 40 ppm.

Table 11.1 Santa Maria Basin (California, USA) comparison of DCEOR field demonstration to baseline field performance (after Wittle and Hill, 2006a, 2006b; Wittle *et al.*, 2008a, 2008b, 2011).

Item	Baseline production	DCEOR production
Production Rate (BOPD)	5	50
Water Cut (%)	45	12
Oil API Gravity	8.1°	10.7°
Gas Production (SCFPD)	1,750–2,000	3,800
Produced Gas Energy Content (BTU/SCF)	1.197	1.730
H_2S Content (ppm)	2,290	2–40
Viscosity (cps@ 100°C)	51,747	21,275

The viscosity, API gravity and PGEC changes are consistent with *in situ Cold Cracking,* with DCEOR breaking down complex hydrocarbon molecules into simpler liquid hydrocarbons and gases. The near complete destruction of H_2S suggests that the DCEOR is electrochemical, rather than temperature related. These results and conclusions are consistent with those of Döring *et al.* (2000, 2001, 2003) for DC electrokinetic soil remediation projects.

Table 11.2 summarizes the DCEOR light (C_1–C_9) hydrocarbon and hydrogen (H_2) production, in comparison to the baseline production. With the exception of methane (C_1), these results are consistent with some type of *in situ Cold Cracking*. The decrease in methane content may be due either to difficulties in capturing this light hydrocarbon, during the baseline production and/or the destruction of methane to hydrogen gas (H_2), which went from non-detect to 1.9%.

Table 11.3 compares the oil production efficiency for DCEOR, cyclic (Huff and Puff) steam injection, and steamflood, for the Santa Maria Basin field demonstration reservoir. DCEOR was approximately 185 times as efficient as "Huff and Puff" and 613 to 1,083 times more effective than steam flood. This reservoir is at a depth of approximately 2,800 ft, significantly deeper than the 2,000 to 2,500 ft practical limit for steam injection. The production efficiencies of Table 11.3 reflect that.

Table 11.2 Santa Maria Basin (California, USA), DCEOR field demonstration changes in produced gas composition (after Wittle and Hill, 2006b; Wittle *et al.*, 2008a, 2008b, 2011).

Hydrocarbon species	Baseline (%)	DCEOR (%)
C_1	85.4	45.89
C_2	2.06	4.22
C_3	2.60	10.02
Iso C_4	0.53	2.65
N-C_4	1.35	6.93
Iso C_5	0.69	3.76
HC_5	0.63	3.42
C_6	0.79	3.99
C_7	0.57	2.57
C_8	0.29	0.56
C_9	0.46	0.19
H_2	-	12.90

Table 11.3 Santa Maria Basin (California, USA), Comparison of DCEOR, cyclic steam injection, and steam flood oil production efficiencies (after Wittle and Hill, 2006a, 2006b; Wittle *et al.*, 2008a, 2008b, 2008c, 2011).

Recovery method	BTU/Incremental Bbl produced
DCEOR	7,014
Cyclic (Huff & Puff) Seam Injection	1,300.000
Steam Flood	4,300,000–7,600,000

11.9.2 Lloydminster Heavy Oil Belt (Alberta, Canada) DCEOR Field Demonstration

The Lloydminster Heavy Oil Belt DCEOR Field demonstrations, including breaks for data analyses and laboratory testing, lasted over 2 years. As with the Santa Maria Basin Field Trial, the Lloydminster Field operator had difficulty with fines production.

The Lloydminster Heavy Oil Belt demonstration reservoir was an 11-m-thick (36.3 ft) unconsolidated, braided stream sand at a depth of approximately 517 m (1,706 ft) BGS, with a 1.5-m (4.95 ft) shale break 3 m (9.9 ft) below the top. The well used for this DCEOR demonstration appeared to have missed the main channel sands of the field and, if possible, was an even worse performer than the well used for the Santa Maria Basin demonstration. Baseline production was approximately 2.32 bbl/day (heavy) oil with a 73% BS&W cut, on pump.

The Lloydminster Heavy Oil Belt demonstration evaluated cyclic DCEOR, as shown in Figure 11.23. Production rates rose with each power cycle.

Table 11.4 summarizes the Lloydminster Heavy Oil Belt field demonstration. Average DCEOR production was 5.78 Bbl/day with a 2.8% water cut, compared with baseline values of 1.32 bbl/day and 31% water cut. This was a 4.38-fold production increase with a 91% water cut reduction. The fluid level in a nearby observation well also rose approximately 60 m, when the DCEOR system was activated. The oil production efficiency was approximately 23.7 BTU/Incremental bbl produced. This DCEOR production efficiency value was comparable to steamflood, for this shallow, heavy oil, field.

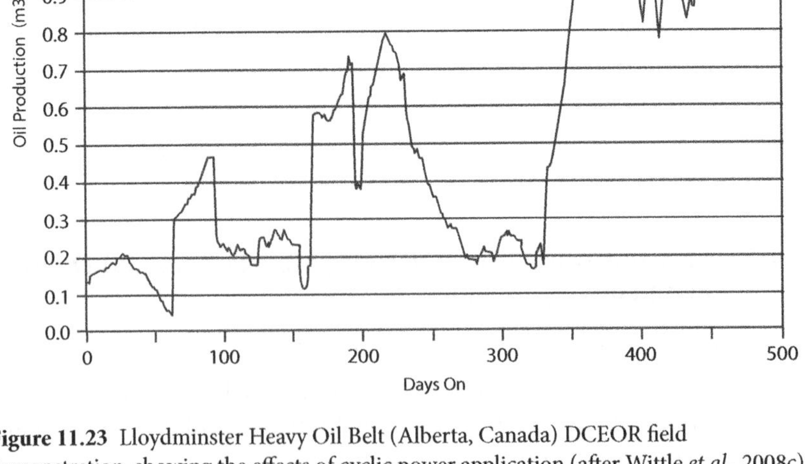

Figure 11.23 Lloydminster Heavy Oil Belt (Alberta, Canada) DCEOR field demonstration, showing the effects of cyclic power application (after Wittle *et al.*, 2008c). Production declines indicate "power off" periods.

Table 11.4 Lloydminster Heavy Oil Belt (Alberta, Canada) comparison of DCEOR Field demonstration to baseline field performance (after Wittle *et al.*, 2008c).

Item	Baseline production	DCEOR production
Production Rate (BOPD)	1.32	5.78
Water Cut (%)	31	2.8
Oil API Gravity	NA	N/A
Gas Production (SCFPD)	Nil	Minimal
Produced Gas Energy Content (BTU/SCF)	-	-
H_2S Content (ppm)	None	None
Viscosity (cps@ 100°C)	10,275	6,230

The Golfo San Jorge Basin DCEOR Field demonstration, was conducted over a period of 51 days (Stainoh, 2011). Golfo San Jorge Basin reservoirs are interbeded fluvial Cenozoic tuffaceous sandstones (Sylwan, 2001 [85]).

The Golfo San Jorge Basin field demonstration reservoir was an 20-m-thick (66 ft) consolidated, braided stream sand at a depth of approximately 820 m (2,706 ft) BGS. The well used for this DCEOR demonstration

Table 11.5 Golfo San Jorge Basin (Santa Cruz, Argentina) DCEOR Field demonstration to baseline field Performance (after Stainoh, 2011).

Item	Baseline production	DCEOR production
Production Rate (BOPD)	11.46	69.79
Water Cut (%)	96	61
Oil API Gravity	15.17°	18.42°
Gas Production (SCFPD)	N/A	N/A
Produced Gas Energy Content (BTU/SCF)	-	-
H_2S Content (ppm)	N/A	N/A
Viscosity (cps@ 50°C)	2,854	640

Golfo San Jorge Basin (Santa Cruz, Argentina), DCEOR Field Demonstration.

was a former field production well converted, without any modifications, to be the DCEOR cathode and production well. Baseline production was approximately 11.46 bbl/day, 15.17° API gravity (Heavy) oil with a 96% BS&W cut, on pump.

Table 11.5 summarizes the Golfo San Jorge Basin DCEOR Field demonstration. Average DCEOR production was 69.79 bbl/day with a 61% BS&W cut, compared to baseline values of 11.46 bbl/day, with a 96% water cut. This represents a 5.09-fold production increase, with a 34% decrease in water cut.

11.10 Produced Fluid Changes

The DCEOR demonstrations at oilfields in the Santa Maria Basin, Lloydminster Heavy Oil Belt and Golfo San Jorge Basin all suggest that DCEOR may change the chemistry of the produced fluids. These results were not expected, at the time, and have been the subject of considerable discussion, since. Increased oil production and reduced water cut were expected, but not the other results. The higher API gravity, increased gas production, and increased BTU content of the produced gasses were initially attributed to *in situ* thermal cracking, due to joule heating of the formation. Joule heating, however, could not explain the dramatic reduction in produced H_2S. The DCEOR field demonstration production wells were also cathodes, for the DCEOR system. Because of this fact, the production of OH^- ions by electrolysis at the cathode was considered as a potential explanation for chemical changes in the produced fluids.

A recently developed low current density electrical soil remediation technology (Döring, 1993, 1996) called Electrochemical Geo-Oxidation (ECGO), by the developer, however, provides insight for yet another potential explanation of the Santa Maria Basin DCEOR results.

Figure 11.24 shows a series of Gas Chromatography–Mass Spectrograph (GCMS) spectra, taken at various times, during *in situ* ECGO remediation a large-scale poly-aromatic hydrocarbon (PAH) contaminated soil site. The GCMS spectra vertical axes are relative to hydrocarbon (HC) species concentration. The horizontal axes are residence times, which are related to the HC species complexity.

The simpler PAH species are vaporized with shorter residence times (left end of the axis), while the more complex species require longer residence times (right end of the axis). Over the course of the site remediation, the GCMS spectra peaks shift from complex to simpler hydrocarbons (i.e., to the left, on the GCMS spectra), suggesting a possible *in situ* break

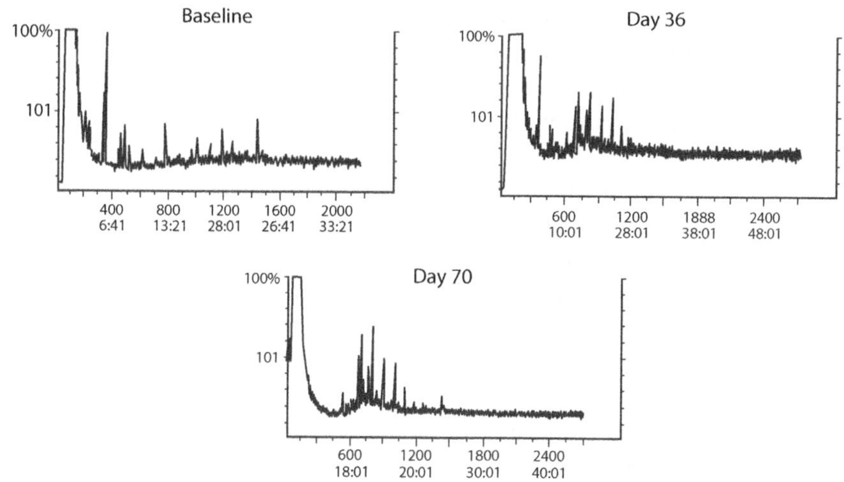

Figure 11.24 EC PAH destruction documented by GCMS changes, with time, during treatment (after Döring *et al.*, 2000). Vertical axes are abundance in ppm; horizontal axes time, in minutes.

down of the complex PAH molecules to simpler hydrocarbon daughter products. The measurement samples were ***not*** collected near the electrodes, but from the body of the contaminated soil site. As a result, the chemical changes in the PAH were ***not*** just occurring at the cathodes or as a result of electrode reactions.

11.11 Laboratory Measurements

The DCEOR demonstrations at oilfields in the Santa Maria Basin, Lloydminster Heavy Oil Belt and Golfo San Jorge Basin all exhibited:

- Increased oil production;
- Reduced water cut;
- Increases in API gravity of the produced crude.

Controlled laboratory tests were developed to further evaluate these field results. Standard petroleum Special Core Analysis Laboratory (SCAL) electrical and flow measurements involve small (1in diameter by 1 to 2 in long) samples housed in Hasler Load Cells, for short term (hours to days) tests. Using these types of tests for investigating the mechanisms involved with DCEOR present the following problems:

- The weeks to months involved in the DCEOR field demonstrations create considerable electrode reaction products at the anodes and cathodes, which could overwhelm any reactions occurring with in the standard petroleum SCAL test sample.
- The very low (μV/m–mV/m) potential gradients involved in the DCEOR field demonstrations are extremely difficult to maintain in a laboratory environment.
- Synthetic surrogate reservoir formation, crude, and brine samples often used for laboratory measurements may not be representative of the actual reservoir material.

To overcome these test design problems, Electro-Petroleum, Inc. (EPI: Wittle and Hill, 2006a, 2006b; Wittle 2008a, 2008b, 2008c, 2011) and P + P Geotechnik GmbH (P + P: Döring, 1993, 1996) used large (0.5 to 1.0 m^3) samples for laboratory testing.

Figure 11.25 and Table 11.6 describe the progress of a large-scale (~1 m^3 volume sample) laboratory test conducted on a California heavy oil field reservoir rock, via Gas Chromatograph–Mass Spectroscopy analyses of samples collected from the test sample at the start of the DCEOR test (red curve of Figure 11.25) and after 74 days (Black curve of Figure 11.25).

The more complex hydrocarbon molecule abundances (right end of the Figure 11.25 spectra) have been reduced, while the simpler molecule abundances (left end of spectra) have increased, as a result of this 74-day

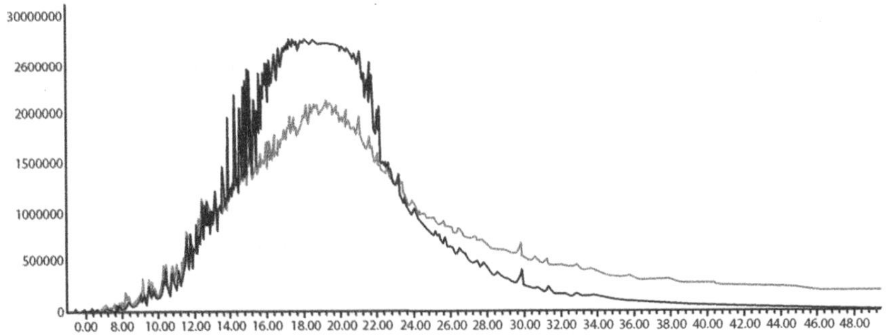

Figure 11.25 Laboratory San Joaquin Valley (California, USA) heavy oil field reservoir rock laboratory DCEOR test (after Wittle and Hill, 2006a, 2006b; Wittle *et al.*, 2008a, 2008b, 2008c, 2011). GCMS Spectra: Vertical axis is abundance, while horizontal axis is retention time (corresponding to HC complexity) in minutes. Red curve, start of test. Black curve, after 74 days of DCEOR.

Table 11.6 California heavy oil field reservoir (rock and crude) laboratory test GCMS results.

GCMS spectra property	Base line sample	74 day sample
Spectra Peak abundance	2,100,000	2,750,000
Spectra Peak Location (min)	19.05	18.40
Spectra Peak Half-Width Abundance	1,050,000	1,375,000
Spectra Peak Half Width (min)	13.2–24.5	14.45–22.9
Spectra Tail	Extensive to 49 min	Insignificant after 30 min

DCEOR treatment. As with the case, for the ECGO PAH remediation, the samples for the spectra in Figure 11.25 ***were not*** collected at the cathode, but from the ~1 m^3 sample.

Figure 11.26 shows viscosity changes during similar large volume DCEOR laboratory tests, involving crudes from two California heavy oil fields. There are significant declines in viscosity (at constant temperature) for both of these crudes, during the course of the tests.

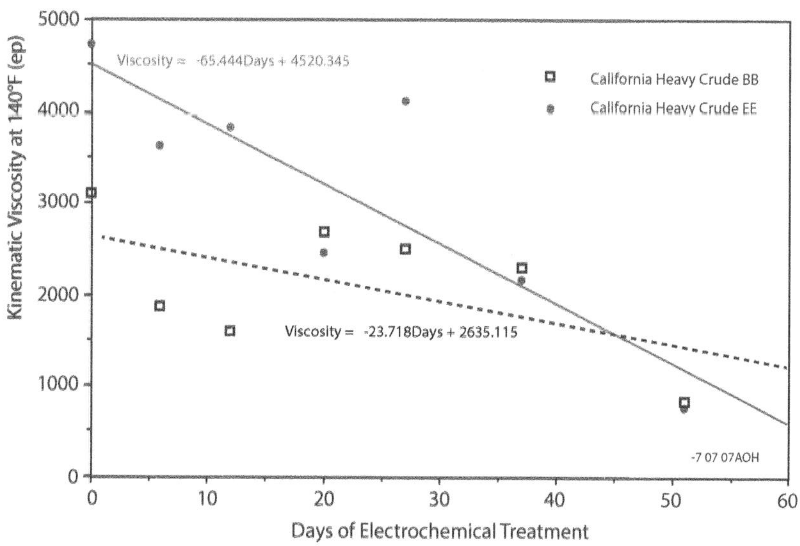

Figure 11.26 California, USA, Heavy Crude viscosity changes during DCEOR laboratory tests (after Wittle *et al.*, 2008a, 2008b, 2008c, 2011).

The results shown in Figures 11.25 and 11.26 are consistent with the Santa Maria Basin Lloydminster, and Golfo San Jorge Basin field DCEOR results as well as the ECGO GCMS spectra of Figure 11.24.

11.11.1 Electrokinetics and Effective Permeability

Figure 11.27 illustrates the effect of DC current application on fluid flow in a micro-darcy fine-grained silty-clay. The effective permeability, in this laboratory test, increases 1 to 1.5 orders of magnitude, when electrical current is applied. This is consistent with the approximately 60 m rise in the fluid level of a monitoring well, during the Lloydminster Heavy Oil Belt, DCEOR field demonstration, and the USC electroosmosis results described earlier and with the results of Chilingar *et al.* (1970).

Several successful laboratory and field tests, using DC electrical currents to introduce bacteria and nutrients into the fine-grained sediments were conducted at the University of Southern California (Chilingar *et al.*, 1997). The process, termed "Electro-Bio-Remediation", utilized DC electrical currents to force bacteria and nutrients into these very low permeability soils and rocks to remediate light hydrocarbon (gasoline and diesel) contamination from spills.

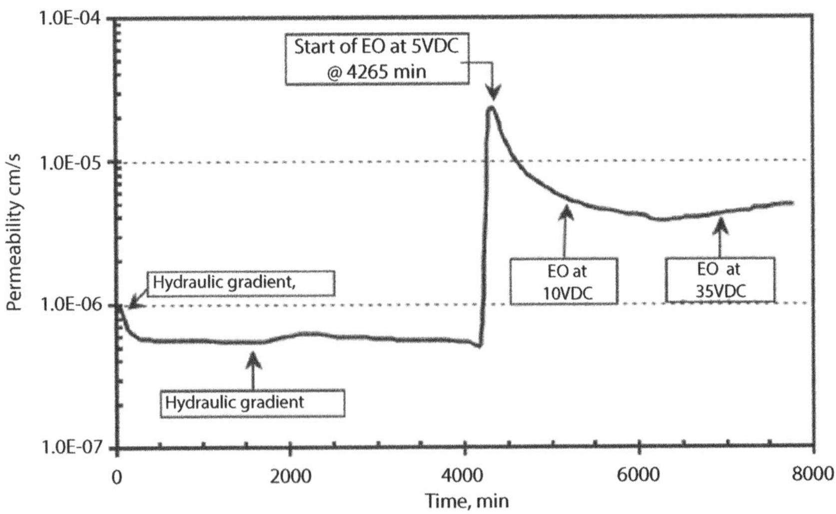

Figure 11.27 Changes in effective permeability, due to the passage of DC current (after, Pamukcu and Pervizpour, 1997 [73]).

11.11.2 Sulfur Sequestration

The dramatic Santa Maria Basin H_2S reduction, during DCEOR, raised questions as to whether these results were due to preferential production, or actual sulfur sequestration. Table 11.7 summarizes some preliminary laboratory results, which suggest that DCEOR moves sulfur from the hydrocarbons to the brines, *in situ*.

11.11.3 Carbonate Reservoir Laboratory Tests

An earlier discussion (above) indicated that carbonate reservoirs, because they tend to be preferentially non-water wet, might behave differently than clastic reservoirs, unless the carbonate reservoir throats were lined with clay minerals. Muhammad Haroun and his students at the Petroleum Institute Abu Dhabi have conducted a series of small-scale (1 in diameter by 1.5 in long core plug) laboratory carbonate DCEOR-assisted water flood experiments (Ansari *et al.*, 2012 [9]; Haroun *et al.*, 2009, 2013 [42, 43]).

The small sample size is subject to the electrode reaction product risks, cited above, in spite of the authors' best attempts, including flushing the sample holder electrode chambers, during the tests. The potential gradients of 2 V/cm are much higher than possible for field operations and may also skew the

Table 11.7 Laboratory evaluation of DCEOR sulfur sequestration studies (after Wittle, 2007).

Oil system sample	Initial % sulfur in oil	Initial % sulfur in electrolyte	Final % sulfur in oil	Final % sulfur in electrolyte
Thriophene in Mineral Oil	1.24	0.00	0.20	2.96
Thriophene in Mineral Oil	1.33	0.01	0.15	1.31
Thriophene in Mineral Oil	2.71	0.00	0.30	1.50
Thrianapthalene in Mineral Oil	3.77	0.00	5.23	0.80
Thrianapthalene in Mineral Oil	6.17	0.07	4.90	0.64
Crude Oil	6.13	0.00	4.57	0.80
Crude Oil	6.13	0.00	4.57	0.80

results. Finally, the aggressive core cleaning and restoration utilized by the authors risks irreversibly altering the electrochemical properties of the cores. Timur (1992 [88]) once attributed the following quote to M. R. J. Wyllie:

> *"The only thing we can say, for certain, about core measurements is that they come from rocks which are no longer part of the reservoir"*

With the above disclaimers:

- Haroun *et al.* (2009) reported increased waterflood and surfactant injection rates with DCEK aided flow.
- Ansari *et al.* (2012 [9]) and Haroun *et al.* (2012) reported increased oil recovery with DCEK assisted waterflood.

11.12 Technology Comparisons

Any emerging technology needs to be compared to both existing and other emerging technologies.

11.12.1 Comparison of DCEOR and Steam Flood Efficiency

Table 11.3 compared DCEOR oil recovery efficiency (energy expended/incremental oil produced) during the Santa Maria Basin, California field demonstration to steam flood and cyclic (Huff and Puff) steam injection, in the same field. For this particular field, with the reservoir at a depth of about 2,800-ft BGS, DCEOR was considerably more efficient than either steam flood or cyclic steam injection.

The Lloydminster Heavy Oil Belt well pay zone was much shallower (517 m or about 1,700-ft BGS). No details on these tests are available, but the DCEOR and steam injection efficiencies for this well were considered to be comparable.

The Santa Maria Basin vs. Lloydminster efficiency comparisons reflect the practical depth range of steam injection EOR.

11.12.2 Comparison of DCEOR and Steam Flood Costs

Table 11.8 compares the costs of a potential 20-acre DCEOR operation to that of a comparable 20 A San Joaquin Valley, California (USA), steam flood operation. This comparison was developed jointly with a representative of a major San Joaquin Valley heavy oil producer, with extensive steam flood experience. The dollar cost figures are in 1998 U$S, not adjusted for inflation.

Table 11.8 Comparison of DCEOR and Steam Flood cost estimates (Wittle and Hill, 2006a, 2006b). All costs are in terms of 1998 U$S, uncorrected for inflation.

Item	Steam flood[1]	DCEOR[2]
Capital, Design, and Installation Costs (U$S)[3]	$3,400,000	$3,125,300
Annual Energy Costs (U$S)[4]	$ 620,800	$ 86,200
Non-Energy O&M Costs (U$S/Bbl)[5]	$2.23	$0.45
Equipment Life Expectancy (yr.)	7	10

[1]Steam flood figures from San Joaquin Valley, California, for 8 × 2.5 A 5-spots (20 A total), with 8 steam injectors and 15 1,500-ft production wells with associated generation, compression, and transmission equipment.
[2]DCEOR figures extrapolated from Santa Maria Basin and Lloydminster Heavy Oil Belt Field Demonstrations, for 6 power converters, and surface anode arrays with 23 subsurface cathodes located in production wells covering a 20 A footprint.
[3]Exclusive of normal production well costs. Steam flood costs are front loaded, whereas DCEOR costs can be incremental, over the project life cycle.
[4]Steam flood: 155,200,000 cu ft/yr natural gas at U$S 0.40/Therm
DCEOR: 862,000KW-hr/yr at U$S 0.10/kW-hr.
[5]Exclusive of lifting costs.

While the capital, design and installation cost estimates, for the two EOR technologies are close, Steam flood installation costs are front-loaded, while the DCEOR system can be added incrementally, spreading its costs over the life of the project. Current approaches, involving intermittent energizing of individual subsurface cathodes, would reduce the number of surface power converters and anode arrays needed. This new approach could significantly reduce DCEOR capital costs. Annual energy costs are much lower for DCEOR, than for steam flood. Non-energy O&M costs for DCEOR are also much lower than those for steam flood. DCEOR equipment is virtually maintenance free, and can be monitored remotely. This also translates into longer equipment life expectancy and lower operating costs than that needed for steam injection technologies.

11.12.3 Comparison of DCEOR to Other EOR Technologies

Table 11.9 compares DCEOR to other emerging, as well as commercial EOR technologies. DCEOR appears to have the following advantages:

- Heat is generated *in situ*, via Joule heating.
- It does not use a working fluid.
- It does not need a significant water supply.

Table 11.9 Heavy Oil EOR technology maturity, asset, and limitation comparison (after Wittle and Hill, 2006a, 2996b; Wittle *et al.*, 2008a, 2008b, 2008c, 2011).

Process limitation	**DCEOR**	**Steam flood**	**CO_2 Flood**	**Fire flood**	**Surfactant flood**	**Co-solvent flood**
Front End Costs	Power Supply Cables Electrodes	Steam Generator Compressor Transmission Lines	C_2 Supply Compressor Transmission Lines	Compressor Transmission Lines	Mixing Tanks Pumps Transmission Lines	Mixing Tanks Pumps Transmission Lines
Uniform Sweep	Yes	No Thief Zones	No Thief Zones	No Thief Zones	No Thief Zones	No Thief Zones
Depth Limitations	No	≤ 2,500 ft	No	No	No	No
Water Demand	Nil	Very High	None	None	High	Variable
Operational Limitations	None	Emission and Water Supply Limits	None	None	Hazardous Chemicals	Hazardous Chemicals
Reservoir and Crude Type	High and Low Permeability	High Permeability Non-Reactive Matrix	High Permeability	High Permeability	High Permeability	High Permeability
Maturity	Field Demonstrated	Commercial	Commercial	Commercial	Pilot Tested	Pilot Tested

- It reduces water cut.
- It is independent of permeability.
- There are no Thief Zone concerns.
- There is no apparent depth limit.
- There are no emissions concerns.
- There are no hazardous chemical concerns.
- It increases apparent permeability.
- Electrokinetics positively influence produced fluids and flow.

The biggest DCEOR limitation, to date, appears to be its limited field application portfolio. This will change with additional field experience.

11.13 Summary

The following conclusions can be drawn, about DCEOR:

- There have been three encouraging field demonstrations, to date.
- It appears to facilitate beneficial chemical changes in the produced fluids.
- It appears to be cost competitive, to steam flood, for shallow reservoirs, and less expensive for deeper reservoirs.
- There are no Tight Zone or Thief Zone problems, with DCEOR.
- There is no water, or other working fluid requirement, with DCEOR.
- It reduces water cut.
- There are no greenhouse gas emission or hazardous liquid problems, with DCEOR.
- Facility installation is incremental, so that capital costs can be spread over the life of the field.
- There appear to be no depth limits to the DCEOR technology.

11.14 Nomenclature

AC	=	Alternating current
Bbl	=	Barrels
BGS	=	Below ground surface
BOPD	=	Barrels oil per day

BTU	=	British thermal unit
BTU/SCF	=	British Thermal Unit/Standard Cubic Foot
COF	=	Cost of finding
COL	=	Cost of lifting
D_i	=	Ionic diffusion coefficient tensor
D_p	=	Particle diffusion coefficient tensor
DC	=	Direct current
$\mathbf{E}$	=	Local electric field vector
ECGO	=	Electrochemical geo-oxidation
DCEK	=	Direct current electrokinetics
DCEOR	=	Electrokinetically Enhanced Oil Recovery
Eh	=	Electron activity
EM	=	Electromagnetic
EOR	=	Enhanced oil recovery
EPI	=	Electro-Petroleum, Inc.
GCMS	=	Gas chromatograph mass spectrograph
GE	=	General electric
I	=	Electrical current
$\mathbf{J}$	=	Electrical current density vector
$\mathbf{J}_i$	=	Generalized flux (vector)
K	=	Thermal conductivity tensor
k	=	Absolute permeability tensor
L_{ij}	=	Generalized coupling coefficient (tensor)
O&M	=	Operations and maintenance
P	=	Circuit element power loss
P_T	=	Total power loss
p_{xyz}	=	Local power loss
PAH	=	Poly-aromatic hydrocarbon
PGEC	=	Produced gas energy content
pH	=	Hydrogen ion activity
R	=	Resistance
RF	=	Radio frequency
SCFGPD	=	Standard cubic feet gas per day
SAGD	=	Steam-assisted gravity drainage
USC	=	University of Southern California
U$S	=	Dollars, US
$\Delta \mathbf{V}$	=	Circuit element voltage drop
∇	=	Vector gradient operator
$\cdot$	=	Dot (inner) vector product
$\oiiint$	=	Volume integral with no limits

$\| \|$	=	Magnitude
μ	=	Fluid viscosity
σ	=	Local electrical conductivity tensor
ρ	=	Local electrical resistivity tensor
$\emptyset$	=	Electrical potential field
$\emptyset_j$	=	Generalized potential field

References

1. Anon., Electro-Osmosis Stabilizes Earth Dam's Tricky Foundation, in: *Engineering News-Record*, pp. 176 (25): 36–37, 40, 45, 1966.
2. Acar, Y.B. and Alashawabkeh, A.N., Principles of Electrokinetic Remediation. *Environ. Sci., Technol.*, 27, 13, 2638–2647, 1993.
3. Y.B. Acar and R.J. Gale, Electrochemical Decontamination of Soils or Slurries. United States Patent No. 5,137,608, 1992.
4. Ace, A.G., *An Investigation of the influence of Applied Electrical Potential Gradients on Flow of Fluids in Porous Media, M. S. Thesis*, Department of Petroleum Engineering, The University of Southern California, 1955.
5. Al Shalabi, E.W., Haroun, M., Ghosh, B., Pamukcu, S., The Stimulation of Sandstone Reservoirs, Using DC Potential. *Pet. Sci. Technol.*, 30, 20, 2137–2147, 2012.
6. Amba, S.A., Chilingar, G.V., Beeson, C.M., Use of Direct Electrical Current for Increasing the Flow Rate of Reservoir Fluids During Petroleum Recovery. *J. Can. Pet. Technol.*, 3, 1, 8–14, 1964a.
7. Amba, S.A., Chilingar, G.V., Beeson, C.M., Application of Electrokinetic Phenomena in Civil and Petroleum Engineering. *Ann. N. Y. Acad. Sci.*, 3, 1, 8–14, 1964b.
8. Amba, S.A., Chilingar, G.V., Beeson, C.M., Use of Direct Electrical Current for Increasing the Flow Rate of Oil and Water in a Porous Medium. *J. Can. Pet. Technol.*, 4, 1, 81–88, 1965.
9. Ansari, A., Haroun, M., Sayed, N.A., Al Kindy, N., Ali, B., Shrestha, R.A., Sarma, H., A New Approach Optimizing Mature Waterfloods with Electrokinetics-Assisted Surfactant Flooding in Abu Dhabi Carbonate Reservoirs. *Soc Pet. Eng. J.*, 2012. SPE-163379.
10. Attanasi, E.D. and Meyer, R.F., Natural Bitumen and Extra-Heavy Oil, in: *2010 Survey of Energy Resources*, J. Trinnaman and A. Clarke (Eds.), pp. 123–150, World Energy Council, 2010.
11. C.W. Bell and C.H. Titus, Electro-Thermal Process for Production of Offshore Oil through Onshore Wells. US Patent No. 3,724,543, 1973.
12. Berry, L.G. and Mason, B., *Mineralogy - Concepts, Descriptions, and Determinations*, W. H. Freeman and Company, San Francisco, 1959.
13. C.W. Bell and C.H. Titus, Electro-Thermal Process for Promoting Oil Recovery. US Patent No. 3,782,465, 1974.

14. C.W. Bell, C.H. Titus, J.K. Wittle, *In Situ* Method for Yielding a Gas from a Subsurface Formation of Hydrocarbon Material. US Patent No. 4,473,114, 1985.
15. G.T. Bell, Electrolytically Promoting the Flow of Oil from a Well. US Patent No. 2,799,641, 1957.
16. Blevins, T.R., Duerksen, J.H., Ault, J.W., Light-Oil Steamflooding An Emerging Technology. *J. Pet. Technol.*, 36) 7, 1115–1122, 1984.
17. Bruell, C.J., Segall, B.A., Walsh, M.T., Electroosmotic Removal of Gasoline Hydrocarbons and TCE from Clay. *J. Environ. Eng., ASCE*, 118, 1, 68–83, 1992.
18. Casagrande, L., Electro-Osmotic Stabilization of Soils. *J. Boston Soc. Civ. Eng.*, 39, 1, 285–317, 1952.
19. Casagrande, L., A Review of Past and Current Work on Electro-Osmotic Stabilization of soils, in: *Harvard Soil Mechanics Series*, vol. 45, Harvard College, Cambridge, 1959.
20. Casagrande, L., Loughney, R.W., Matich, M.A.J., Electro-Osmotic Stabilization of a High Slope in Loose Saturated Silt. *Proc., 5th. ICSMFE*, vol. 2, pp. 555–461, 1961.
21. Chelidze, T.L., The Effect of Surface Polarization on the Electrical Properties of Rocks in Variable Fields. *Izv., Earth Phys.*, 10, 80–87, 1969.
22. Chilingar, G.V., Chang, K.S., Davis, J.E., Farhangi, H.J., Adamson, L.G., Sawabini, S., Possible Use of Direct Electrical Current for Augmenting Reservoir Energy During Petroleum Production. *Compass*, 45, 4, 272–285, 1968.
23. Chilingar, G.V., El-Nassir, A., Stevens, R.G., Effect of Electrical Current on Permeability of Sandstone Cores. *J. Pet. Technol.*, 22, 7, 830–836, 1970.
24. Chilingar, G.V., Loo, W.W., Khilyuk, L.F., Katz, S.A., Electrobioremediation of Soils contaminated with Hydrocarbons and Metals. *Energy Sources*, 19, 129–146, 1997.
25. Chute, F.S. and Vrmeulen, F.E., Present and Potential Applications of Electromagnetic Heating in the *In-Situ* Recovery of Oil. *OSTRA J. Res.*, 4, 19–33, 1988.
26. Chute, F.S., Vrmeulen, F.E., Stevens, L.G., Study of the Technical and Economic Feasibility of an Electric Preheat Process for *In-Situ* Recovery from Athabasca Oil Sands. *OSTRA J. Res.*, 3, 139–154, 1987.
27. Cohen, R.M. and Mercer, J.W., *DNAPL Site Evaluation*, C. K. Smoley/CRC Press, Boca Raton, 1992.
28. Conway, B.E., *Electrochemical Supercapacitors – Scientific Fundamentals and Technological Applications*, Kluwer Academic/Plenum Publishers, New York, 1999.
29. F.I. Crowson and W.G. Gill, Method and Apparatus for Secondary Recovery of Oil. US Patent No. 3,605,888, 1971.
30. F. Döring, Verfahren und Einrichtung zur Beseitigung von Schadstoffen, insbesondere im Erdbodenbereich. European Union Patent No. 0 578 925 A1, 1993.

31. F. Döring, Verfahren zur Abtoetung von Mikrooganismen und/order Mineralisierung von Organischen Substanzen im Boden Und im Grundwasser mittens elektrischen Stromes. European Union Patent No. 0 792 796 A1, 1996.
32. Döring, F., Iovenitti, J.L., Döring, N., Satre, R., II, McIlvride, W.A., Hill, D.G., ElectroChemical Remediation Technologies for Soil and Ground Water. *Proc. EREM 2000 International Conference on Nuclear and Hazardous Waste Management*, Chattanooga, pp. 24–28, September, 2000.
33. Döring, F., Iovenitti, J.L., Döring, N., Satre, R., II, McIlvride, W.A., Hill, D.G., ElectroChemical Remediation Technologies for Soil and Ground Water Remediation. *EREM 2001, 3rd Symposium on Electrokinetic Remediation*, Karlsruhe, pp. 18–20, April, 2001.
34. Döring, R., Pamucku, S., Wittle, K., Zanko, L., *The Electrochemical Remediation of Petroleum Hydrocarbons: Theory and Results, from a Soil Recycling Plant*, Electro-Petroleum, Inc., Wayne, PA, 2003.
35. S.R. Feris, Well Treating Process Using Electro Osmosis. US Patent No. 3417823, 1968.
36. Ghazanfari, E., Miroshnik, A., Shrestha, R., Pamukcu, S., Wittle, K., Investigation of Underlying electrochemical Processes in Electrically Enhanced Oil Recovery. *SPE Journal, Rev.*, 2013.
37. Ghazanfari, E., Shrestha, A.R., Miroshnik, A., Pamukcu, S., Electrically Assisted Liquid Hydrocarbon Transport in Porous Media. *Electrochem. Acta*, 86, 185–191, 2012.
38. W.G. Gill, Method and Apparatus for Secondary Recovery of Oil. US Patent No. 3,507,330, 1970.
39. W.G. Gill, Electrical Method and Apparatus for the Recovery of Oil. US Patent No. 3,542,066, 1972.
40. Gray., D.H. and Mitchell, J.K., Fundamental Aspects of Electro-Osmosis in Soils. *J. Soil Mech. Found. Div., ASCE*, 93, SM6, 209–236, 1967.
41. Gray, D.H. and Schlocker, J., Electrochemical Alteration of Clay Soils. *Clays Miner.*, 17, 309–322, 1969.
42. Haroun, M., Ansari, A.A., Wittle, J.K., Shrestha, R.A., Chilingar, G.V., Electrokinetic Assisted Waterflooding in Abu Dhabi Carbonate Reservoirs. *EREM2013*, pp. 23–26, June, 2013.
43. Haroun, M., Chilingar, G.V., Pamukcu, S., Wittle, J.K., Belhaj, H.A., Al Bloushi, M.N., Optimizing Electroosmotic Flow Potential for Electrically Enhanced Oil Recovery (EEOR®) in Carbonate Rock Formations of Abu Dhabi Based on Rock Properties and Composition. *International Petroleum Technology Conference*, 2009, IPTC 13812.
44. Hill, D.G., A Laboratory Investigation of Electrical Anisotropy in Precambrian Rocks. *Geophysics*, 37, 6, 1022–1038, 1972.
45. Hill, D.G., Electro-Osmotic Aided Remediation for Non-Aqueous Liquid Contamination of Fine-Grained Aquitards, in: *Weiss Associates*, Emeryville, CA, August, 1994.

46. Hill, D.G., Enhanced Restoration of Contaminated Aquitards, Using Electro-Osmotic Aided Remediation, in: *Weiss Associates, Presentation to California Environmental Protection Agency Department of Toxic Substances Control*, May 7, 1997.
47. Hill, D.G., Enhanced Restoration of Contaminated Aquitards, Using Electro-Osmotic Aided Remediation, in: *Environmental Management and Technology Expo*, pp. 17–19, Long Beach, CA, November, 1998.
48. Hill, D.G., Geophysical Basis for Electro-Chemical Geo-Oxidation Mechanisms, in: *Weiss Associates*, Emeryville, CA, 1999.
49. Hill, D.G., Iovenitti, J.L., Cooke, M.A., Enhanced Restoration of Contaminated Aquitards Using Electro-Osmotic Aided Remediation. *HazWaste World SuperFund XVIII*, Washington, DC, pp. 2–4, December 1997.
50. Hill, D.G., Wittle, J.K., Fricker, D.J., Chilingar, G.V., Moving Goo: Direct Electric Current Oil Recovery (EEOR)—A New Approach to Enhancing Oil Production. *Sacramento Petroleum Assoc.*, November 17, 2010.
51. Howell, B.F. and Licastro, P.H., Dielectric Behavior of Rocks and Minerals, American Mineralogist. 46, 269–288, 1961.
52. Iovenitti, J.L., McIlvride, W.A., Döring, F., Hill, D.G., Döring, R., Electrochemical Remediation Technologies for Organic and Metal Contaminants in Soil, Sediment, and Ground Water; Presentation of Case Histories. *Air and Waste Management Association Annual Conference and Exhibition*, 2003.
53. Kasevich, R.S., Price, S.L., Faust, D.L., Fontaine, M.F., Pilot Testing of a Radio Frequency Heating System for Enhanced Oil Recovery From Diatomaceous Earth. *SPE Annual Technical Conference and Exhibition*, 1994.
54. Katsube, T.J., The Electrical Polarization Mechanism Model for Moist Rocks. *Geo. Surv. Canada*, Paper 75-1C: 353–360, 1975.
55. Keller, G.V. and Licastro, P.H., *Dielectric constant and Electrical Resistivity of Natural-State Cores*, pp. USGS Bulletin 1052–H: 257–285, 1959.
56. Komarov, S.G. and Shmarova, V.P., Membrane Potential of Clays, in: *Prikladnaya Geofizika*, vol. 31, pp. 288–293, 1961.
57. Lageman, R. and Pool, W., and Seffinga, G., 1989, Electro-Reclamation: Theory and Practice. *J. Chem. Ind.*, 9, 585–590.
58. Lorenz, P.B., Phenomenology of Electro-Osmosis and Streaming Potential. *J. Phys. Chem.*, 56, 775–778, 1952.
59. Marshall, D.J. and Madden, T.R., Induced Polarization, A Study of its Causes. *Geophysics*, 24, 4, 790–816, 1959.
60. Mayper, V., The Normal Effect - Part I, in: *Overvoltage Research and Geophysical Applications*, J.R. Wait (Ed.), Pergamon Press, New York, 1959a.
61. Mayper, V., The Normal Effect - Part II, in: *Overvoltage Research and Geophysical Applications*, J.R. Wait (Ed.), Pergamon Press, New York, 1959b.
62. McGee, B.C.W., Electro-Thermal Pilot in the Athabasca Oil Sands: Theory versus Performance. *World Oil*, 11, 47–54, 2008.
63. McGee, B.C.W., Vermeulen, F.E., Yu, L., Field Test of Electrical Heating with Horizontal and Vertical Wells. *J. Can. Pet., Technol.*, 38, 3, 46–53, 1999.

64. Meyer, R.F., Attanasi, E.D., Philip, A., Freeman, P.A., Heavy Oil and Natural Bitumen Resources in Geological Basins of the World, in: *Open File Report 2007-1084*, U.S. Geological Survey, 2007.
65. Mitchell, J.K., In-Place Treatment of Foundation Soils. *J. Soil Mech. Found. Div., ASCE*, 96, SM1, 73–110, 1970.
66. Mitchell, J.K., *Fundamentals of Soil Behaviour*, John Wiley and Sons, Inc., New York, 1993.
67. Nourbehecht, B. and Madden, T.R., *Irreversible Thermodynamics in Inhomogeneous Media and Geoelectric Applications*, Massachusetts Institute of Technology, Cambridge, 1963.
68. Onsager, L., Reciprocal Relations in Irreversible Processes I. *Phys. Rev.*, 37, 405–426, 1931a.
69. Onsager, L., Reciprocal Relations in Irreversible Processes II. *Phys. Rev.*, 38, 2265–2279, December, 1931b.
70. Onsager, L., The Motions of Ions: Principles and Concepts (Nobel Laureate Lecture). *Science*, 166, 3911, 1359–1364, 1969.
71. Pamukcu, S., Electrokinetic Removal of Coal Tar Constituents from Contaminated Soils, in: *EPRI TR-103320, Electric Power Research Institute*, 1994.
72. Pamukcu, S., Electro-Chemical Technologies for *In-Situ* Restoration of Contaminated Subsurface Soils. *Electron. J. Geotech. Eng.*, 1997. http://geotech.okstate.edu/ejge/ppr97031/.
73. Pamukcu, S. and Pervizpour, M., *Electroosmotic Aided NAPL Remediation of Contaminated Soil: Stanford Linear Accelerator Site - Final Report to Weiss Associates*, Emeryville, Ca, November, 1997.
74. Rahner, D. and Röhrs, J., Electrochemically Induced Reactions in Soils – A New Approach to the *in-Situ* Remediation of Contamination Soils? Part 1: The Microconductor Principle. *Electrochem. Acta*, 47, 1395–1414, 2002.
75. Röhrs, J., Ludwig, G., Rahner, D., Electrochemically Induced Reactions in Soils – A New Approach to the *in-Situ* Remediation of Contamination Soils? Part 2: Remediation Experiments with a Natural Soil Containing Highly Chlorinated Hydrocarbons. *Electrochem. Acta*, 47, 1405–1403, 2002.
76. Rokityansky, I.I., The Nature of Induced Polarization of Ion-Conducting Soils. *Izv. Geophysics Ser.*, 1055–1060, 1959.
77. Sahni, A., Kumar, M., Knapp, R.B., Electromagnetic Heating Methods for Heavy Oil Reservoirs. *SPE/AAPG/SEG Western Regional Meeting, Paper SPE-62550*, Long Beach, CA, pp. 19–23, June, 2000.
78. Sanberg, C.L., Hale, A., Kovscek, A., History and Application of Resistance Electrical Heaters in Downhole Oil Field Applications. *SPE/AAPG/SEG Western Regional Meeting, Paper SPE-165323*, Monterey, CA, pp. 19–25, March, 2013.
79. E. Sarapuu, Method of Underground Electrolinking and Electrocarbonization of Mineral Fuels. US Patent No. 2,795,279, 1957.
80. Scott, J.H., Carroll, R.D., Cunningham, D.R., Dielectric Constant and Electrical Conductivity Measurements of Moist Rock. *J. Geophys Res.*, 72, 20, 5101–5115, 1967.

81. Schlumberger, C., *Étude sur la Prospection électrique du sous-sol*, Gauthier-Villars, Paris, 1920.
82. Smythe, W.R., *Static and Dynamic Electricity*, John Wiley & Sons, Inc., New York, 1968.
83. *Energy Resources Conservation Board*, Edmonton, Alberta, 2011, Staff, Total E&P Canada Ltd., Surface Steam Release of May 18, 2006.
84. Stainoh, M., *QOR Test Report: Sur Rio Deseado Este Oilfield*, Roach SA, Buenos Aires, 2011.
85. Sylwan, C.A., Geology of the Golfo San Jorge Basin, Argentina. *J. Iber. Geol.*, 27, 123–157, 2001.
86. Sumner, J.S., *Principles of Induced Polarization for Geophysical Exploration*, Elsevier Scientific Publishing Company, New York, 1976.
87. Tchillingarian, G.V., Possible Utilization of Electrophoretic Phenomenon for Separation of Fine Sediments into Grades. *J. Sediment. Petrol.*, 221), 29–32, 1952.
88. Timur, A., personal communication, 1992.
89. Tikhomolova, K.P., *Electro-Osmosis*, Ellis Horwood, New York, 1993.
90. C.H. Titus, J.K. Wittle, C.W. Bell, Apparatus for Passing Electrical Current Through an Underground Formation. US Patent No. 4,495,990, 1985.
91. Vacquier, V., Holmes, C.R., Kintzinger, P.R., Lavergne, M., Prospecting for Ground Water by Induced Polarization. *Geophysics*, 22, 3, 660–687, 1957.
92. H.J. Vinegar and R.M. Bass, *In Situ* Recovery from a Hydrocarbon Containing Formation Using Conductor-in-Conductor Heat Sources with an Electrically conductive Material in the Overburden. US Patent No. 7,165,165, 2007.
93. H.J. Vinegar, I.E. Berchenko, E.P. de Rouffignac, T.D. Fowler, R.C. Ryan, S.L. Wellington, E. Zhang, *In Situ* Thermal Processing of an Oil Reservoir. US Patent No. 7077199, 2006a.
94. H.J. Vinegar, J.M. Coles, E.P. de Rouffignac, B. Hunsucker, J.L. Menotiti, C.A. Pratt, S.L. Wellington, Installation and Use of Removable Heaters in a Hydrocarbon Containing Formation. US Patent No. 7,156,176, 2007.
95. H.J. Vinegar, P. Veenstra, S.P. Giles, C.L. Sandburg, F.H.K. Rambow, M.G. Picha, E. Zhang, G. Beer, F.G. Carl, T. Bai, D.S. Kim, M.D. Fairbanks, G.P. Sanz, *In Situ* Thermal Processes for Subsurface Formations. US Patent No. 7121342, 2006b.
96. Wait, J.R. (Ed.), *Overvoltage Research and Geophysical Applications*, Pergamon Press, New York, 1959.
97. Ward, S.H. and Fraser, D.C., Conduction of Electricity in Rocks, in: *Mining Geophysics, Vol. II*, D.A. Hansan, W.E. Heinrichs, R.C. Holmer, R. e. MacDougall, J.S. Sumner, S.H. Ward (Eds.), pp. 198–223, Society of Exploration Geophysicists, 1967.
98. Waxman, M.L. and Smitts, L.J.M., Electrical Conductivities of Oil-Bearing Shaly Sands. *SPE J. AIME Trans.*, 243, 1, 107–122, 1968.
99. J.K. Wittle and C.W. Bell, Electrochemical Process for Effecting Redox-Enhanced Oil Recovery. US Patent No. 6,877,556 B2, 2005.

100. J.K. Wittle and C.W. Bell, Method for Enhancing Oil Production Using Electricity. US Patent No. 7,325,604 B2, 2008.
101. Wittle, J.K. and Hill, D.G., Use of Direct Current Electrical Stimulation for Heavy Oil Production, in: *Society of Petroleum Engineers Applied Technology Workshop - Technologies for Thermal Heavy Oil and Bitumen Recovery and Production*, pp. 14–15, Calgary, March, 2006a.
102. Wittle, J.K. and Hill, D.G., Direct Current Electrical Stimulation - A New Approach to Enhancing Heavy Oil Production". *First World Heavy Oil Conference*, pp. 12–15, November 2006b.
103. Wittle, J.K., Hill, D.G., Chilingar, G.V., *Direct Current Electrical Enhanced Oil Recovery in Heavy-Oil Reservoirs to Improve Recovery, Reduce Water Cut, and Reduce H2S Production While Increasing API Gravity*, pp. SPE-114012, Society of Petroleum Engineers, 2008a.
104. Wittle, J.K., Hill, D.G., Chilingar, G.V., Electro-Enhanced Oil Recovery (EEOR) Using Direct Current. *Oil Sands and Heavy Oil Technologies Conference*, Calgary, 2008b.
105. Wittle, J.K., Hill, D.G., Chilingar, G.V., Direct Current Stimulation for Heavy Oil Production. *Second World Heavy Oil Conference, Edmonton*, pp. Paper No. 2008-374, 2008c.
106. Wittle, J.K., Hill, D.G., Chilingar, G.V., Direct Current Oil Recovery (EEOR) - A New Approach to Oil Production. *ENERG Source Part A*, 33, 9, 805–822, 2011.
107. Wittle, J.K. and Pamukcu, S., *Electrokinetic Treatment of Contaminated Soils, Sludges, and Lagoons, DOE/CH-9206*, U. S. Department of Energy, 1993.
108. P.E. Workman, Method of Recovering and Increasing the Production of Oil. US Patent No. 1,784,214, 1930.

Addendum

Improving Injectivity of Liquids Into Tight Rocks

"In as much as in many procedures proposed in this book involved injection of some liquids (e.g., acid)...," this problem of injectivity can be solved by using technology proposed by Chilingar *et al.* (references [1] through [4]). The stimulation of carbonate reservoirs is mainly achieved by acidizing treatments (Figure A.1). Acids may be injected into the pores and pre-existing fractures or at hydraulic fracturing rates depending upon the results desired. The acid dissolves the carbonate (limestones/dolomites), enlarging the pores. This gives rise to an increase in permeability.

The main problem in acidizing is the fact that the radial distance the acid will penetrate until being spent is short, especially in tight carbonates. In the case of matrix acidizing, with enlargement of pores or pre-existing fractures:

(1) the specific surface area decreases,
(2) the velocity decrease,
(3) spending time increase, and
(4) the penetration distance increase.

In the case of the uniform penetration of acid, the reaction rate declines uniformly with deceasing acid concentration. The weight of carbonate dissolved per increment of distance penetrated declines uniformly until the acid is completely spent. With stronger acid, the spending time decreases. On assuming a homogeneous formation and that the volume of acid injected is equal to the pore volume invaded [$q_i t = \pi\Phi h\ (r_a^2 - r_w^2)$], the radial distance the acid will penetrate until being spent, r_a (ft), is equal to

$$r_a = \sqrt{\frac{0.0936 q_i t}{\pi \phi h} + r_w^2}$$

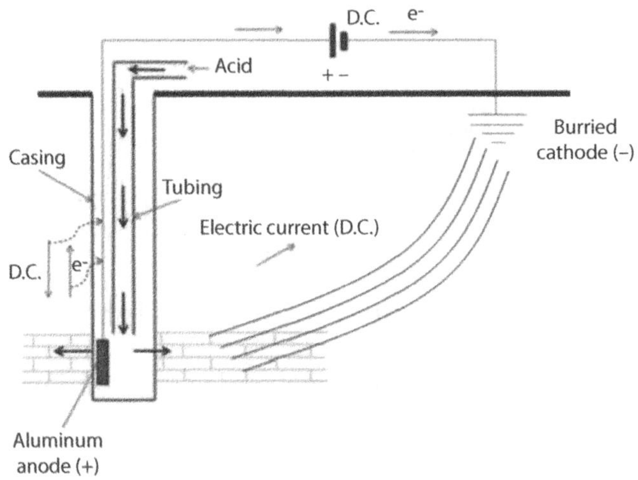

Figure A.1 Electrodes arrangement for acidizing operation.

where

q_i = acid injection rate (bbl/min);
t = spending time (sec);
Φ = fractional porosity;
h = formation thickness (ft);
r_w = wellbore radius (ft).

In the case of uniform penetration of acid, the reaction rate declines uniformly with decreasing acid concentration. The weight of carbonate dissolved per increment of distance penetrated declines uniformly until the acid is completely spent.

References

1. Chilingar, G.V., and Haroun, M. *Electrokinetics for Petroleum and Environmental Engineers*. Scrivener Publishing (Wiley) New Jersey, 240 pp, 2014.
2. Chilingarian, G.V., Mazzullo, S.V., and Rieke, H.H. *Carbonate Reservoir Characterization. A Geologic-Engineering Analysis, Part I.* Elsevier, Amsterdam, 1992.
3. Chilingar, G.V., Shin, S., Haroun, M., Albannay, A., Wittle, K., Meshkati, N., and Pamukcu, S. Improving Acidizing Operations. *J. Sustainable Energy Engineering*., 1(1), 1–5, 2013.
4. Chilingar, G.V., El-Nassir, A., Stevens, R.G. Effect of direct electrical current on permeability of sandstone cores. *J. Petrol. Tech.* **22**, 830–836, 1970.

Nomenclature

AE	Acoustic Emission
ADC	Analog-to-Digital Converter
ACF	Autocorrelation Functions
BT	Billion Tons
HydroVibroFrac	Hydraulic Fracturing along with Vibrowave Treatment
GRVP	Deep Chemical-Wave Reservoir Treatment
IMASH RAN	Institute of Machine Science of the Russian Academy of Sciences
INNT	Institute of Advanced Petroleum Technologies
KVT	Coiled Tubing Wave Technologies
OGVP	Elimination of the Water and Gas Zone Isolation Issues
RANS	Russian Academy of Natural Sciences
SAE	Seismic Acoustic Emission
SE	Seismic Emission
SLOE	Seismic Location of Emission Sources
SLOB	Seismic Side Scanner
VDHV	Integrated Vibrowave, Pressure-Drawdown and Chemical Near-Wellbore Zone Treatment
VPV	Vibrowave and Foam Treatment
VST	VibroSeismic Treatment

Symbols

A	- oscillatory downhole pressure amplitude, Pa
$A(t)$	- autocorrelation function of the acoustic-signal time series, dimensionless
C	- sound speed in the medium, m/s
C_m	- phase velocity of the m-th wave, m/s
C_∞	- high-frequency limit of the sound velocity, m/s
D_c	- Hausdorff correlation dimensionality, dimensionless
$\bar{d}$	- effective (average or median) diameter of the medium pore channels, μm
E	- acoustic-signal pulse energy, conditional units of the V^2 (square voltage) dimension
E_ς	- accumulated energy of acoustic signals, conditional units of V^2 (square voltage) dimension
$\frac{dE}{dt}$	- velocity of the acoustic emission energy release, conditional units V^2/s
E_k	- kinetic energy, J
$\boldsymbol{E}$	- oscillatory energy flow density in the given point of the medium, W/m^2
E_1, E_2	- elastic moduli for the wellbore tubular materials and cement column, respectively, Pa
e	- = 2,7182…
f	- oscillation frequency, Hz
F	- force, *N*
$F_{аД}$	- adhesion force, Pa
$F_{Д}$	- dynamic pressure force (velocity head or drag force), Pa
f_B, f_H	- relative medium permeabilities for water and oil, respectively, fractions
G	- shear modulus, Pa
G	- shear modulus for "dry" porous medium, Pa
$G_\alpha(t, r)$	- Green function of wave differential equation
G_0	- gas-oil ratio, m^3/m^3
g	- gravity acceleration, m/s^2

$H(z)$	- piecewise constant Heaviside function, 0 or 1
h	- fracture, channel width, reservoir thickness, m
i	- complex number imaginary unity
I	- flow of the matter through a pore (channel), m^3/s
J	- intensity of oscillations in a wave, W/m^2
K	- matrix bulk modulus (elasticity), Pa
K_y	- relative limit of vibration strength
K_y^0	- relative limit of vibration strength with zero amplitude of loading cycle
k	- porous medium permeability, μm^2
k_0	- dimensionless coefficient of velocity field irregularity near the droplet frontal surface
$k_В$, $k_Н$	- relative permeabilities for water and oil, respectively, μm^2
$k_О$, $k_З$	- formation water permeability during liquid production and injection, μm^2
k_+, k_-	- dimensionless liquid drag coefficient subject to asymmetric flow in opposite directions
M	- dimensionless phase mobility parameter during liquid flow
m	- (ϕ in English language literature) porosity, fraction
m	- mass of particle, body, medium volume, kg
N	- oscillatory emission power, W
n	- wellbore shot density, number of shots per unit of wellbore perforation interval length, N/m
N_k	- perforated-channel oscillatory-emission power, W
P	- pressure, Pa
P_s	- static pressure in the well during cement slurry setting, Pa
P^0	- pressure in the wellbore liquid, Pa
p_k	- capillary pressure, Pa
p_0	- bubblepoint pressure, Pa
$p_{пл}$	- formation pressure, Pa
Q	- fluid volumetric flow rate, injectivity, well production rate, m^3/s
R	- particle speed return coefficient after collision with reflecting plane
R^0	- limit of cement-column uniaxial-compression vibration strength, Pa
R_p	- cement-column ultimate strength, Pa
Re	- viscous liquid Reynolds number, ratio of inertia forces to viscous friction forces
r	- distance, radius, m

r_k	- channel radius, m
$r_{Эф}$	- effective radius of medium pore channel, m
$r_В{}^*r_Н{}^*$	- effective radii at two-phase flow for water and oil, respectively, m
S	- surface, m^2
S_k	- channel cross-section area, m^2
sgn x	- piecewise-constant signum (sign) function of actual × argument
t	- time, seconds
$\boldsymbol{T}$	- constant force, N
T	- medium temperature, °C
$\boldsymbol{u}$	- solid-phase displacement in Bios equations to describe elastic wave propagation in fluid-filled media
$\boldsymbol{V}$	- liquid-phase displacement in Bios equations to describe elastic wave propagation in fluid-filled media
V	- velocity, velocity amplitude, m/s
$V_i(t)$	- phase trajectories (integrated curves) of a non-linear dynamic system in a space of m dimension
x	- coordinate x, m
y	- coordinate y, m
Z	- coordination number of porous medium structure
α	- coefficient of oscillation decay in a wave, 1/m
β	- liquid compressibility, Pa^{-1}
β_i	- i-th phase compressibility for a multiphase medium, Pa^{-1}
$\Gamma(z)$	- (Gamma) Euler Gamma function
Δp	- pressure drop, Pa
$\Delta P^*(r)$	- dimensionless pressure at distance r from the wellbore axis
$\Delta\sigma_\tau$, $\Delta\sigma_r$	- variation in tangential and radial stresses on the inner surface of the cement column of the well, respectively, Pa
δ	- coefficient of oscillation decay in a wave, 1/m
$\delta(x)$	- Dirac delta function
η	- fluid viscosity, mPa*c
θ	- dimensionless coefficient of suspended bridging particles state during flow process
λ	- dimensionless instantaneous friction (sliding friction) coefficient at particle collision
$\lambda_П$	- effective reservoir thermal conductivity, W/(m*K)
μm^2	- permeability (1 darcy ≈ 0.9869 μm^2) (1 D = 0.9869 mkm^2 = 0.9869 $\times 10^{-12}$ m^2 =1,000 mD)
μ	- fluid dynamic viscosity, mPa*s
$\mu_В$, $\mu_Н$	- water and oil viscosities, respectively, mPa*s

$\mu_O, \mu_З$	- fluid viscosities during reservoir production and injection into reservoir, respectively, mPa*s
μ_-, μ_+	- dimensionless flow rates subject to asymmetric flows in opposite directions
μ_1, μ_2	- dimensionless Poisson ratios for the wellbore tubular materials and cement column
ν	- fluid kinematic-viscosity coefficient, m^2/s
ξ	- oscillatory displacement of medium particles, m
$\ddot{\xi}$	- oscillatory acceleration of medium particles, m/s^2
π	- mathematical constant, ratio of circle circumference to its diameter
ρ	- medium density, kg/m^3
ρ_f	- fluid density, kg/m^3
$\rho_В, \rho_Н$	- water and oil viscosities, respectively, kg/m^3
ρ_C	- water and oil mixture density, kg/m^3
σ	- interfacial tension, N/m
$\sigma_{ТВ}, \sigma_{ТН}$	- interfacial solid-water and solid-oil tensions, respectively, N/m
σ_*	- maximum stress of the cement-column vibration loading cycle, Pa
τ	- tangential (shear) stress, Pa
τ_0	- ultimate shear stress, Pa
τ_r	- relaxation time of physical processes, s
$\varphi_\alpha(t)$	- singular function Abel core
$\varphi_r(r)$	- numerical function of medium pore distribution by radii
$\varphi_\alpha(\zeta) * f_\alpha(\zeta)$	- convolution of functions
Φ	- rapidly changing alternating force in the field of oscillations, N
χ	- reservoir medium piezoconductivity, m^2/s
ω	- circular (angular) frequency, rad/s (radian per second)

Subscripts and superscripts

0	- ultimate lower
о	- process of liquid production from the reservoir
з	- process of liquid injection into the reservoir
в (w)	- water
н (p)	- oil
пл	- reservoir
эф	- effective

i,j	- indices
r	- radial
τ	- tangential
s	- static
div	- differential operator (divergence)
grad	- vector that shows the direction of growth of a certain value φ, grad $\varphi = \nabla \varphi$ exp(x) – e^x exponent
$K!$	- factorial of number K
Δ	- difference
∇^2	- scalar product of nabla (Hamiltonian) operator by itself, to show the Laplace operator Δ in the Cartesian coordinate system
Δ	- Laplace operator symbol

Classifications of the reservoirs on the basis of permeability

1 D = 0.9869 mkm^2 = 0.9869 $\times 10^{-12}$ m^2 =1,000 mD

Low	0 to 100 mD
Medium	100 to 500 mD
High	> 500 mD

Very Good	> 1 mkm^2 (μm^2)
Good	0.1 – 1.0 mkm^2 (μm^2)
Medium	0.01 – 0.1 mkm^2 (μm^2)
Low	0.001 – 0.01 mkm^2 (μm^2)
Very Low	< 0.001 mkm^2 (μm^2)

About the Authors

Professor George Varos Chilingar

Professor George V. Chilingar(ian) is Professor Emeritus of Petroleum, Civil and Environmental Engineering at the University of Southern California (USC), Los Angeles, California. He received his Bachelor's and Master's Degrees in Petroleum Engineering at the University of Southern California (USC) in 1949 and 1950, respectively. While he was working on his Ph.D. in Geology at USC, he spent 4 years in the U.S. Air Force, where he made a scientific breakthrough in jet fuel and was given the position of Chief of the Petroleum and Chemicals Laboratory at Wright Patterson Air Force Base, Ohio. By 1956, he was awarded a Ph.D. in Geology minor in Petroleum Engineering. After his term serving in the U.S. Air Force, he returned to USC as a member of the School of Engineering faculty. For over 60 years, Professor George V. Chilingar taught various courses in Petroleum Engineering, Petroleum Geology, and Environmental Engineering at USC. In 2008, he was honored with the Lifetime Achievement Award from Viterbi School of Engineering at USC. Professor Chilingar is Academician, USC International Ambassador; Member of the Russian Academy of Sciences; Founder and Past President of the Russian Academy of Natural Sciences USA Branch; Honorary Professor of Gubkin University, Russia; Honorary Consul of Honduras in Los Angeles, CA, since 1983; and Certified Petroleum Geologist, American Association of Petroleum Geologists and State of California. The results of his investigation are presented in over 500 research articles and 73 books in the fields of Petroleum and Environmental Engineering and Petroleum Geology. With many of his books being translated into Russian, Chinese, and Farsi. He has numerous patents, especially in the field of his inventions: EEOR and Acoustic Vibrational Enhanced Oil Recovery. Professor Chilingar is recipient of over 100 medals, awards, and diplomas from various governments, universities, and organizations, including Lomonosov's medal of Russian Academy of Sciences: Order of Homayun Gold Medal from the Shah of Iran, and White Elephant Medal from King of Thailand. In 2021, Professor Chilingar was given the Society of Petroleum Engineers (SPE) Honorary Membership award in Dubai for outstanding service to SPE and distinguished scientific and engineering achievements. Professor Chilingar founded several journals including (1) Energy Sources and (2) Journal of Petroleum Science and Engineering. Professor Chilingar is with Global Pioneer Oil Company, Los Angeles, CA, USA. Please find Professor Chilingar's email [GeorgeV.Chilingar@gmail.com].

Professor Kazem Majid Sadeghi

Professor Kazem Majid Sadeghi has a Bachelor of Science in Chemistry from University of California, Santa Barbara (1985), Master of Science in Environmental Engineering (1987) and Engineer Degree in Civil Engineering from University of Southern California (1989). In addition, he has a degree of Doctor of Philosophy in Geography from University of California, Santa Barbara (2015). Professor Sadeghi has been researching and teaching many years at University of California, Santa Barbara and California State Polytechnic University, Pomona. He has over 30 years of civil and environmental engineering and consulting experience, including hazardous waste management, pollution prevention assessments, design of industrial wastewater pretreatment facilities and gas collection/treatment systems, treatment of carbonaceous materials, and soil remediation. As an engineer, he has worked in private and government jobs and has been a teacher at university level. He has been responsible for civil and environmental engineering tasks for many projects. In addition, he has been working in the Green Stormwater Infrastructure and capture projects, reduce pollutants, and augment water supply projects. He is knowledgeable in water quality, wastewater, and hazardous waste regulations and has represented industries in regulatory negotiations, the preparation of various civil and environmental engineering designs, compliance reports, as well as local, state, and federal permits. He assisted in development of stormwater pollution prevention plans and Environmental Impact Statement (EIS) reports. He also has 38 reviewed journal publications, 3 U.S. patents, 1 Canadian patent, 1 International patent, and 22 presentations at international conferences. Professor Sadeghi is with Global Pioneer Oil Company, Los Angeles, CA, USA. Please find Professor Sadeghi's email [kmsadeghi007@gmail.com].

Professor Oleg Leonidovich Kuznetsov is a graduate from Moscow Geological-Prospecting Institute S. Ordzhonikidze (1962). He received a degree of Doctor of Technical Sciences in Geophysics (1975). Upon graduation he worked at Institute of Geology and Mining of Fossil Fuels of the Academy of Sciences (1962-1970) and All-Union Institute of Nuclear Geophysics and Geochemistry (1972-1986). He currently works in the All-Russia Institute of Geosystem (since 1986) and Professor of M. V. Lomonosov Moscow State University (since 1986). In addition, he is Professor at Dubna State University (since 1994) working on research development and teaching. Professor Kuznetsov is President of Russia's Academy of Natural Sciences; Honorary member of the Society of Hungarian Geophysicists; Member of the Academy of Creativity; International Academy of Sciences; New York Academy of Sciences; Eastern Academy of Oil and Gas; Honored Worker of Science and Technology of Russia; Laureate of the USSR State Prize. Over the years of scientific and practical activities set the stage oil geoacoustics, concept and project global geoinformation system GEOS (1984-1987) and Foundations of Nonlinear Geophysics. He is the author of a numbers papers and books on applied geophysical technology. He is the author of monographs, including: ***Transformation and the Interaction of Geophysical Fields in the Lithosphere*** (1994); ***Geoinformatics*** (1992); ***Geophysics at the Turn of the XXI Century***. Please find the Professor Kuznetsov email [olk@uni-dubna.ru].

Index

Also of Interest

Check out these other related titles from Scrivener Publishing

From the Same Author:

ACOUSTIC AND VIBRATIONAL ENHANCED OIL RECOVERY, By George V. Chilingar, Kazem Majid Sadeghi, and Oleg Leonidovich Kuznetsov, ISBN: 9781119760153. Written by one of the most renowned petroleum engineers in history and his esteemed co-author, this new and groundbreaking method for enhanced oil recovery revolutionizes a changing industry. *NOW AVAILABLE!*

THE EVOLUTION OF EARTH'S CLIMATE, By John O. Robertson, George V. Chilingar, O. G. Sorokhtin, N. O. Sorokhtin, and W. Long, ISBN: 9781119407065. Written by a group of the most experienced and well-known environmental engineers in the world, from a unique perspective, this volume explores the hot-button issue of climate change, its causes, and the future of the planet. *NOW AVAILABLE!*

ENVIRONMENTAL ASPECTS OF OIL AND GAS PRODUCTION, By John O. Robertson, George V. Chilingar, ISBN: 9781119117377. Written by two of the most experienced and well-known petroleum engineers in the world, this volume explores the environmental aspects of oil and gas production, one of the hottest and most contested issues of our time in the energy sector. *NOW AVAILABLE!*

ELECTROKINETICS FOR PETROLEUM AND ENVIRONMENTAL ENGINEERS, By George V. Chilingar and Mohammed Haroun, ISBN: 9781118842690. Written by a pioneer in electrokinetics and one of the most well-known petroleum engineers in the world, this volume is the first volume ever written that explores the practical uses in petroleum and environmental engineering of this technology. *NOW AVAILABLE!*

Fundamentals of the Petrophysics of Oil and Gas Reservoirs, by Buryakovsky, Chilingar, Rieke, and Shin. ISBN 9781118344477. The most comprehensive book ever written on the basics of petrophysics for oil and gas reservoirs. *NOW AVAILABLE!*

Petroleum Accumulation Zones on Continental Margins, by Grigorenko, Chilingar, Sobolev, Andiyeva, and Zhukova. ISBN 9781118385074. Some of the best-known petroleum engineers in the world have come together to produce one of the first comprehensive publications on the detailed (zonal) forecast of offshore petroleum potential, a must-have for any petroleum engineer or engineering student. *NOW AVAILABLE!*

Mechanics of Fluid Flow, by Basniev, Dmitriev, and Chilingar, ISBN 9781118385067. The mechanics of fluid flow is one of the most important fundamental engineering disciplines explaining both natural phenomena and human-induced processes. A group of some of the best-known petroleum engineers in the world give a thorough understanding of this important discipline, central to the operations of the oil and gas industry. *NOW AVAILABLE!*

Other Related Titles available from Scrivener Publishing:

Measurement While Drilling (MWD) Signal Analysis, Optimization, and Design, by Wilson C. ChinYinao Su, Limin Sheng, Lin Li, Hailong Bian and Rong Shi, ISBN9781118831687. The only book explaining modern MWD technology, to include hardware design, signal processing and telemetry, offering unique approaches to high-data-rate well logging. *NOW AVAILABLE!*

Formation Testing: Pressure Transient and Formation Analysis, by Wilson C. Chin, Yanmin Zhou, Yongren Feng, Qiang Yu, and Lixin Zhao, ISBN 9781118831137. This is the only book available to the reservoir or petroleum engineer covering formation testing algorithms for wireline and LWD reservoir analysis that are developed for transient pressure, contamination modeling, permeability, and pore pressure prediction. *NOW AVAILABLE!*

Electromagnetic Well Logging, by Wilson C. Chin, ISBN 9781118831038. Mathematically rigorous, computationally fast, and easy to use, this new approach to electromagnetic well logging does not bear the limitations of existing methods and gives the reservoir engineer a new dimension to MWD/LWD interpretation and tool design. *NOW AVAILABLE!*

Oil Spill Risk Management: Modeling Gulf of Mexico Circulation and Oil Dispersal, By David Dietrich, Malcolm J. Bowman, Konstantin Korotenko, and Hamish Bowman, ISBN: 9781118290385. This book describes and applies state-of-the-art software designed to help balance cost and profit estimates against risk in the petrochemical industry using oil extracted from ocean bottom deposits. *SEPTEMBER 2014*

Bioremediation of Petroleum and Petroleum Products, by James Speight and Karuna Arjoon, ISBN 9780470938492. With petroleum-related spills, explosions, and health issues in the headlines almost every day, the issue of remediation of petroleum and petroleum products is taking on increasing importance, for the survival of our environment, our planet, and our future. This book is the first of its kind to explore this difficult issue from an engineering and scientific point of view and offer solutions and reasonable courses of action. *NOW AVAILABLE!*

Sustainable Resource Development, by Gary Zatzman, ISBN 9781118290392. Taking a new, fresh look at how the energy industry and we, as a planet, are developing our energy resources, this book looks at what is right and wrong about energy resource development. *NOW AVAILABLE!*

An Introduction to Petroleum Technology, Economics, and Politics, by James Speight, ISBN 9781118012994. The perfect primer for anyone wishing to learn about the petroleum industry, for the layperson or the engineer. *NOW AVAILABLE!*

Ethics in Engineering, by James Speight and Russell Foote, ISBN 9780470626023. Covers the most thought-provoking ethical questions in engineering. *NOW AVAILABLE!*

Zero-Waste Engineering, by Rafiqul Islam, ISBN 9780470626047. In this controversial new volume, the author explores the question of zero-waste engineering and how it can be done, efficiently and profitably. *NOW AVAILABLE!*

Formulas and Calculations for Drilling Engineers, by Robello Samuel, ISBN 9780470625996. The most comprehensive coverage of solutions for daily drilling problems ever published. *NOW AVAILABLE!*

Emergency Response Management for Offshore Oil Spills, by Nicholas P. Cheremisinoff, PhD, and Anton Davletshin, ISBN 9780470927120. The first book to examine the Deepwater Horizon disaster and offer processes for safety and environmental protection. *NOW AVAILABLE!*

Advanced Petroleum Reservoir Simulation, by M.R. Islam, S.H. Mousavizadegan, Shabbir Mustafiz, and Jamal H. Abou-Kassem, ISBN 9780470625811. The state of the art in petroleum reservoir simulation. *NOW AVAILABLE!*

Energy Storage: A New Approach, by Ralph Zito, ISBN 9780470625910. Exploring the potential of reversible concentrations cells, the author of this groundbreaking volume reveals new technologies to solve the global crisis of energy storage. *NOW AVAILABLE!*

Printed and bound by CPI Group (UK) Ltd, Croydon, CR0 4YY
17/09/2026
14964246-0004